DATE DUE FOR RETURN

This book may be recalled
before the above date

NEW PERSPECTIVES IN QUANTUM FIELD THEORIES

Edited by
J Abad, M Asorey & A Cruz

NEW PERSPECTIVES IN QUANTUM FIELD THEORIES

Proceedings of the
XVIth GIFT International Seminar on Theoretical Physics

3-8 June 1985
Jaca, Huesca, Spain

World Scientific

Published by

World Scientific Publishing Co Pte Ltd.
P. O. Box 128, Farrer Road, Singapore 9128

Library of Congress Cataloging-in-Publication data is available.

NEW PERSPECTIVES IN QUANTUM FIELD THEORIES

ISBN 9971-50-048-5
9971-50-057-4 pbk.

Printed in Singapore by Fu Loong Lithographer Pte Ltd.

PREFACE

This volume contains the proceedings of the XVI GIFT International Seminar on Theoretical Physics, which was held at Jaca, Spain, from 3 to 8 June 1985, and was organized by two of the editors (J. A. and M. A.) and Professor J. L. Alonso.

The Seminar was devoted to New Perspectives in Quantum Field Theories, especially in the topics: Anomalies, Lattice Gauge Theory and Quantum Gravity. The lecturers invited are leading experts in the subjects.

The level was intended for advanced graduate students and recent Ph.D's. The organizers express their gratitude to the lecturers for their very beautiful and stimulating lectures.

The Seminar was funded by the Instituto de Estudios Nucleares, Comisión Asesora de Investigación Científica y Tecnica and the University of Zaragoza, the Rector and Vice Rector of which, Dr. Camarena and Dr. Aporta, we wish to most heartily, as well as the Department of Theoretical Physics and its Director, L. J. Boya, for their invaluable support.

J. Abad
M. Asorey
A. Cruz

UNIVERSIDAD
DE
ZARAGOZA

CONTENTS

NEW PERSPECTIVES IN QUANTUM FIELD THEORIES

NUMERICAL SIMULATION OF QUANTUM FIELD THEORIES

by

BERND A. BERG

Supercomputer Computations Research Institute

Florida State University

Tallahassee, Florida 32306

ABSTRACT

A pedagogical introduction to Monte Carlo simulations of quantum field theories is given. Literature on numerical results on Quantum Chromodynamics is summarized. Possibly beyond quantum field theory a recent exploratory simulation of discrete quantum gravity is summarized.

I. INTRODUCTION

Advanced computer technology has made numerical simulations a powerful tool not only for physics, but quite general in more or less all branches of science. In these lectures I report on simulations of quantum field theory (QFT). Most work relies on applying the Monte Carlo technique [1-3] and within recent years our qualitative understanding of quantum field theories has considerably improved due to numerical results.

In section II, corresponding to my first lecture, I will give an introduction to the MC method. The major emphasis is on illustrating the flexibility in setting up algorithms according to the requirements of balance or detailed balance. The hope is to provide enough information to enable the interested non-expert reader to sett up own algorithms adapted to his or her particular problems.

Section III and IV correspond to my second and third lecture. In section III a report on QCD results is given. I will be very brief, as my lecture was mainly covering material as published elsewhere [4] under the title "Status Report on Lattice QCD". More detailed reports and reviews on the subject can be found in the literature, for instance Ref.[5-8]. In the final section IV I summarize my own recent exploratory work simulating Euclidean quantum gravity. This goes to some extend beyond quantum field theory, because it is not at all understood whether quantum field theoretical concepts are still valid at a scale of the order of the Planck length. T.D. Lee and collaborators [9], for instance, emphasize the possiblity of a fundamental length.

II. THE MONTE CARLO METHOD

The Monte Carlo (MC) method is quite old in statistical physics. Originally it was introduced by Metropolis et. al. [1]. I will give an introduction emphasizing practical aspects for actually doing calculations. For a review the interested reader is referred to the article of Binder [2].

II.1 Importance Sampling

The typical task in statistical mechanics is to calculate expectation values

$$\bar{O} = \langle O \rangle = Z^{-1} \int \prod_i d\Phi_i \; O(K) \; \exp(-\beta \, S(K)) \qquad (II.1)$$

Here the <u>partition function</u> Z normalizes $\langle 1 \rangle = 1$. S is the action and a <u>configuration</u> K of the system is defined by

$$K = \{ \Phi_1, \ldots , \Phi_N \}. \qquad (II.2)$$

For simplicity I consider a discrete system. The configurations can be labelled by integers: K_n $(n=1,2,\ldots,N,\ldots)$ and equation (1) becomes

$$O = \langle O \rangle = Z^{-1} \sum_n O(K_n) \; \exp(-\beta S(K_n)). \qquad (II.3)$$

A simple discrete system is for instance an Ising model on a finite or infinite lattice. In this case $\Phi_i = \pm 1$.

We would now like to attempt a numerical evaluation of the partition function (3). A naive sampling is certainly impractical. Already, for the case of a 2d Ising model on a 30x30 lattice, 2^{900} configurations would contribute. The large number of configurations suggests a statistical treatment. Generating configurations randomly with a uniform distribution would, however, be completely inefficient. Most of the configurations have a very small Boltzmann factor $e^{-\beta S(K_n)}$ and their contribution to the sum (3) is neglegible.

Importance Sampling generates configurations K_n with probability

$$P(K_n) = \text{const } e^{-\beta S_n} , \; S_n = S(K_n). \qquad \text{(II.4)}$$

The constant is fixed by the normalization $\sum_n P(K_n) = 1$.

Expectation values become arithmetic averages:

$$\overline{O} = \langle O \rangle = \lim_{N \to \infty} \frac{1}{N} \sum_{n=1}^{N} O(K_n). \qquad \text{(II.5)}$$

Configurations with equilibrium distribution (4) can be generated by a Markov process. The elements of the Markov chain are the configurations. These configurations are generated sequentially, each configuration from the previous one alone. The transition probability of creating the configuration K_n in one step from K_m is given by $W_{mn} = W(n \to m) \geq 0$ and is required to satisfy the following properties:

i) Ergodicity:

$$e^{-\beta S_m} > 0 \quad \text{and} \quad e^{-\beta S_n} > 0 \quad \text{imply:}$$

an integer number $k > 0$ exists such that $(W^k)_{mn} > 0$.

ii) Normalization: $\sum_m W_{mn} = 1$.

iii) Balance:

$$\sum_m W_{nm} e^{-\beta S_m} = e^{-\beta S_n}$$

Balance means: The Boltzmann weight factors form an eigenvector of the matrix $W = (W_{nm})$.

The Boltzmann ensemble E_B is defined to be the ensemble with probability distribution

$$P_B^n = \text{const } e^{-\beta S_n} \quad , \sum_n P_B^n = 1 \qquad \text{(II.6)}$$

for the configurations K_n. Under the conditions i), ii) and iii) the Boltzmann ensemble is the only <u>equilibrium ensemble</u> E_{eq} (i.e. satisfying $W P_{eq} = P_{eq}$) of the Markov process.

To prove*) this claim we have to define a distance between ensembles. Suppose we have two ensembles E and E', each of which is a collection of many configurations. Denote the probability for configuration K_n in E or E' by P^n or P'^n respectively. Then we define the distance between E and E' as

$$|| E - E' || = \sum_n | P_n - P'_n | , \qquad \text{(II.7)}$$

where the sum is over all possible configurations. Now suppose that E' resulted from the application of transition probability W to ensemble E. We can compare the distance of E' from the Boltzman ensemble with the distance of E from the Boltzmann ensemble:

---

*) Here I follow closely Ref.[3], p.130.

$$|| E' - E_B || = \sum_n | \sum_m W_{nm}(P^m - P_B{}^m) |$$

$$\leq \sum_n \sum_m | W_{nm} (P^m - P_B{}^m) | \quad \text{(triangle inequality)}$$

$$= \sum_m | P^m - P_B{}^m | = || E - E_B || . \qquad \text{(II.8)}$$

(The last step is valid because of $\sum_n W_{nm} = 1$ and $W_{nm} \geq 0$.)

We conclude that the algorithm reduces the distance of an ensemble from the equilibrium Boltzmann ensemble. The inequality (8) is strict if W_{nm} never vanishes. This is easily seen: through the normalization $\sum P^n = \sum P_B{}^n = 1$ it follows that $P^k - P_B{}^k > 0$ implies $P^\ell - P_B{}^\ell < 0$ for some appropriate ℓ. Conseqently the contributions in $\sum_m W_{nm} (P^m - P_B{}^m)$ flip sign. Ergodicity implies $(W^k)_{nm} > 0$ for sufficiently large k. As $(W^k)_{nm}$ can be used in equation (8) as well as W_{nm} it follows that the distance to the Boltzmann ensemble will be reduced by applying the Markov process sufficiently often.

There are many ways to construct a Markov process satisfying i), ii) and iii). In practice most MC algorithms are based on products of steps, each satisfying the condition of <u>detailed balance</u>:

$$\text{iii')} \qquad W_{nm}\, e^{-\beta S_m} = W_{mn}\, e^{-\beta S_n}$$

Using the normalization $\sum_m W_{mn} = 1$ detailed balance immediately implies balance iii).

Detailed balance still does not uniquely fix the transition probabilities W_{nm}. The Metropolis algorithm is a popular choice because of its computational simplicity in practical applications. The Metropolis algorithm chooses:

$$W_{mn} = \begin{cases} 1 & \text{for} \quad S_m < S_n \\ e^{-\beta(S_m - S_n)} & \text{for} \quad S_m > S_n, \end{cases} \tag{II.9}$$

and W_{mn}/W_{nm} satisfies detailed balance.

There is an amazing flexibility in the choice of transition probabilities. In sections II.2 and II.3 I illustrate this fact by discussing in detail two examples.

II.2 Example 1: 2d O(3) σ-Model

The 2d O(3) σ-model is considered to be interesting because of its analogies with the 4d Yang-Mills theory [10]. Vacuum expectation values are calculated with respect to the partition function

$$Z = \int \prod_j ds_j \; e^{-\beta S(\{s_j\})} \quad . \tag{II.10}$$

The spins s_j are defined on the sites of a 2d torus with periodic boundary conditions. They are normalized to $(s_j)^2=1$ and the measure ds_j is defined by

$$\int ds_j = \frac{1}{4\pi} \int_{-1}^{+1} d\cos\theta_j \int_0^{2\pi} d\phi_j \; . \tag{II.11}$$

The action is

$$S = \sum_{\ell} \left(1 - s_{i(\ell)} s_{j(\ell)} \right) , \tag{II.12}$$

where the sum goes over all links ℓ of the lattice and $i(\ell), j(\ell)$ are the endpoints of link ℓ.

Each spin interacts only with its nearest neighbours. We would like to change a typical single spin s as depicted in Figure 1.

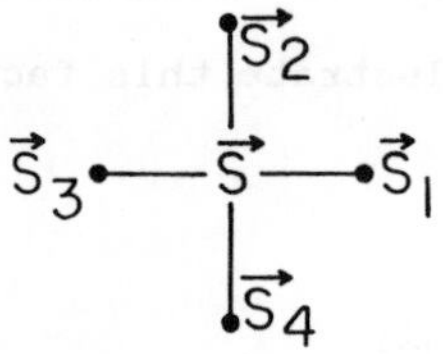

Figure 1

The sum of its neighbours is

$$S = \sum_{j=1}^{4} s_j$$

Spin s contributes $2 - sS$ to the action and the Metropolis algorithm yields the transition probabilites

$$W(s \to s') = \begin{cases} 1 & \text{for} \quad s'S > sS \\ e^{-\beta(sS - s'S)} & \text{for} \quad s'S < sS \end{cases} \tag{II.13}$$

Repeating the Metropolis algorithm again and again for the same spin s leads (as has been proven in the previous section) to the equilibrium distribution of this spin, which reads

$$P(s';S) = \text{const } e^{\beta s'S}$$

and

$$\int P(s';S)\, ds' = 1. \tag{II.14}$$

One would of course prefer to choose s' directly with the probability

$$W(s \to s') = P(s';S) = \text{const } e^{\beta s'S} \tag{II.15}$$

The algorithm, which creates this one variable limiting distribution directly, is called <u>heat bath</u> method. The heat bath algorithm is efficient, if the action is sufficiently simple to allow an explicit calculation of the probability $P(s';S)$.

Setting it up is an easy task in case of the O(3) σ-model. Let

$$\alpha = \sphericalangle(s',S), \quad x = \cos\alpha \quad \text{and} \quad \hat{S} = \beta S.$$

The Boltzmann weight becomes $\exp(x\hat{S})$ and the normalization constant follows from

$$\int_{-1}^{+1} e^{x\hat{S}}\, dx = \frac{2}{\hat{S}} \sinh(\hat{S}) .$$

Therefore the probability is

$$P(s';S) = P(x;\hat{S}) = \frac{\hat{S}}{2\sinh(\hat{S})} e^{x\hat{S}} . \qquad \text{(II.16)}$$

Computers are equipped with efficient pseudo random number generators, which generate uniformly distributed pseudo random numbers in the range $0 \leq x_\omega \leq 1$. We have to convert each proposed pseudo random number x_ω into a choice of $x = \cos\alpha$. This is done by interpreting x_ω as the shaded area of Figure 2:

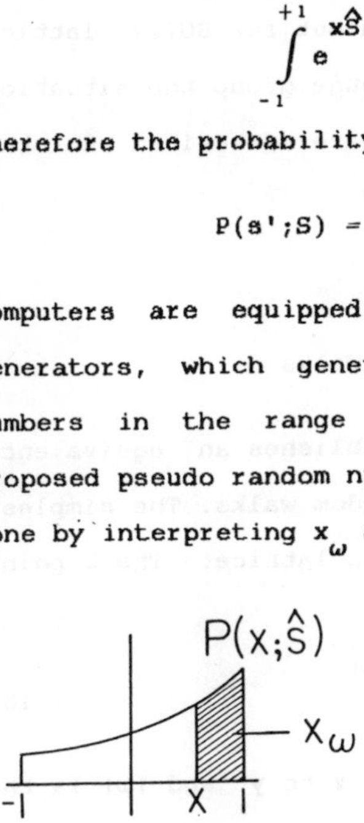

Figure 2

$$x_\omega = \int_{x}^{1} P(x';\hat{S})\, dx'$$

and it follows

$$x = \cos\alpha = \frac{1}{\hat{S}} \ln[\ \exp(\hat{S}) + x_\omega \exp(+\hat{S}) - x_\omega \exp(-\hat{S})\ ]. \qquad \text{(II.17)}$$

Next we have to fix the direction of s' in the plane orthogonal to S. This is done by choosing a random angle β, uniformly distributed in the range $0 \leq \beta \leq 2\pi$. $x = \cos\alpha$ and β completely

determine s' with respect to S. Before finally storing s' in the computer memory, we have to calculate the coordinates of s with respect to a cartesian coordinate system, which is globally defined for the whole lattice. The components of s with respect to this coordinate system are then kept in the computer memory.
A heat bath like method can also be worked out for SU(2) lattice gauge theory [11]. In case of the SU(3) gauge group the situation is more difficult. Various algorithms are described in the literature [12].

II.3 Example 2: Random Walks

The Symanzik polymer expansion [13] establishes an equivalence between field theories and interacting random walks. The simplest case is a free field theory on a hypercubic lattice. The 2-point function can be written as

$$G(x,y) = \sum_{\omega:x\to y} e^{-\beta|\omega|} \tag{II.18}$$

The sum goes over all random walks ω from x to y and $|\omega|$ is the length of walk ω. A critical point β_c exists such that the continuum free field theory is recovered in the limit $\beta\to\beta_c+0$.
I would like to describe in some detail a MC procedure [14] for simulating the partition function (18), because it illustrates the creative possiblities that one has in setting up transition probabilities. Two models on hypercubic d-dimensional lattices are considered. Model 1 allows spikes, see Figure 3, and in

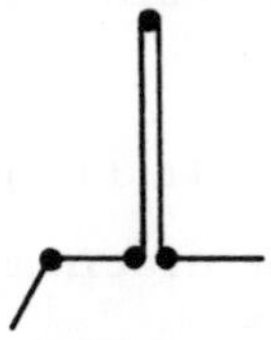

model 2 spikes are forbidden. An easy entropy argument gives $\beta_c = \beta_c^1 = \ln(2d)$ for model 1 and $\beta_c = \beta_c^2 = \ln(2d-1)$ for model 2.

Figure 3

The basic idea of the MC procedure described now is to shift single links orhtogonal to their own direction. This allows to implement constraints such as keeping the starting point x and the final point y fixed. We would like to shift a randomly choosen link. For the transition probabilities the change in length is crucial. One has therefore to classify all possible cases, which lead to different changes in the length of the random walk. Altogether there will be four distinct cases (including some subcases). Each case gets its own probabilities assigned. With the help of the next four Figures this is outlined now.

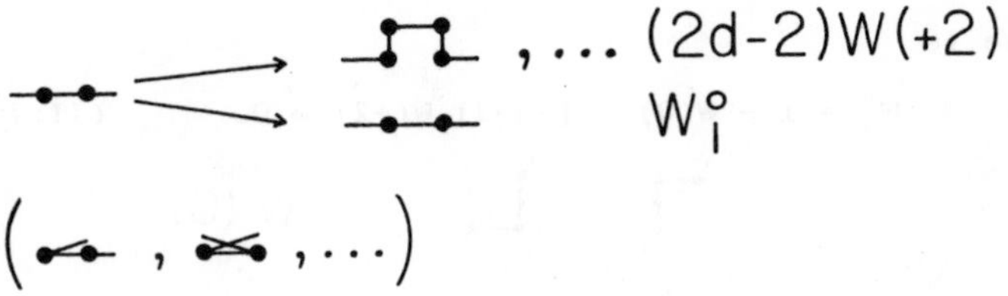

Figure 4.1

<u>Case 1)</u> The situation of Figure 4.1 allows (2d-2) shifts, each increasing the length of the path $|\omega|$ by +2 to $|\omega|+2$. We assign to each of these shifts the transition probability W(+2). The

probability that nothing happens (this means we reject doing a shift) is

$$W_1^0 = 1 - (2d-2)\, W(+2) \geqslant 0. \qquad (II.19.1)$$

We call W_1^0 a 'zero-shift probability'. Each of the four distinct cases of our classification will have its own zero-shift probability. (Subcases with identical shift probabilities are indicated in parenthesis in Figure 4.1 and subsequent figures.)

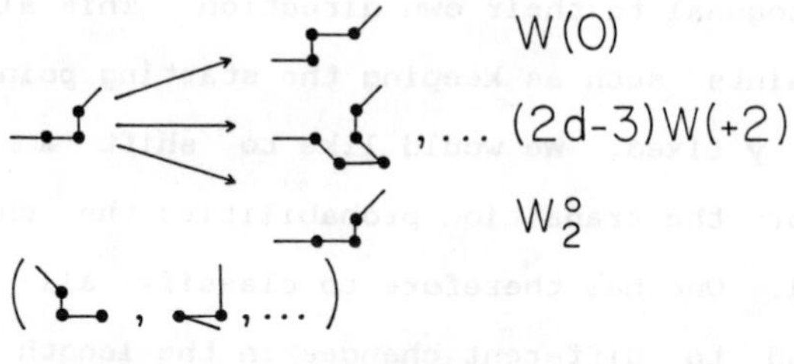

Figure 4.2

Case 2) Figure 4.2 allows one shift, which does not change the length $|\omega|$ of the walk ω. This shift becomes the transition probability W(0) assigned. Furthermore (2d-3) shifts with probability W(+2) are possible. Hence the zero shift probability in this case is

$$W_2^0 = 1 - W(0) - (2d-3)\, W(+2) \geqslant 0. \qquad (II.19.2)$$

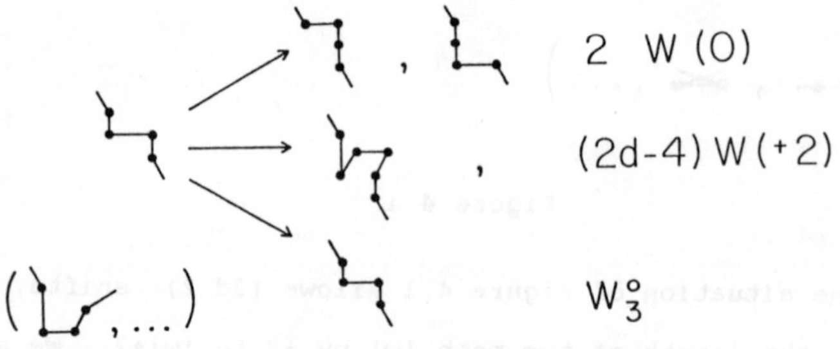

Figure 4.3

Case 3) There are two shifts which keep $|\omega|$ constant and (2d-4) shifts increasing $|\omega|$ by +2. Consequently the zero-shift probability is

$$W_3^0 = 1 - 2\,W(0) - (2d-4)\,W(+2) \geq 0. \qquad \text{(II.19.3)}$$

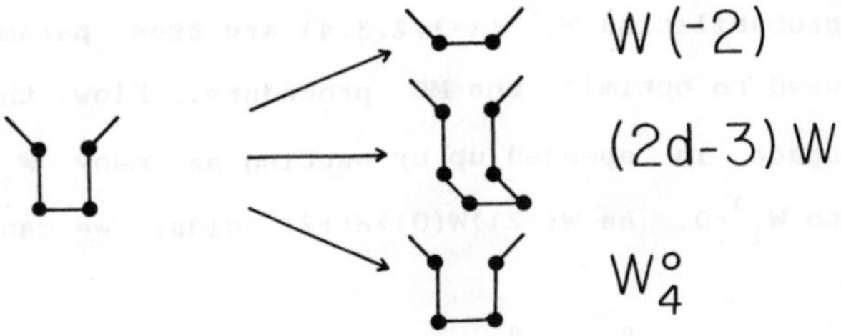

Figure 4.4

Case 4) In this (final) case there is one shift allowing to shrink the length by -2 from $|\omega|$ to $|\omega|-2$. This shift becomes the probability W(-2) assigned. The zero-shift probability becomes:

$$W_4^0 = 1 - w(-2) - (2d-3)\,W(+2) \geq 0. \qquad \text{(II.19.4)}$$

For $d \geq 3$ all walks $\omega: x \to y$ can be constructed by carrying out subsequent shifts. This implies ergodicity.

Let us now set up detailed balance. The probability for randomly hitting a certain link is $|\omega|^{-1}$. Therefore detailed balance iii') reads:

$$\frac{W(n)}{|\omega|}\, e^{-\beta|\omega|} = \frac{W(-n)}{|\omega|+n}\, e^{-\beta(|\omega|+n)} \quad , \ (n=0,2) \qquad \text{(II.19.5)}$$

We get rid of the ugly factor $\frac{|\omega|}{|\omega|+n}$ by simulating instead of (18) the modified partition function

$$G(x,y) = \sum_{\omega:x\to y} |\omega|\ e^{-\beta|\omega|} \qquad \text{(II.20)}$$

and detailed balance becomes

$$\frac{W(n)}{W(-n)} = e^{-\beta n} \qquad (n=0,2). \qquad \text{(II.21)}$$

The zero-shift probabilities W_i^0 (i=1,2,3,4) are free parameters which can be used to optimize the MC procedure. Flow through configuration space is speeded up by setting as many W_i^0 as possible equal to $W_i^0=0$. As W(-2)>W(0)>W(+2) holds, we can only choose

$$W_3^0 = W_4^0 = 0 \quad . \qquad \text{(II.22)}$$

We are left with three equations (19.3, 19.4 and 21) determining uniquely the three unknown transition probabilies. Introducing $\kappa=e^{-2\beta}$ the solutions read:

$$W(+2) = \frac{\kappa}{1+(2d-3)\kappa} \quad , \qquad \text{(II.23.1)}$$

$$W(0) = \frac{1}{2}\,\frac{1+\kappa}{1+(2d-3)\kappa} \quad , \qquad \text{(II.23.2)}$$

$$W(-2) = \frac{1}{1+(2d-3)\kappa} \quad , \qquad \text{(II.23.3)}$$

The possible bottleneck of the procedure is W(+2), despite the fact that in accordance with equations (19.4) and (21) W(+2) is optimized.

In model 2 we have to reject transitions which would generate spikes. Such a rejction does not affect detailed balance, but adds only to the zero-shift probabilites. As a net result the speed of moving through configuration space slows down.

Let me emphasize a pitfall. The procedure as outlined so far is wrong for model 1, because detailed balance is violated in this case. The problem is caused by spikes: The transition indicated in Figure 5 has no inverse! In practice a way out is to discard

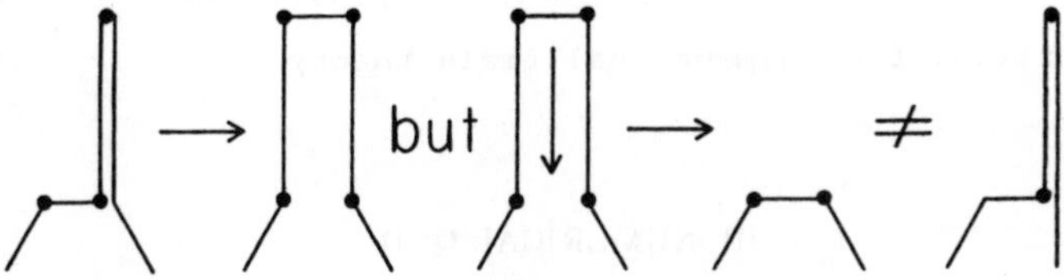

Figure 5

all transitions without inverse. Again this adds only to the zero-shift probabilities without affecting detailed balance. A more efficient but complicated solution would be to add new transitions.

By comparing with exact results, we can convince ourselves that the Monte Carlo procedure realy works. This is demonstrated in Figure 6, Ref. [14], where MC results for the average lenght of a walk are plotted. The full line is the analytic result. There are a large number of numerical investigations on random walks using a variety of different methods. The emphasis is not only on field theory but as well on polymer physics. The literature is best traced back from Ref.[15].

The MC method of section II.2 has also been worked out for random surfaces. This is of some interest, because a well-known conjecture is that gauge theories have a simple formulation in terms of interacting random surfaces. For a review on the subject see Ref.[17]. Very recent numerical results [18] yield an anomalous dimension for the two point function in a lattice surface model, which is the analog to random walks without spikes. In case the model is well-defined this result implies a non-trivial underlying field theory and random surfaces without spikes may be a good starting point for trying an existence proof for a non-trivial four dimensional field theory.

III. NUMERICAL QCD

To answer the question "Whether and to what extent is QCD the theory of strong interactions?" requires non-perturbative methods, because it is at present-day energies impossible to disentangle perturbative and non-pertrubative QCD predictions. Basic assumptions of QCD like

- confinement
- spontaneous chiral symmetry breaking
- dynamical mass generation

are of non-perturbative nature. Monte Carlo calculations allow in principle to obtain quantitative non-perturbative results. Unfortunately statistical and systematical errors are a severe problem. For this reason the major gain up to now is a better qualitative understanding of what is going on in dynamical QCD. Various reviews and summaries of the subject exist [4-8] and here I will only give an introduction to the literature on various subtopics.

III.1 String Tension

Historically it was, after the pioneering work by Wilson [19], the paper by Creutz [11] on the SU(2) string tension which tremendously stimulated the interest in numerical simulations of lattice QCD. Numerically, however, his string tension estimate did not survive later criticism.

Estimates for the square root of the string tension, $\sqrt{\kappa}$, are best given in units of Λ_L:

$$\sqrt{\kappa} = \text{const } \Lambda_L \tag{III.1}$$

In these units the original result of Creutz for the SU(2) string tension reads $\sqrt{\kappa} = (78 \pm 12)\Lambda_L$, whereas the most recent SU(2) estimates (for instance Ref.[20,21]) are of order $\sqrt{\kappa} = (54 \pm 5)\Lambda_L$.

In the case of the SU(3) gauge group the improvement is even more drastic. The first estimates [22] were of of order $\sqrt{\kappa} \approx 190\Lambda_L$, whereas the most recent calculations (for instance Ref.[23]) give $\sqrt{\kappa} \approx 90\Lambda_L$.

As always in lattice gauge theory an open problem is to include the effect of virtual quark pairs. The first step would be testing better methods for including the fermion determinant.

III.2 Glueballs

Glueballs are the mass spectrum states of pure lattice gauge theories. For a detailed review of lattice glueball calculations, see Ref.[24]. So far all calculations agree in finding

$$m(0^{++}) = \text{const } \Lambda_L \tag{III.2}$$

to be the lowest glueball (mass gap). Only for this state is an unambiguous cross-over from strong coupling to [asymptotic(?)] scaling seen. Recent estimates are of order $m(0^{++}) = (190 \pm 30)\Lambda_L$ for the SU(2) gauge group [21,25] and of order $m(0^{++}) = (270 \pm 30)\Lambda_L$ for the SU(3) gauge group [26,27]. Most results rely on a Monte Carlo variational method as first tested in Ref.[28]. Unfortunately correlations disappear into statistical noise at the rather short distance $t \geq 3$. Source method calculations (see [27] and references given there) obtain signals up to a much larger distance, but have other disadvantages: a) The upper bound property of the obtained mass gap results is lost and b) the mean value of the internal energy enters as an additional new free parameter.

Investigating correlations between non-local operators (adjoint Polyakov loops in the investigated [21] SU(2) case) gives a dramatic improvement for appropriate lattices and β -values. It is presently under investigation, whether similar results can be obtained for the SU(3) gauge group.

Various results on exited glueball states

$$0^{PC},\ 1^{PC},\ 2^{PC} \text{ and } 3^{PC} \text{ with } P = \pm \text{ and } C = \pm$$

do also exist. They rely almost exclusivly on the MC variational method and are plagued by large statistical noise. As the signal disappears at very small distances, it is even difficult to decide whether the masses of these states are indeed high (in units of m(0^{++})) or wether just the trial wave functions are bad. More details can be found in Ref.[24,29].

III.3 Finite Temperature Phase Transition

A finite physical temperature is introduced on the lattice by taking (at fixed β) the time direction extension N_T of the Euclidean lattice to be finite, while sending $N_X = N_Y = N_Z \mapsto \infty$ An order parameter is the vacuum expectation value $< L_T >$ of the Polyakov loop. The Polyakov loop is a Wilson loop which is closed by means of periodic boundary conditions in the time direction. The critical temperature T_c separates the confinement phase ($< L_T >= 0$) from a supposed plasma phase of free quarks and gluons ($< L_T > \neq 0$). Polyakov loops are suitable for MC calculations [30]. In the case of the SU(3) gauge group one finds a first order phase transition and recent estimates [31] give

$$T_c \approx 55\ \Lambda_L. \tag{III.3}$$

Alternatively, one may study singularities in thermodynamic variables like the average plaquette action, the gluonic energy density, etc. [31]. This is particularly useful, if one studies the influence of fermions on the finite temperature phase transition. Much exploratory work has already been done on this subject (for partial references see [33]). The investigations agree on finding a rapid crossover in the region of the suspected finite temperature phase transition. Unfortunately the question, whether this is still a first order phase transition or a second order phase transition or no phase transition, is a difficult quantitative one and controversial. Including light quarks raises the critical temperature to the order of magnitude $T_c \approx 100\ \Lambda_L$. In physical units this may well be above 400 Mev, causing some problem seeing signatures of the phase transition in heavy ion collisions. Future work may concentrate on applying improved fermion methods to the problem.

III.4 Monte Carlo Renormalization Group

String tension and deconfinement temperature exhibit an overshooting of asymptotic scaling in the region $\beta \approx 6$. It would be of interest to know whether this behaviour is universal and already reflecting scaling in the non-asymptotic sense. MC renormalization group investigations [34,19] aim to follow the scaling to larger values of β and may therefore link the region of present MC calculations to the really asymptotic one.
Recently various groups [35] concentrated on SU(3) gauge theory. Combined results look are displayed in Figure 6.

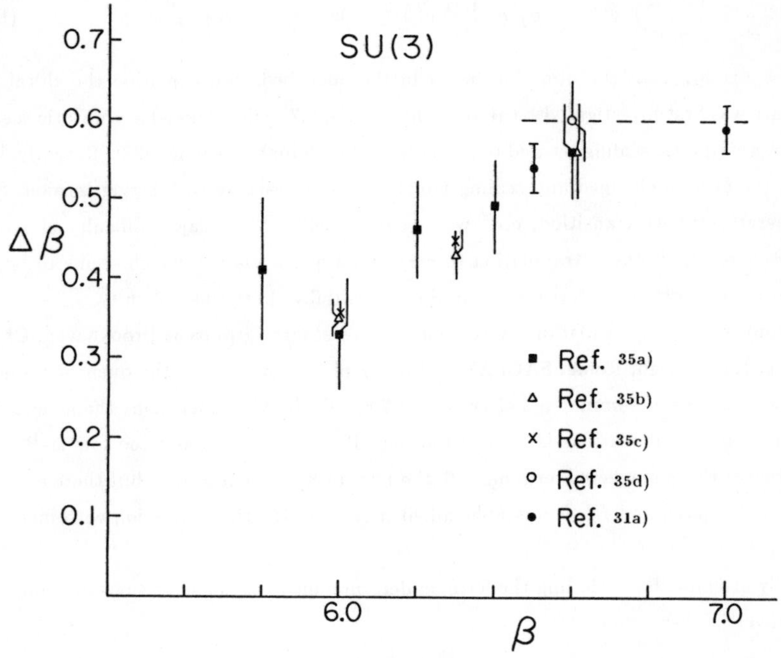

Figure 6

$\Delta\beta$ is the shift in the fundamental plaquette coupling of the Wilson action corresponding to a change of scale by a factor of two. Estimates from string tension and deconfinement temperature are in agreement with the MC renormalization group results. Adding more results from other authors somewhat spoils the picture. This may be either due to numerical inaccuracy or due to non-universal behaviour in the region $\beta \approx 6$.

III.5 Chiral Symmetry Breaking and Hadronic Spectrum

Chiral symmetry breaking is of central importance for understanding the QCD spectrum (see, for instance, Ref.[36]) One has

$$< \bar{\psi}\psi >= n_f \ \alpha^{-4/11} \ a(\beta)^3 \ < \bar{\psi}\psi >_{inv}, \quad \alpha = \frac{g^2}{(4\pi)} \tag{III.4}$$

and $< \bar{\psi}\psi >_{inv}$ has the dimesion $[m^3]$. In the quenched approximation the chiral condensate was first investigated by Hamber and Parisi [37], who claimed asymptotic scaling. Later, asymptotic scaling seemed to be confirmed with high precission [38]. Recently, however, the fashion changed and scaling violations in accordance with string tension, finite temperature phase transition, etc., were reported [40]. The major difficulty of reliable $< \bar{\psi}\psi >$ results is the extrapolation to very small quark masses, which seems to be ambiguous. All hadronic spectrum calculations essentially share this difficulty.

Hadronic spectrum calculations were done by several large groups at Brookhaven, CERN, DESY, Edingburgh, Rome, SACLAY and other places. Results in the quenched approximation have been summarized elsewhere [4,7,8]. With Wilson fermions the general tendency of these calculations is toward too high P/ρ mass ratio and too low Δ-P mass difference. Using staggered fermions [39] the P/ρ mass ratio improves, but there are problems with scaling and F_π is unstable, albeit moving in the right direction with increasing β.

Finally attempts for including the fermion determinant in such calculations are subject to controversial discussions [41].

In conclusion the obtained MC results are consistent with QCD being the theory of strong interactions. Precise results will, however, require very serious large scale experiments and considerable effort on inventing and testing new methods.

IV. DISCRETE EUCLIDEAN QUANTUM GRAVITY

It is well-known that classical gravity is inconsistent with quantum principles. A simple argument (here I follow closely the presentation in the last paper of Ref.[9]) relies on the uncertainty relation. Let us consider two points A and B, separated by a spatial distance $\triangle x$ and a time difference $\triangle t$ such that

$$\triangle t < \triangle x < l_p \sim 10^{-33}\text{cm}$$

$\triangle t < \triangle x$ insures us that A and B are outside each other's light cone and local quantum field theory states that experiments at A and B can be done independently of each other. Yet, based on the uncertainty principle, we expect a fluctuation of energy $\triangle E \sim \frac{1}{\triangle t} > \frac{1}{l_p}$, and the gravitational field associated with such a fluctuation $\triangle E$ is very strong at small distances. Indeed, its Schwarzshild radius, i.e. the black hole radius, is

$$R \sim G\triangle E > \frac{G}{l_p},$$

where $G = l_p^2$ is Newton's gravitational constant. Thus we find $R > \triangle x$ and it is clearly impossible that classical gravity still applies to the points A and B, because there is no classical recovery from a black hole. Also it is quite unreasonable that measurements at A and B could be viewed as independent and the concept of locality very likely breaks down at such distances. Classical space-time does not exist anymore and we have quantum fluctuations in space and time.

Little is known, however, about quantum gravity. A major reason is that the gravitational coupling is dimensionful and "naive" quantization of the classical action leads to a non-renormalizable theory. A fundamental length [9] would, however, provide a natural ultraviolett cut-off and may remedy these problems. But the classical action is also unbounded, which seems to be a further obstacle for a successful quantization and in particular for any functional integral treatment. On the other hand the validity of classical gravity is well-established over many orders of magnitude in length scale. In view of the fact that experimantal guidance about gravitational quantum effects does not exist it seems to be a formidable task to construct a better action out of thin air. The aim of my numerical investigation [42] is not to solve quantum gravity, but to get a better hold on the diffuculties arising in attempting to quantize the classical Einstein action in its discrete

Regge version [43]. A first result of my MC simulation is that entropy might cure the problem of the unboundedness of the action.

On a Regge skeleton space the Einstein action becomes [43,9]

$$S = \sum_t \alpha_t A_t \tag{IV.1}$$

The sum goes over all triangles of the 4d skeleton space. A_t is the area of triangle t and α_t the associated deficit angle. Following Ref. [44] the theory is formulated in the Euclidean. This is not really satisfactory, because the relation to the Minkowskian theory is not well understood. For my exploratory MC study the hope is that qualitative features of quantum gravity are not destroyed by using the Euclidean formulation and the particular measure as defined below.

We would like to calculate vacuum expectation values with respect to the partition function

$$Z = \int_V D[space]\; e^{m_p^2 S}. \tag{IV.2}$$

Here m_p is (up to a constant which is irrelevant in our context) the Planck mass and the sign of the action is taken to give negative modes and zero modes for small fluctuations around the original flat space configuration [45,9]. The measure $D[space]$ is choosen to be

$$D[space] = \prod_l d\log(x_l) \prod_p F_\theta(x_{p_1}, ..., x_{p_{10}}) \tag{IV.3}$$

Here x_l is the length of link l, $\prod_l$ extends over all links l and $\prod_p$ over all pentahedra p (= 4-simplexes) of the Regge skeleton. The function $F_\theta(x_1, ..., x_{10})$ takes care of the pentahedra constraints. $F_\theta = 1$ if a flat pentrahedron can be constructed from the ordered links of length $x_1, ..., x_{10}$, and otherwise $F_\theta = 0$. The choosen measure is the simplest [42] allowing simulations within a fixed finite volume V. For the sake of having a tractable model this is sufficient. However, to simulate real gravity one should try to derive the discrete measure from the continuum measure [44] along similar lines as Friedberg and Lee [9] did when deriving the Regge action from the continuum action.

To set the length scale one needs an analog of the lattice spacing a in normal lattice theories. This is done by defining the expectation value of one length to be constant. For instance:

$$l_0 = (v_0)^{\frac{1}{4}} \quad \text{with} \quad v_0 =< v_p >, \quad V = N_p v_0 \tag{IV.4}$$

v_p is the volume of pentahedron p, and l_0 is the introduced length scale. The choice of l_0 is satisfactory if other obvious choices are equivalent, such as

$$l'_0 = < x_l > \quad \text{and} \quad l''_0 = \sqrt{< A_t >} \quad \text{etc.} \tag{IV.5}$$

Equivalent means $l'_0 = c' l_0$ and $l''_0 = c'' l_0$ in the limit of an infinite number of links. In other words it means having canonical dimensions.

MC calculations were carried out for the hypercubic model [45] and a simplicial variant of it. This limitation to regular structures is entirely due to the increased technical problems of simulating a random lattice [46], which ideally should be used, because it may be of fundamental importance for discrete gravity [9]. The regular models used can be considered as qualitative approximations to the random lattice. The results for both models considered are very similar and in this representation I will entirely refer to the hypercubic model. For the numerical simulations of Ref.[42] systems of size $N = 2^4$ and $N = 3^4$ at $m_p^2 = 0, \pm 0.3$ were used. Measurements were performed after each sweep. Meanwhile new results at various m_p^2 values were obtained [47].

$m_p^2 = 0$ results are entirely due to the entropy of the measure (3). Finite size effects are small and canonical dimensions (5) are strongly supported. Negative expectation values are obtained for the average action $< S >$ and in the infinite system limit

$$< S > \approx -.280 \tag{IV.6}$$

is indicated.

Of major interest is analyzing $< S >$ as a function of m_p^2. Unfortunately $m_p^2 \neq 0$ slows down the computational speed by a factor of order ≈ 40. Therefore only the values $m_p^2 = \pm 0.3$ were investigated with high statistics in Ref.[42]. Now the more detailed investigation [47] collects high statistics at varying m_p^2-values. Figure 7 is representative of these new results. The system size is 2^4. The upper curve is the expectation value $(< S > +1)10$, where the number 1 has been added for convenience to allow a logarithmic scale. The lower curve gives the expectation value $< A_t >$ of triangles and is closely correlated with $< S >$. (In contrast the average deficit angle is strongly anticorrelated. It stays negative and takes its largest negaitve values for the largest positive action values.) A phase transition between an entropy dominated and an energy dominated region is supported by the response of $< S >$ and $< A_t >$ to changes of m_p^2. In the entropy dominated region the action density $< S >$ stays stable and finite, whereas in the energy dominated region the action density

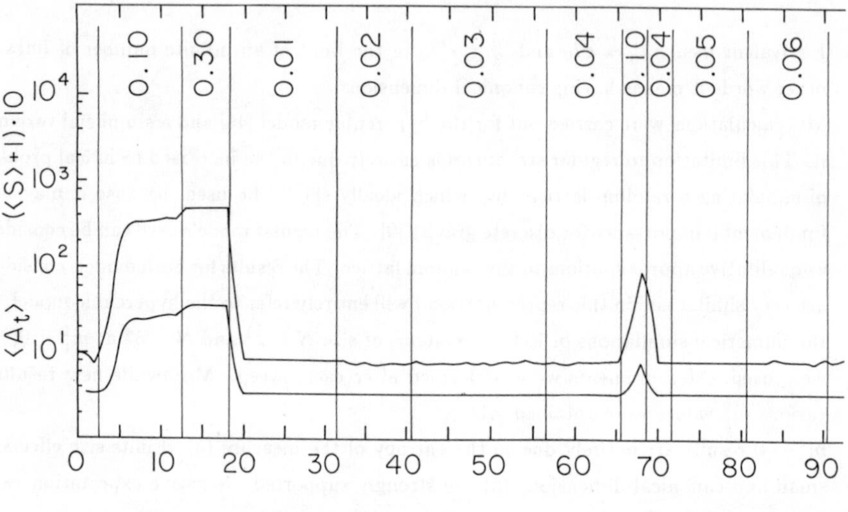

⟨A_t⟩ ⟨S⟩+1) 10
10^4
10^3
10^2
10^1
0.10
0.30
0.01
0.02
0.03
0.04
0.10
0.04
0.05
0.06
0 10 20 30 40 50 60 70 80 90
0.07
0.08
0.07
0.06
0.05
0.04
0.03
0.02
0.02
0.01
100 110 120 130 140 150 160 180 170 190

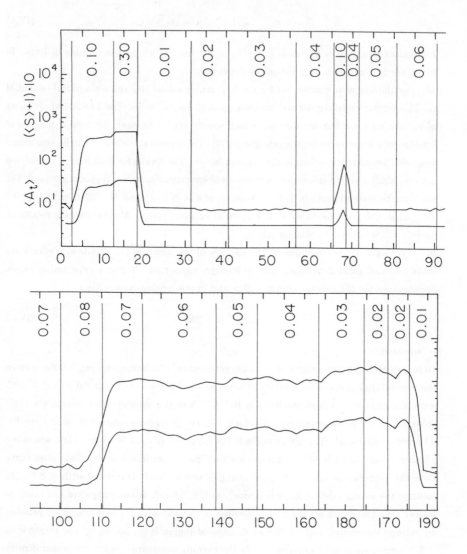

Figure 7

ends up in metastable states. In the latter region careful variations of m_p^2 seem to allow $< S >$ an arbitrarily large value, reflecting the unboundedness of the Regge-Einstein action. Many metastable states are discovered, preventing really approaching the limit of an infinite action density. The situation reminds one somewhat of structures encountered in spin glass models.
A crude estimate of the critical m_p^2 is

$$0.01 < m_{p_c}^2 < 0.08 \tag{IV.7}$$

This estimate is based on observing $< S >$ approach a (large) metastable value for $\beta \geq 0.08$ and decreasing from there down to stable values by taking $\beta \leq 0.01$. The bounds are rather bad due to problems with metastability. But the ability to come down from high action densities to low action densities without setting $m_p^2 = 0$ is strong evidence for the two phase structure. At a not too high action value the program found also at $m_p^2 = 0.04$ back to the entropy dominated region. It is amazing to note that within the numerical uncertainties

$$< S >_c \approx 0.$$

Some consistency checks were performed for finite size effects on 3^4 lattices. CPU time became, however, a serious problem and therefore the finite size investigation remained rather inconclusive. Despite the small size of the investigated system, I would like to argue that the indicated phase transition is a real effect surviving the large system limit. Each hypercube of the gravity model has already a rich substructure of 24 pentahedra and the exhibited relevance of entropy for the small sized lattice certainly demonstrates that entropy is not a neglegible factor for more reasonably sized lattice.
In conclusion, quantum gravity is an exciting, although speculative, topic for numerical simulations. New qualitative insight into the physics of gravitational quantum fluctuations may be gained.

Acknowledgements:
I would like to thank the organizers of the GIFT seminar for creating a very pleasant meeting. I am indebted to T.D. Lee for encouragement and a useful discussion on quantum gravity. Finally I appreciated Dennis Duke's comments on the manuscript.

References

(1) S.N. Metropolis, A.W. Rosenbluth, M.N. Rosenbluth, A.H. Teller and E. Teller, J. Chem. Phys. 21 (1953) 1087

(2) K. Binder, in "Phase Transitions and Critical Phenomena", C. Domb and M.S. Green, eds., Vol.5B (Academic Press, NY 1976)

(3) M. Creutz, "Quarks and Gluons on a Lattice", Cambridge University Press 1983

(4) B. Berg, Nucl. Phys. A434 (1985) 151c

(5) J.B. Kogut, Rev. Mod. Phys. 55 (1983) 775

(6) J.M. Drouffe and J.B. Zuber, Phys. Reports 102 (1983) 1

(7) M. Creutz, L. Jacobs and C. Rebbi, Phys. Reports 95 (1983) 201

(8) P. Hasenfratz, Ericé Lectures 1984;
A. Hasenfratz and P. Hasenfratz, Preprint, FSU-SCRI-85-2

(9) R. Friedberg and T.D. Lee, Nucl. Phys. B242 (1984) 145;
G. Feinberg, R. Friedberg, T.D. Lee and H.C. Ren, Nucl. Phys. B245 (1985) 343;
T.D. Lee, "Difference Equations as the Basis of Fundamental Physics Theories", Preprint, CU-TP-297;

(10) A. Migdal, Sov. Phys.-JETP 42 (1976) 413, 743;
L. Kadanoff, Ann. Phys. 100 (1976) 359;
A.A. Belavin and A.M. Polyakov, JETP Lett. 2 (1975) 245

(11) M. Creutz, Phys. Rev. D21 (1980) 313

(12) E. Pietarinen, Nucl. Phys. B190[FS3] (1981) 349;
P. Lisboa and C. Michael, Phys. Lett. 113B (1982) 303;
N. Cabibbo and E. Marinari, Phys. Lett. 119B (1982) 387;
D. Barkai, K.J.M. Moriarty and C. Rebbi, Comp. Phys. Comm. 32 (1984) 1;
A.D. Kennedy, J. Kuti, S. Meyer and J. Pendleton, J. Comp. Phys. (to be published)

(13) K. Symanzik, Varenna Lectures 1969

(14) B. Berg and D. Foerster, Phys. Lett. 106B (1981) 323

(15) A. Berretti and A. D. Sokal, Preprint, Courant Institute of Mathematical Research, May 1985

(16) B. Berg and A. Billoire, Phys. Lett. 297B (1984) 297;
B. Berg, A. Billoire and D. Foerster, Nucl. Phys. B251 (1985) 665

(17) J. Fröhlich, Cargèse Lectures 1983

(18) B. Baumann and B. Berg, Preprint, DESY-85-078

(19) K. Wilson Phys. Rev. D10 (1974) 2445;
Ericé Lectures 1975;
Cargèse Lectures 1979

(20) A. Billoire and E. Marinari, Phys. Lett. 128B (1984) 399;

(21) B. Berg and A. Billoire, Preprint, DESY-85-082

(22) M. Creutz, Phys. Rev. Lett. 45 (1980) 313;
E. Pietarinen, Nucl. Phys. B190 [FS3] (1981) 349;

(23) G. Parisi, R. Petronzio and F. Rapuano, Phys. Lett. 128B (1983) 418;
D. Barkai, K. J. M. Moriarty and C. Rebbi, Phys. Rev. D30 (1984) 1293;
S. Otto and J. Stack, Phys. Rev. Lett. 52 (1984) 2328;

(24) B. Berg, Cargèse Lectures 1983

(25) B. Berg, A. Billoire, S. Meyer, and C. Panagiotakopoulos, Commun. Math. Phys. 97 (1985) 31;

(26) B. Berg and A. Billoire, Nucl. Phys. B221 (1983) 109; B226 (1983) 405;

(27) Ph. de Forcrand, G. Schierholz, H. Schneider and M. Teper, Phys. Lett. 148B (1984) 140

(28) B. Berg, A. Billoire and C. Rebbi, Ann. Phys. (NY) 142 (1982) 185; 146 (1983) 470;
M. Falcioni, E. Marinari, M. L. Paciello, G. Parisi, F. Rapuano, B. Taglienti and Zhang Yi-Cheng, Phys. Lett. 110B (1982) 295;
K. Ishikawa, G. Schierholz and M. Teper, Phys. Lett. 110B (1982) 399

(29) H. Kamenzki and B. Berg, Phys. Rev. D, to appear

(30) L.D. McLerran and B. Svetitsky, Phys. Lett. 98B (1981) 195;
J. Kuti, J. Polonyi and K. Szlachanyi, Phys. Lett. 98B (1981) 301

(31) a) T. Celik, J. Engels and H. Satz, Z. Phys. C22 (1983) 301;
b) A. D. Kennedy, J. Kuti, S. Meyer and B. J. Pendelton, Phys. Rev. Lett. 54 (1985) 87

(32) J. Engels, F. Karsch, I. Montvay and H. Satz, Nucl. Phys. B205 [FS5] (1982) 545

(33) P. Hasenfratz, F. Karsch and I. O. Stamatescu, Phys. Lett. 133B (1983) 221;
T. Celik, J. Engels and H. Satz, Phys. Lett. 133B (1983) 427;
J. Polonyi, H.W. Wyld, J. B. Kogut, J. Shigemitsu and D. K. Sinclair, Phys. Rev. Lett. 53 (1984) 644;
Nucl. Phys. B251 [FS13] (1985) 333

(34) S.K. Ma, Phys. Rev. Lett. 37 (1976) 461;
L.P. Kadanoff, Rev. Mod. Phys. 49 (1977) 267;
R.H. Swendsen, Phys. Rev. Lett. 42 (1979) 859;
R.H. Swendsen, in Real Space Renormalization, Topics in Current Physics, Vol. 30, edited by Th. W.

Burkhardt and J.M.J. van Leeuwen (Springer, Berlin, 1982)

(35) a) A. Hasenfratz, P. Hasenfratz, U. Heller and F. Karsch, Phys. Lett. 143B (1984) 193;
b) K.C. Bowler, A. Hasenfratz, P. Hasemfratz, U. Heller, F. Karsch, R. D. Kenway, I. Montvay, G. S. Pawley and D. J. Wallace, Nucl. Phys. B257 [FS14] (1985) 155l;
c) K.C. Bowler, F. Gutbrod, P. Hasenfratz, U. Heller, F. Karsch, R.D. Kenway, I. Montvay, G.S. Powley, J. Smit and D.J. Wallace, to be published;
d) R. Gupta and A. Patel, Preprint, CALT-68-1142 (1984);
R. Gupta, G. Guralnik, A. Patel, T. Warnock and C. Zemach, Preprint, UCSD-10p10-245, June 1985

(36) J. Gasser and H. Leutwyler, Physics Reports 87 (1982) 77

(37) H. Hamber and G. Parisi, Phys. Rev. Lett. 47 (1981) 1792

(38) I. M. Barbour, P. Gibbs, J. P. Gilchrist, G. Schierholz and M. Teper, Phys. Lett. 136B (1984) 80

(39) H. Hamber, Phys. Rev. D31 (1985) 586;
I. M. Barbour, Talk given at "Advances in Lattice Gauge Theorie", Tallahassee, April 1985

(40) K. Kogut and L. Susskind, Phys. Rev. D11 (1975) 395;
H. S. Sharatchandra, H. J. Thun and P. Weisz, Nucl. Phys. B220 [FS8] (1983) 447;
H. Kluberg-Stern, A. Morel, O. Napoly and B. Petersson, Nucl. Phys. B220 [FS8] (1983) 447;

(41) D. Weingarten, Nucl. Phys. B257 [FS14] (1985) 629

(42) B. Berg, Phys. Rev. Lett. 55 (1985) 904

(43) T. Regge, Nuovo Cimento 19 (1961) 558

(44) S.W. Hawking, in "General Relativity", edited by S.W. Hawking and W. Israel (Cambridge Univ. Press, England, 1979)

(45) M. Rocek and R.M. Williams, Z. Phys. C21 (1984) 171

(46) N.H. Christ, R. Friedberg and T.D. Lee, Nucl. Phys. B202 (1982) 89; and B210 [FS6] (1982) 337

(47) B. Berg, work in preparation

Applications of Topological and Differential Geometric Methods to Anomalies in Quantum Field Theory †

Paul Ginsparg

Lyman Laboratory of Physics

Harvard University

Cambridge, MA 02138

1. **Gauge theories**
   A. Abelian (axial) anomaly
   B. Non-abelian (gauge) anomaly
   C. Formalism: differential forms, index theorems, etc.
   D. Topological interpretation of gauge anomalies

2. **Gravitational theories**
   A. Formalism: vielbeins, curved space index theorems, etc.
   B. Gravitational anomalies
   C. Green-Schwarz superstring
   D. Axionic interpretation

3. **Sigma-models**
   A. Formalism: coset spaces
   B. Isometry anomalies
   C. Holonomy anomalies

Introduction

My intent in these three lectures is to give a pedagogical account of recent developments in the study of anomalies in quantum field theory. I hope to exemplify in a reasonably self-contained manner the many guises in which the same basic anomaly may appear. My approach here will follow rather closely the development in [1]–[3]. This is not to say that these references are either the first or the last word on any of the topics treated here, only that they follow the approach with which I am most comfortable, and I refer you to them for further references to the original and contemporary literature.

---

† supported in part by NSF contract PHY-82-15249

In general, by an anomaly we mean that some symmetry present in the classical action of a theory is not preserved by the full quantum theory. Historically, the original axial anomaly[4] emerged through consideration of a fermion triangle diagram with one axial current and two vector currents. Imposing current conservation and Bose symmetry in the vector channels, one finds that the axial current is not conserved, instead we have $\partial_\mu j_5^\mu \sim F\tilde{F}$, and chiral symmetry is broken in the presence of gauge fields coupled to the conserved vector currents. This breakdown forms the basis of our understanding of π^0 decay and the resolution of the $U(1)$ problem[5]. Replacing the vector currents by energy-momentum tensors in the triangle diagram and imposing energy-momentum conservation in both energy-momentum channels similarly leads[6] to a non-conserved axial current, and a breakdown of chirality in the presence of an external gravitational field.

The anomalous triangle diagram with gauge fields coupled to $V-A$ currents at each vertex also has important consequences. Unless the anomalies cancel when summed over different fermion species, the non-conservation of the $V-A$ currents constitutes a breakdown in gauge invariance, rendering the theory inconsistent. The conditions for anomaly cancellation then provide useful constraints for model building[7][8]. Replacing the gauge currents by energy-momentum tensors leads[9] to a so-called gravitational anomaly, and if uncancelled gives a theory with a quantum breakdown in the conservation of energy-momentum. Anomalies have also proven useful in providing constraints on the possible spectra of composite models by requiring certain matching conditions on Green's functions of flavor currents[10].

The anomaly encountered in each instance above is related to a chiral, i.e. left-right, asymmetry of the theory. Since nature appears to be chirally asymmetric at some fundamental level, the philosophy we take here is that firming our understanding of anomalies in general and the way certain theories manage to avoid them in particular should prove useful in the long run for understanding physics.

1. Gauge theories

A. We begin by describing the relation between the standard axial anomaly and the Atiyah-Singer index theorem[11]. It is most easily exhibited by means of Fujikawa's method[12] for calculating anomalies. For a massless Dirac fermion ψ defined on a $2n$-dimensional flat euclidean space with coordinates x^μ and coupled to an external gauge field A_μ with gauge group G, the fermion effective action functional $\Gamma_{\rm eff}[A]$ is obtained by performing the functional integral over fermion fields

$$e^{-\Gamma_{\rm eff}[A]} = \int_{\psi,\overline{\psi}} e^{-\int d^{2n}x\, \overline{\psi} \not{D} \psi} \tag{1.1}$$

where $\not{D} = D_\mu \gamma^\mu = (\partial_\mu + A_\mu)\gamma^\mu$ (Our dirac matrices are taken to be hermitian, $\gamma^\mu = \gamma^{\mu\dagger}$, with $\{\gamma^\mu, \gamma^\nu\} = 2\delta^{\mu\nu}$. The gauge field $A_\mu = A^a_\mu \lambda^a$, on the other hand, is anti-hermitian with λ^a anti-hermitian generators of the lie algebra of G. In the next section, we shall generalize the Dirac operator to curved manifolds. It is sufficient for our purposes in the present section, however, to take spacetime to be any compact manifold whose intrinsic geometry plays no role, for example S^{2n}). The classical action $\int \overline{\psi} \not{D} \psi$ is invariant under global axial rotations of the Fermi fields $\psi \to e^{i\alpha\gamma_5}\psi$, $\overline{\psi} \to \overline{\psi} e^{i\alpha\gamma_5}$, where $\gamma_5 \equiv i^n \prod_1^{2n} \gamma^\mu$ (the phase factor is chosen so that γ_5 is hermitian, with $(\gamma_5)^2 = +1$). To define the functional integral in (1.1), we expand $\psi = \sum a_n \psi_n$ and $\overline{\psi} = \sum \overline{b}_n \psi_n^\dagger$ in terms of the eigenfunctions ψ_n of $i\not{D}$ (satisfying $i\not{D}\psi_n = \lambda_n \psi_n$, with λ_n real by hermiticity of $i\not{D}$; and normalized to $(\psi_m, \psi_n) = \int dx\, \psi_m^\dagger \psi_n = \delta_{mn}$), where the a_n's and $\overline{b}_n$'s are anticommuting variables. The measure is then defined as $\int_{\psi,\overline{\psi}} \to \int \prod_n d\overline{b}_n\, da_n$. If we make a change of variables

$$\psi \to \psi + i\alpha(x)\gamma_5 \psi, \quad \overline{\psi} \to \overline{\psi} + \overline{\psi}\, i\gamma_5 \alpha(x), \tag{1.2}$$

then the exponent in the integrand changes by

$$\int d^{2n}x\, \overline{\psi} \not{D} \psi \to \int d^{2n}x\, \overline{\psi} \not{D} \psi - i \int d^{2n}x\, \alpha(x)\, \partial_\mu j_5^\mu \quad (j_5^\mu = \overline{\psi} \gamma^\mu \gamma_5 \psi). \tag{1.3}$$

Expanding in α and taking into account the invariance of $\Gamma_{\rm eff}[A]$ under a change of variables gives naive Ward identities implying conservation of the axial current, $\partial_\mu j_5^\mu = 0$, at the quantum level. We must, however, also take into account the change in the measure of the functional integral, i.e. the jacobian of the change of variables which, if not identically unity, gives an additional anomalous term. To compute this jacobian for the a_n's, we expand $\psi' \equiv (1+i\alpha(x)\gamma_5)\psi = \sum a'_n \psi_n$. The matrix C_{nm} relating the a'_n's to the a_n's via

$$\begin{aligned} a'_n &= \sum_m C_{nm}\, a_m \\ &= (\psi_n, (1+i\alpha\gamma_5)\psi) = \sum_m (\psi_n, (1+i\alpha\gamma_5)\psi_m)\, a_m \end{aligned}$$

is then given by $C_{nm} = \delta_{nm} + (\psi_n, i\alpha(x)\gamma_5\psi_m)$. The jacobian for the change in the a_n's for infinitesimal α is thus

$$\begin{aligned} \det C_{nm} &= e^{\operatorname{tr}\log C_{nm}} \\ &\approx e^{\operatorname{tr}(\psi_n, i\alpha\gamma_5\psi_m)} = e^{\sum_n (\psi_n, i\alpha\gamma_5\psi_n)} . \end{aligned}$$

The contribution from the change in the $\bar{b}_n$'s turns out to be identical with the *same* sign, so the measure is multiplied by an overall factor of the inverse determinant

$$e^{-2i\int dx\, \alpha(x) \sum_n \psi_n^\dagger(x)\gamma_5\psi_n(x)} \equiv e^{-2i\int \alpha(x)\mathcal{A}(x)} \tag{1.4}$$

under the change of variables (1.2). From (1.3), we then identify

$$\partial_\mu j_5^\mu = -2\mathcal{A}(x). \tag{1.5}$$

The trace in (1.4) is divergent, but may be regularized with a gaussian cut-off:

$$\int dx\, \alpha(x)\mathcal{A}(x) \equiv \lim_{M\to\infty} \int dx\, \alpha(x) \sum_n \psi_n^\dagger(x)\gamma_5\psi_n(x)\, e^{-\lambda_n^2/M^2} \tag{1.6}$$

In the limit in which $\alpha(x)$ becomes a constant independent of x, we can use (1.6) to generate a proof of the index theorem for the Dirac operator, or conversely, use the index theorem to obtain $\mathcal{A}(x)$. The exponent in (1.4) is then proportional to

$$\int dx\, \mathcal{A}(x) = \int \sum_n \psi_n^\dagger \gamma_5 \psi_n\, e^{-\lambda_n^2/M^2}\Big|_{M\to\infty} = \int \operatorname{tr}\gamma_5 e^{-(i\not{D}/M)^2}\Big|_{M\to\infty},$$

which only gets contributions from eigenfunctions ψ_n with eigenvalue $\lambda_n = 0$. This is because for $\lambda_n \neq 0$, $\gamma_5\psi_n$ has eigenvalue $-\lambda_n$ and is thus orthogonal to ψ_n. This leaves only the contribution from the zero mode eigenfunctions of γ_5

$$\begin{aligned} &\int \operatorname{tr}\gamma_5 e^{-(i\not{D}/M)^2}\Big|_{M\to\infty} = \\ &\int dx \left(\sum_{i=1}^{n_+} \psi_{i+}^\dagger(x)\psi_{i+}(x) - \sum_{i=1}^{n_-} \psi_{i-}^\dagger(x)\psi_{i-}(x) \right) = n_+ - n_-, \end{aligned}$$

where n_+ and n_- are the numbers of positive and negative chirality zero modes of the Dirac operator. To provide an integral expression for this quantity, we need to expand the exponent and pick out the piece which survives the $M \to \infty$ limit. Since the γ-matrix trace requires at least $2n$ γ-matrices to compensate the γ_5, the first non-vanishing contribution comes from pulling down exactly n factors of F from $\not{D}^2 = D_\mu D^\mu + F_{\mu\nu}\gamma^\mu\gamma^\nu$ in the exponent. This leading contribution is finite and moreover is the only contribution which survives the $M \to \infty$ limit, giving the result

$$\int dx\, \mathcal{A}(x) = n_+ - n_- = \frac{i^n}{2^{2n}\,\pi^n\, n!}\,\epsilon^{\alpha_1\beta_1\ldots\alpha_n\beta_n}\,\mathrm{tr}\,\overbrace{F_{\alpha_1\beta_1}\ldots F_{\alpha_n\beta_n}}^{n \text{ times}}\,. \tag{1.7}$$

From (1.5) we find that this also implies

$$\partial_\mu j_5^\mu = -2\mathcal{A}(x) = -2\frac{i^n}{2^{2n}\,\pi^n\, n!}\,\epsilon^{\alpha_1\beta_1\ldots\alpha_n\beta_n}\,\mathrm{tr}\,F_{\alpha_1\beta_1}\ldots F_{\alpha_n\beta_n}\,, \tag{1.8}$$

generalizing the familiar 4-dimensional result

$$\partial_\mu j_5^\mu = \frac{1}{16\,\pi^2}\epsilon^{\mu\nu\alpha\beta}\,\mathrm{tr}\,F_{\mu\nu}F_{\alpha\beta} = \frac{1}{8\,\pi^2}\mathrm{tr}\,F\tilde{F}. \tag{1.9}$$

B. Now we turn to consider anomalies in gauge currents in theories with gauge fields coupled to Weyl fermions transforming in some representation r of the gauge group G under gauge transformations. The effective action $\Gamma_r[A]$ is defined by

$$e^{-\Gamma_r[A]} = \int_{\psi,\overline{\psi}} e^{-\int d^{2n}x\,\overline{\psi}\,\not{D}_+\psi}$$
$$\not{D}_+ = \gamma^\mu(\partial_\mu + A_\mu)\tfrac{1}{2}(1+\gamma_5). \tag{1.10}$$

The gauge currents are $\overline{\psi}\gamma^\mu\lambda^a\frac{1}{2}(1+\gamma_5)\psi$ where the λ^a's are the (anti-hermitian) generators of G in the representation r. Our conventions are such that under an infinitesimal gauge transformation parametrized by $v = v^a\lambda^a$, we have

$$A_\mu \to A_\mu + D_\mu v, \quad D_\mu v = \partial_\mu v + [A_\mu, v] \tag{1.11}$$

so that the effective action changes by

$$\begin{aligned}\Gamma_r[A] &\to \Gamma_r[A + Dv]\\ &= \Gamma_r[A] + \int dx\ \mathrm{tr}\left[Dv\,\frac{\delta}{\delta A}\Gamma_r[A]\right]\\ &= \Gamma_r[A] - \int dx\ \mathrm{tr}\left[v\,D\frac{\delta}{\delta A}\Gamma_r[A]\right].\end{aligned} \tag{1.12}$$

But since $\delta\Gamma_r[A]/\delta A^a_\mu = \langle\overline{\psi}\gamma^\mu\lambda^a\frac{1}{2}(1+\gamma_5)\psi\rangle_A \equiv j^{\mu a}$ is the expectation value of the gauge current in the presence of the background gauge fields, we learn that the condition for gauge invariance, $\Gamma_r[A] = \Gamma_r[A + Dv]$, is equivalent to the condition

$$0 = D_\mu \frac{\delta\Gamma_r[A]}{\delta A_\mu} = D_\mu j^\mu, \tag{1.13}$$

i.e. that the gauge current have vanishing divergence (we are using a matrix notation in which $j^\mu = j^{\mu a}\lambda^a$, and the covariant derivative D_μ is defined as in (1.11)).

General arguments allow us to determine when the effective action $\Gamma_r[A]$ defined by (1.10) is potentially gauge non-invariant[9]. The real part of Γ_r, $\mathrm{Re}\,\Gamma_r = \frac{1}{2}(\Gamma_r + \Gamma_r^*) = \frac{1}{2}(\Gamma_r + \Gamma_{\bar{r}}) = \frac{1}{2}\Gamma_{r+\bar{r}}$, always admits a gauge invariant definition since it is equal to half the effective action $\Gamma_{r+\bar{r}}$ of a theory with additional Weyl fermions in the complex conjugate representation $\bar{r}$ of G, which admits a gauge invariant definition in terms of, for example, a Pauli-Villars regulator, since its representation content $r+\bar{r}$ forms a real representation of G. We thus learn that only complex representations give a gauge non-invariant $\Gamma_r[A]$ and only the imaginary part of $\Gamma_r[A]$ may pick up an anomalous variation under gauge transformations.

The identification of the functional integral in (1.10) with a determinant

$$\int_{\psi,\overline{\psi}} e^{-\int dx\,\overline{\psi}\,\not{D}_+\psi} \stackrel{?}{=} \det \not{D}_+ \tag{1.14}$$

is problematic because the operator $\not{D}_+ = \not{D}\frac{1}{2}(1+\gamma_5)$ maps positive to negative chirality spinors and consequently does not have a well-defined eigenvalue problem: $i\not{D}_+\Phi_+ = \lambda\Phi_+$ is meaningless.

Instead we define

$$e^{-\Gamma_r[A]} = \int_{\psi,\overline{\psi}} e^{-\int\overline{\psi}\,\hat{D}\psi} = \det\hat{D}, \tag{1.15}$$

where the operator $\hat{D}$ is defined by

$$\hat{D} = \not{\partial} + \not{A}\tfrac{1}{2}(1+\gamma_5) = \not{\partial}_- + \not{D}_+ = \begin{pmatrix} & \not{\partial}_- \\ \not{D}_+ & \end{pmatrix} \tag{1.16}$$

(on the far right $\hat{D}$ is represented in a γ_5-diagonal basis). This definition is suggested by considering a Pauli-Villars prescription (requiring opposite statistic regulator fermions) and satisfies certain necessary consistency requirements. The operator $\hat{D}$ acts on Dirac fermions, taking the combined space of positive plus negative chirality spinors into itself, so it has a well-posed eigenvalue problem. It has gauge couplings only to the positive chirality pieces, however, so the gauge theory defined by (1.15) is identical to that following from (1.10), up to an overall constant independent of the gauge field. $\det \hat{D}$ thus generates the same perturbative expansion as does $\not{D}_+$ for small gauge fields. Furthermore, since $i\not{\partial}_-$ has no non-trivial zero modes, $i\hat{D}$ has only positive chirality zero modes which coincide with those of $i\not{D}_+$. The chirality selection rules in the presence of instantons are thus the same for $i\hat{D}$ and $i\not{D}_+$.

Although not hermitian, $i\hat{D}$ is an elliptic operator and its eigenvalues λ_n, satisfying $i\hat{D}\psi_n = \lambda_n\psi_n$, are in general complex. They are not necessarily gauge invariant but can be used to define a gaussian regulator for the Fujikawa prescription to follow. Although there is no reason for individual eigenvalues of $i\hat{D}$ to be gauge invariant, the absolute value of the product of the eigenvalues is gauge invariant, as follows from considering

$$
\begin{aligned}
|\det \hat{D}|^2 = \det \hat{D}\,\det \hat{D}^\dagger &= \det \begin{pmatrix} \not{\partial}_-\not{\partial}_+ & \\ & \not{D}_+\not{D}_- \end{pmatrix} \\
&= \det(\not{\partial}_-\not{\partial}_+)\det(\not{D}_+\not{D}_-).
\end{aligned}
\tag{1.17}
$$

But the ordinary Dirac operator $\not{D} = \begin{pmatrix} & \not{D}_- \\ \not{D}_+ & \end{pmatrix}$ satisfies

$$
(\det \not{D})^2 = \det \begin{pmatrix} \not{D}_-\not{D}_+ & \\ & \not{D}_+\not{D}_- \end{pmatrix} = (\det \not{D}_+\not{D}_-)^2,
$$

which admits, as previously mentioned, a gauge invariant regulator. (1.17) then implies, up to an irrelevant constant, that $|\det \hat{D}| \sim (\det \not{D})^{1/2}$ is gauge invariant, thereby satisfying the constraint that only the imaginary part of the effective action should have an anomalous variation under gauge transformations.

Finally, to see how $\Gamma_{\rm r}[A]$ defined in (1.15) varies under gauge transformations, we again use a variant of Fujikawa's method. The integrand of (1.15) is left unchanged by an infinitesimal gauge transformation $A \to A + Dv$ and a compensating change of fermion variables $\psi \to \psi - v\psi_+$, $\overline{\psi} \to \overline{\psi} + \overline{\psi}\_v$. The fermion functional integral (1.15) is defined by expanding ψ and $\overline{\psi}$ in terms of separate right and left eigenfunctions ψ_n and $\chi_n^\dagger$ of $i\hat{D}$, satisfying $i\hat{D}\psi_n = \lambda_n\psi_n$, $(i\hat{D})^\dagger\chi_n = \lambda_n^*\chi_n$, and $\int dx\,\chi_n^\dagger\psi_m = \delta_{nm}$. In terms of $\psi = \sum_n a_n\psi_n$, $\overline{\psi} = \sum_n \chi_n^\dagger \bar{b}_n$, the fermion measure is defined as $\prod_n d\bar{b}_n da_n$. Orthogonality of $\chi_n^\dagger$ and ψ_m shows that the action satisfies $\int \overline{\psi}\hat{D}\psi = \sum_n(-i\lambda_n)\bar{b}_n a_n$, and (1.15) is thus formally equivalent to $\det i\hat{D} = \prod_n(-i\lambda_n)$ (this infinite product of eigenvalues may be given meaning by means of either ζ-function or Pauli-Villars regularization). The anomalous variation under gauge transformations comes from the Jacobian factor contributed by the change in the fermion measure $\prod_n d\bar{b}_n da_n$ under the change of variables corresponding to the above infinitesimal gauge transformation. This anomalous contribution can be written as (see [1] for more details)

$$\lim_{M\to\infty}\int dx \lim_{x\to y} \mathrm{tr}\, v(x)\gamma_5 e^{-(i\hat{D}_x)^2/M^2}\,\delta(x-y) = \int \mathrm{tr}\left[v D_\mu \frac{\delta\Gamma_{\mathrm{r}}}{\delta A_\mu}\right], \qquad (1.18)$$

with the trace defined in terms of eigenfunctions of $i\hat{D}$ and regulated with its eigenvalues. The trace may be rewritten as

$$\begin{aligned}\mathrm{tr}\left(v\,\gamma_5\, e^{\hat{D}^2/M^2}\right) &= \mathrm{tr}\left[v\left(\tfrac{1}{2}(1+\gamma_5)-\tfrac{1}{2}(1-\gamma_5)\right)e^{(\partial\!\!\!/_-D\!\!\!\!/_+ + D\!\!\!\!/_+\partial\!\!\!/_-)/M^2}\right]\\ &= \mathrm{tr}\left[v\tfrac{1}{2}(1+\gamma_5)\,e^{\partial\!\!\!/ D\!\!\!\!/ /M^2}\right] - \mathrm{tr}\left[v\tfrac{1}{2}(1-\gamma_5)\,e^{D\!\!\!\!/ \partial\!\!\!/ /M^2}\right],\end{aligned}$$

whose evaluation is then tedious but straightforward in a plane wave basis. For $2n = 4$ dimensions, it gives the familiar result[7]

$$\begin{aligned}\delta_v\Gamma_{\mathrm{r}}[A] = \Gamma_{\mathrm{r}}[A+Dv]-\Gamma_{\mathrm{r}}[A] &= -\int d^4x\,\mathrm{tr}\,[vD_\mu j^\mu]\\ &= -\frac{1}{24\pi^2}\int d^4x\,\mathrm{tr}\left[v\,\epsilon^{\lambda\mu\alpha\beta}\,\partial_\lambda(A_\mu\partial_\alpha A_\beta+\frac{1}{2}A_\mu A_\alpha A_\beta)\right] \qquad (1.19)\\ &= -\frac{1}{24\pi^2}\int \mathrm{tr}\left[v(x)\,d(A\,dA+\frac{1}{2}A^3)\right]\end{aligned}$$

(with the last line written in differential form notation to be described in a moment). (1.18) can also be evaluated explicitly for any higher dimensional spacetime. The result agrees with that of Feynman diagram techniques and also with the result of our topological analysis to be detailed shortly.

We reiterate that it is the gauge non-invariance of the regulator, due to the gauge non-invariance of the eigenvalues of $i\hat{D}$, which leads to the anomaly in gauge invariance in this approach. Before moving on to our topological analysis of the gauge anomaly, we first pause to introduce some mathematical apparatus.

C. On a general manifold M coordinatized by coordinates x^μ in some neighborhood, the notion of a vector at a point is realized by a linear differential operator $a^\mu(\partial/\partial x^\mu)$ which acts on functions $f\colon M \to R$ defined in the vicinity of the point.

Using $\partial/\partial x^\mu$ as the basis vectors for the tangent space, we then define a set of dual basis covectors dx^μ for the cotangent space (linear operators on vectors) satisfying $(dx^\mu, \partial/\partial x^\nu) = \delta^\mu_\nu$. Combining the dx^μ's antisymmetrically via the wedge product gives a convenient set of bases for the spaces of totally antisymmetric cotensor fields

$$\begin{gathered} dx^\mu \wedge dx^\nu = dx^\mu \otimes dx^\nu - dx^\nu \otimes dx^\mu \\ dx^\mu \wedge dx^\nu \wedge dx^\lambda = dx^\mu \otimes dx^\nu \otimes dx^\lambda \pm \text{permutations}, \\ \vdots \end{gathered}$$

called n-forms in n^{th} rank. We then write the gauge field $A_\mu = A^a_\mu\, \lambda^a$ as a matrix-valued 1-form

$$A = A_\mu\, dx^\mu,$$

called the connection 1-form. The field strength tensor $F_{\mu\nu} = \partial_\mu A_\nu - \partial_\nu A_\mu + [A_\mu, A_\nu]$ is written as a 2-form

$$F = \frac{1}{2} F_{\mu\nu}\, dx^\mu \wedge dx^\nu = dA + A^2 \tag{1.20}$$

(unless otherwise noted, multiplication of forms is always assumed to be the antisymmetrized wedge product), called the curvature 2-form. Here the exterior derivative, generally taking n-forms to $(n+1)$-forms, is defined by $d = (\partial/\partial x^\mu) dx^\mu \wedge$, and generates a minus sign when moved through forms of odd degree. In this notation, the Bianchi identity is derived by taking the exterior derivative of the curvature, $dF = d^2 A + dA\, A - A\, dA = FA - AF$, and can be written

$$DF \equiv dF + [A, F] = 0. \tag{1.21}$$

Our conventions for gauge transformations $g(x)\colon M_{2n} \to G$ take the form

$$\begin{aligned} A \to A^g &= g^{-1}(A + d)g \\ F \to F^g &= dA^g + (A^g)^2 = g^{-1} F g. \end{aligned} \tag{1.22a}$$

For infinitesimal gauge transformations $g \approx 1 + v$ where $v = v^a \lambda^a$, we have the corresponding infinitesimal transformations

$$\begin{aligned} A &\to A + dv + [A, v] = A + Dv \\ F &\to F - [v, F]. \end{aligned} \tag{1.22b}$$

On occasion we shall need the bracket of a matrix-valued p-form $\omega = \omega^a\, \lambda^a$ with a matrix-valued q-form $\eta = \eta^a\, \lambda^a$, defined as

$$[\omega, \eta] \equiv \omega \wedge \eta - (-1)^{pq} \eta \wedge \omega = \omega^a \wedge \eta^b\, [\lambda^a, \lambda^b].$$

Another bit of notation we shall need to introduce is the Hodge dual, defined in d-dimensions to take p-forms to $(d-p)$-forms according to

$$*dx^{\mu_1} \wedge \ldots \wedge dx^{\mu_p} = \frac{1}{(d-p)!} \epsilon^{\mu_1 \ldots \mu_p}{}_{\mu_{p+1} \ldots \mu_d}\, dx^{\mu_{p+1}} \wedge \ldots \wedge dx^{\mu_d}. \tag{1.23}$$

For d=4, for example, we have $*dx^1 = dx^2 \wedge dx^3 \wedge dx^4$, etc. Finally, the volume form is $dx^1 \wedge \ldots \wedge dx^d$ and integrals of d-forms $\alpha = a(x)\, dx^1 \wedge \ldots \wedge dx^d$ over (coordinate patches of) a d-dimensional manifold are defined by

$$\int \alpha = \int d^d x\, a(x).$$

We can now restate the index theorem for the Dirac operator $D\!\!\!\!/ = \gamma^\mu(\partial_\mu + A_\mu)$ on a $2n$-dimensional manifold, as obtained from (1.7), in the form

$$\operatorname{ind} D\!\!\!\!/ = n_+ - n_- = \int \frac{i^n}{(2\pi)^n\, n!} \operatorname{tr} F^n. \tag{1.24}$$

We point out that the integrand together with its normalization is just the form of degree $2n$ in the formal expansion of the Chern character $\operatorname{ch}(F) \equiv \operatorname{tr} e^{iF/2\pi}$. In terms of forms, we can also write (1.8) as

$$\partial_\mu j_5^\mu = -2\mathcal{A}(x) = -2 \frac{i^n}{(2\pi)^n\, n!} \operatorname{tr} *F^n. \tag{1.25}$$

Let us illustrate how the index of the Dirac operator can come out to be nonzero. A $2n$-dimensional manifold M is ordinarily specified in terms of a covering by coordinate patches $\{\mathcal{U}_i\}$, each diffeomorphic to R^{2n}. The gauge field A is defined separately on each patch with the requirement that the gauge fields A_i and A_j on different patches $\mathcal{U}_i$ and $\mathcal{U}_j$ be related by

$$A_i = g_{ij}^{-1}(A_j + d) g_{ij} \tag{1.26}$$

where $g_{ij}\colon \mathcal{U}_i \cap \mathcal{U}_j \to G$ is a gauge transformation defined on the overlap between patches. The topological information is thereby encoded in the transition functions g_{ij}. In the case of the sphere S^{2n}, for example, the simplest covering consists of two coordinate patches, the upper and lower hemispheres $\mathcal{D}_+$ and $\mathcal{D}_-$, intersecting on an equatorial S^{2n-1}. Letting $A_\pm$ denote the gauge field on $\mathcal{D}_\pm$, we see that the gauge field configuration is characterized by a single transition function g_{+-}, which in turn is a map from S^{2n-1} into G. Maps from a sphere S^m into a space K which are continuously deformable into one another are called homotopic, and the equivalence classes of such maps form a group $\pi_m K$, called the mth homotopy group of K. We have thus learned that a gauge field configuration on S^{2n}, representing a connection on a principal bundle over M, can be topologically nontrivial if and only if the gauge group G satisfies $\pi_{2n-1}G \neq 0$. Homotopy groups of compact Lie groups are frequently easy to determine and in any event have been extensively tabulated (see, for example, [13] for all such groups up to the π_{15}). We note that all semi-simple compact Lie groups have $\pi_3 G = Z$, the group of integers, giving the familiar instanton bundles on four dimensional spacetimes characterized by an integer m (the equatorial S^3 on S^4 plays the role of the S^3 at infinity on R^4). We also point out that all $SU(N)$'s have $\pi_{2n-1}SU(N) = Z$ for $n \leq N$, where the integer counts the number of times a representative map wraps around a basic $(2n-1)$-sphere in $SU(N)$. This is easily understood by recalling that as topological spaces, we have $SU(N)/SU(N-1) = S^{2N-1}$ and $SU(2) = S^3$, so $SU(N)$ looks like a twisted product of odd spheres from S^3 up to S^{2N-1}.

The $2n$-form $\operatorname{tr} F^n$ appearing in the index theorem (1.24) is invariant under the gauge transformations (1.22). It is also a closed form, since

$$
\begin{aligned}
d \operatorname{tr} F^n &= n \operatorname{tr} dF\, F^{n-1} = -n \operatorname{tr}[A, F] F^{n-1} \\
&= -n \operatorname{tr} A F^n + n \operatorname{tr} F A F^{n-1} = 0
\end{aligned}
$$

by the Bianchi identity (1.21) (in the last line we have used the fact that F is an even degree form to cyclically permute the trace with no change in sign). By Poincaré's lemma, any such closed form can at least locally be expressed as an exact form. In this case, we write

$$
\operatorname{tr} F^n = dQ_{2n-1}(A), \tag{1.27a}
$$

where

$$
Q_{2n-1}(A) = n \int_0^1 \delta t \operatorname{tr}[A F_t^{n-1}], \quad F_t = t dA + t^2 A^2 \tag{1.27b}
$$

is known as the Chern-Simons $(2n-1)$-form. (1.27) is easily verified by noting that $n\, d \operatorname{tr}[A F_t^{n-1}] = (\partial/\partial t) \operatorname{tr} F_t^n$ and that $\operatorname{tr} F_t^n |_0^1 = \operatorname{tr} F^n$.

Now we are in a position to evaluate the index $\operatorname{ind} \not{D}$ for the instanton bundles over S^4 mentioned above. The integral of $\operatorname{tr} F^2$ over the manifold can be divided into integrals over upper and lower hemispheres $\mathcal{D}_\pm$. Using $\operatorname{tr} F^2 = dQ_3(A)$, where $Q_3(A) = \operatorname{tr}(A\, dA + \frac{2}{3} A^3)$, together with Stokes theorem gives

$$
\begin{aligned}
\operatorname{ind} \not{D} &= \frac{i^2}{8\pi^2} \int_{S^3} [Q_3(A_+) - Q_3(A_-)] \\
&= \frac{i^2}{8\pi^2} \int_{S^3} Q_3(g_{+-}^{-1}\, dg_{+-}) = \frac{1}{24\pi^2} \int_{S^3} \operatorname{tr}(g_{+-}^{-1}\, dg_{+-})^3
\end{aligned} \tag{1.28}
$$

(the relative sign of the Q_3's comes from the opposite orientation of the boundary S^3 with respect to $\mathcal{D}_+$ and $\mathcal{D}_-$). The right hand side of this expression is known to result in an integer proportional to the homotopy class (winding number) of the map $g_{+-}: S^3 \to G$. The above expression thus makes it manifest that the index is only sensitive to the topological class of the transition function on the boundary.

D. The anomalous change in the effective action defined in (1.15) under the infinitesimal gauge transformation (1.22*b*) can be computed either by perturbation theory[4][7][14], by Fujikawa's method[12][1], or by topological arguments[15][1]. Straightforwardly evaluating (1.18), for example, gives

$$\delta_v \Gamma_r[A] = -\int_{S^{2n}} \operatorname{tr} v D_\mu j^\mu = -N \int_{S^{2n}} Q^1_{2n}(v, A) \qquad (1.29a)$$

$$N = \frac{i^{n+2}}{(2\pi)^n (n+1)!}$$

$$Q^1_{2n}(v, A) \equiv n(n+1) \int_0^1 \delta t\,(1-t)\, \operatorname{str}\left[v, d\left(A, F_t^{n-1}\right)\right], \qquad (1.29b)$$

where str denotes the trace symmetrized in its arguments and F_t is as in (1.27*b*). In general, the symmetrized trace of matrix-valued forms $\Sigma_i = \Sigma_i^a \lambda^a$ of arbitrary degree is defined by

$$\operatorname{str}(\Sigma_1, \Sigma_2, \ldots, \Sigma_n) \equiv \Sigma_1^{a_1} \wedge \cdots \wedge \Sigma_n^{a_n} \sum_{\text{perm}} \frac{1}{n!} \operatorname{tr}\left(\lambda^{a_{p(1)}} \cdots \lambda^{a_{p(n)}}\right) \qquad (1.30)$$

(i.e. we pull out the form part and symmetrize in the matrix part).

We now point out that the anomaly polynomial $Q^1_{2n}(v, A)$ above may be determined algebraically[15][1] from the index density

$$\frac{i^{n+1}}{(2\pi)^{n+1}(n+1)!} \operatorname{tr} F^{n+1} = \frac{N}{2\pi i} dQ_{2n+1}(A)$$

for a two higher (i.e. $(2n+2)$-) dimensional index theorem as follows. The gauge variation $\delta_v Q_{2n+1}(A)$ is closed, since $d\delta_v Q_{2n+1}(A) = \delta_v dQ_{2n+1}(A) = \delta_v \operatorname{tr} F^{n+1} = 0$. Thus $\delta_v Q_{2n+1}(A)$ can be written as the total derivative of something, which turns out to be the anomaly polynomial $Q^1_{2n}(v, A)$ of (1.29*b*):

$$\delta_v Q_{2n+1}(A) = d\, Q^1_{2n}(v, A). \qquad (1.31a)$$

Another useful characterization of the anomaly polynomial $Q^1_{2n}(v, A)$ is given by writing the Chern-Simons form as a formal polynomial $Q_{2n+1}(A, F)$ in the gauge field A and field strength F (rather than in A and dA, for example). Then $Q^1_{2n}(\hat{v}, A)$ is also easily verified to be the equal to the variation

$$Q^1_{2n}(\hat{v}, A) \approx Q_{2n+1}(A+\hat{v}, F) - Q_{2n+1}(A, F), \tag{1.31b}$$

to first order in the anticommuting form $\hat{v}$. The relations (1.29) and (1.31) suggest an intimate relation between the gauge anomaly and a higher dimensional index theorem which we now proceed to elucidate.

To do this, we continue to work with a spacetime manifold S^{2n}, gauge group G, and a reference gauge field A chosen, as is sufficient for our purposes here, to give a Dirac operator $\not{D}(A)$ with no zero modes. We then consider a one-parameter family of gauge transformations $g(\theta, x)$ satisfying the boundary conditions $g(0, x) = g(2\pi, x) = 1$. For simply connected gauge groups, the topological classes of maps $g(\theta, x)$: $S^1 \times S^{2n} \to G$ are classified by the $(2n+1)^{\text{st}}$ homotopy group $\pi_{2n+1}G$ since the boundary conditions make $S^1 \times S^{2n}$ look topologically like S^{2n+1}. As a function of the one-parameter family of gauge transformed configurations

$$A^{g(\theta)} = g^{-1}(\theta)(A+d)g(\theta), \tag{1.32}$$

our previous arguments show that $\det \hat{D}(A^{g(\theta)})$ satisfies

$$e^{-\Gamma_{\text{eff}}[A^{g(\theta)}]} = \det \hat{D}(A^{g(\theta)}) = [\det \not{D}(A)]^{1/2}\, e^{iw(A,\theta)}, \tag{1.33}$$

since only the phase may change as we move around the loop $g(\theta)$. We next extend the one-parameter family of gauge fields $A^{g(\theta)}$ to a two-parameter family interpolating between $A = 0$ and $A^{g(\theta)}$,

$$A^{t,\theta} = tA^{g(\theta)}, \tag{1.34}$$

which can be considered as parametrized by a disc with radial coordinate t ranging from 0 to 1, and angular coordinate θ. Since its modulus is constant and non-vanishing on the boundary of this disc, the function $\det \hat{D}(A^{t,\theta})$ restricted to the boundary gives a map from S^1 to S^1 (the image a circle in the complex plane), characterized by an integer

$$m = \frac{1}{2\pi}\int_0^{2\pi} d\theta\, \frac{\partial w(A,\theta)}{\partial\theta} \tag{1.35}$$

which counts the net winding number of the phase of $\det \hat{D}(A^{t,\theta})$ as we move around the boundary of the (t, θ) disc. Our strategy is now to exhibit a situation in which $m \neq 0$ is forced for purely topological reasons, showing in particular that the phase of the determinant must have an anomalous variation under the local gauge transformations around the boundary.

We first relate the winding number m above to the local behavior of the smallest eigenvalues of $\det \hat{D}(A^{t,\theta})$ in the interior of the disc. The determinant is allowed to vanish in the interior, of course, since there the gauge field $A^{t,\theta}$ is no longer a gauge transformation of the original zero-free reference field A. These zeroes generically occur at isolated points and the winding number of the phase of $\det \hat{D}(A^{t,\theta})$ can be determined in the immediate vicinity of each of these points. It follows from continuity in deforming and combining contours that the winding number of $\det \hat{D}(A^{t,\theta})$ around the boundary of the (t, θ) disc is equal to the signed sum of local winding numbers of the phase around contours surrounding points in the interior of the disc at which the determinant vanishes. Furthermore, by regarding the determinant as a regularized product of complex eigenvalues, we see that its winding number around any such contour is equal to the sum of the winding numbers of the individual eigenvalues and moreover receives a net contribution only from those eigenvalues which vanish in the interior of the contour. The net winding number m of the phase around the boundary is thus determined as the sum of the winding numbers of the smallest eigenvalues of $\hat{D}(A^{t,\theta})$ at the points where their vanishing produces zeroes of the determinant. This reduces the problem of calculating m to characterizing the behavior of these eigenvalues locally about their zeroes.

The next step is less straightforward and requires an adiabatic approximation to put interior zeroes of $\det \hat{D}(A^{t,\theta})$ into one-to-one correspondence with zero modes of a specific $(2n+2)$-dimensional Dirac operator $\not{D}_{2n+2}$. The winding number at these interior zeroes then turns out to be equal to the chirality $\chi = \pm 1$ of the associated $(2n+2)$-dimensional zero mode. By the argument of the previous paragraph, this means that the winding number of the $2n$-dimensional Weyl determinant around the boundary of the (t, θ) disc is equal to the number of positive chirality minus the number of negative chirality zero modes of our $\not{D}_{2n+2}$, i.e.

$$\operatorname{ind} \not{D}_{2n+2} = n_+ - n_- = m = \frac{1}{2\pi} \int_0^{2\pi} d\theta \, \frac{\partial w(A, \theta)}{\partial \theta}. \tag{1.36}$$

We shall refer the reader to [1] for further details on the adiabatic approximation and treat more fully only the construction of the $(2n+2)$-dimensional gauge field. This will then allow us to use the integral formula (1.24) for the index of $\not{D}_{2n+2}$ to identify, up to a total derivative, the anomalous local variation $\partial w / \partial \theta$ of the phase.

The specific $(2n+2)$-dimensional Dirac operator required above is defined with respect to a G-bundle over $S^{2n} \times S^2$ constructed as follows. The upper patch is taken to be the spacetime manifold S^{2n} crossed with the (t, θ) disc, the lower patch is S^{2n} crossed with some (trivial) (s, θ) disc, and the transition function on the $S^{2n} \times S^1$ boundary between the patches is simply the function $g(\theta, x)$. The $(2n+2)$-dimensional gauge field $\mathcal{A}$ we take to be

$$\mathcal{A}(x,t,\theta) = tg^{-1}(A + d + d_\theta)g = A^{t,\theta} + tg^{-1}\, d_\theta g \tag{1.37a}$$

$(d = dx^\mu \frac{\partial}{\partial x^\mu},\ d_\theta = d\theta \frac{\partial}{\partial \theta})$ on the upper patch, and

$$\mathcal{A}(x,s,\theta) = A \tag{1.37b}$$

on the lower patch. We note that (1.37a) at the $t = 1$ boundary is equal to (1.37b) gauge transformed by $g(\theta, x)$ as it should be. The index theorem (1.24) then allows us to write

$$\operatorname{ind} \not{D}_{2n+2}(\mathcal{A}) = \frac{i^{n+1}}{(2\pi)^{n+1}(n+1)!} \int_{S^2 \times S^{2n}} \operatorname{tr} \mathcal{F}^{n+1} \tag{1.38}$$

where $\mathcal{F} = (d + d_\theta + d_t)\mathcal{A} + \mathcal{A}^2$ is the $(2n+2)$-dimensional gauge field strength. The integrand above reduces to the difference of Chern-Simons forms $Q_{2n+1}(\mathcal{A})|_{t=1} - Q_{2n+1}(\mathcal{A})|_{s=1}$ for the gauge fields (1.37a, b) evaluated at the boundary between the patches. Using manipulations identical to those preceding (1.28), it is then straightforward to write $\operatorname{ind} \not{D}_{2n+2}$ as

$$\operatorname{ind} \not{D}_{2n+2} = \frac{(-1)^n\, i^{n+1}\, n!}{(2\pi)^{n+1}(2n+1)!} \int_{S^1 \times S^{2n}} \operatorname{tr} \left(g^{-1}(d + d_\theta + d_t)g\right)^{2n+1},$$

where the right hand side again is proportional to the number of times the map $g(\theta, x)$ wraps around a basic $(2n+1)$-sphere in G.

To identify the local index density $id_\theta w(A, \theta)$, we first note that the $Q_{2n+1}(\mathcal{A})$ term evaluated at $s = 1$ gives no contribution to the integral over $S^{2n} \times S^1$ since from (1.37b) it has no $d\theta$ component. (1.38) thus reduces to

$$\operatorname{ind} \not{D}_{2n+2} = \frac{i^{n+1}}{(2\pi)^{n+1}(n+1)!} \int_{S^1 \times S^{2n}} Q_{2n+1}(A^{g(\theta)} + \hat{v}) \tag{1.39}$$

The infinitesimal gauge parameter $\hat{v} = g^{-1} d_\theta g = d\theta\, g^{-1} \partial_\theta g = d\theta\, v$ appears as a 1-form and is the only quantity in the integrand above which carries a $d\theta$ to saturate the integral over S^1. Using $d_\theta \hat{v} = -\hat{v}^2$, we see that $(d + d_\theta)(A^{g(\theta)} + \hat{v}) + (A^{g(\theta)} +$

$\hat{v})^2 = F^{g(\theta)}$ so that we may use the relation (1.31b) satisfied by Q^1_{2n}. The result is that only the term linear in $\hat{v}$, $Q^1_{2n}(\hat{v}, A)$, contributes to the anomaly

$$d\theta \int_{S^{2n}} \operatorname{tr} v D_\mu j^\mu = i d_\theta w(A,\theta) = \frac{i^{n+2}}{(2\pi)^n (n+1)!} \int_{S^{2n}} Q^1_{2n}(\hat{v}, A^{g(\theta)}). \qquad (1.40)$$

For $2n = 4$ dimensions, this gives the result

$$*\operatorname{tr} v D_\mu j^\mu = \frac{1}{(2\pi)^2\, 3!} Q^1_4(v, A^{g(\theta)}) = \frac{1}{24\pi^2} \operatorname{tr}\left[v\, d(A^{g(\theta)} dA^{g(\theta)} + \frac{1}{2}(A^{g(\theta)})^3\right], \quad (1.41)$$

in agreement with (1.19) at $\theta = 0$.

We have shown, then, that the presence of a non-abelian gauge anomaly is forced by the non-trivial topology of the gauge group. In $2n = 4$ dimensions, for example, the existence of a global topologically induced gauge anomaly requires $\pi_5 \mathrm{G} = Z$, as is fulfilled by all $SU(N)$, $N \geq 3$. Even if there is no such *global* topological obstruction, however, there may nonetheless be a local anomalous variation of $\Gamma_{\text{eff}}[A]$ (with the same local form (1.29)) which cannot be eliminated through the addition of local counterterms to the action. A simple example of this is provided by the theory of a single Weyl fermion coupled to a $U(1)$ gauge field on a spacetime S^{2n}, known through perturbation theory to be anomalous despite the vanishing of all $\pi_i U(1)$ for $i > 1$. Another example is provided by an $SU(3)$ gauge theory in 6-dimensions, again anomalous in perturbation theory (having an anomaly proportional to the symmetrized trace of *four* generators), but having no integrated anomaly by the specific construction given here because $\pi_7 SU(3) = 0$. The anomaly for such theories would show up as a global topological obstruction, however, on less topologically trivial spacetimes such as $(S^1)^{2n}$. For a $U(1)$ gauge theory in 4 spacetime dimensions, for example, we could base our adiabatic argument on the 6-dimensional gauge field configuration $\mathcal{A} = \theta_1 d\theta_2 + \theta_3 d\theta_5 + \theta_6 d\theta_4$, where $\theta_1 \ldots \theta_4$ represent the spacetime $(S^1)^4$ and θ_5, θ_6 represent the two parameter manifold, now topologically $S^1 \times S^1$ rather than S^2. This construction generalizes to a maximal torus in any group to give a topologically induced anomaly in all cases where there is a perturbative anomaly (i.e. when the symmetrized trace of $n+1$ group generators is non-vanishing). A simple universality argument then uniquely fixes the form of the local perturbative variation for arbitrary groups and spacetimes since local perturbative methods always give the same result independent of the actual presence or absence of globally non-trivial topology. The result, however, is always required to take a functional form capable of exploiting non-trivial topology if and when it arises.

In closing here, we note that our results can also be interpreted in terms of the non-trivial topology of group orbit space. The gauge degrees of freedom of the theory (1.1) are not parametrized by the full space of gauge fields $\mathcal{A}$, but rather by the gauge orbit space $\mathcal{A}/\mathcal{G}$, where $\mathcal{G}$ is the group of small gauge transformations. The full quantum theory will thus be consistent only when $\Gamma_{\text{eff}}[A]$ can be consistently restricted to $\mathcal{A}/\mathcal{G}$. This is equivalent to requiring that the fermion effective action be invariant under $\mathcal{G}$, i.e. that $\Gamma_{\text{eff}}[A^g] = \Gamma_{\text{eff}}[A]$, where $A^g = g^{-1}(A+d)g$. A topological obstruction to consistently restricting $\Gamma_{\text{eff}}[A]$ from $\mathcal{A}$ to $\mathcal{A}/\mathcal{G}$ was identified here as the net winding of the phase of the complex number $\exp(-\Gamma_{\text{eff}}[A])$ under loops of gauge transformations, i.e. under closed $\mathcal{G}$-orbits. This winding number, if non-zero, acts to prevent the identification of $\mathcal{G}$-orbits in $\mathcal{A}$ to points in $\mathcal{A}/\mathcal{G}$. The possible presence of this obstruction, related to the topological nontriviality of $\mathcal{A}/\mathcal{G}$, was identified by probing the space $\mathcal{A}/\mathcal{G}$ with non-contractible 2-spheres, thereby determining whether or not the functional $\exp(-\Gamma_{\text{eff}}[A])$ may only exist as a section of a nontrivial line bundle over $\mathcal{A}/\mathcal{G}$. For a dynamical gauge theory, we need to insist that the functional exist as a section of a trivial bundle and hence that there be no anomaly. This is because the gauge field dynamics are obtained in principle by performing the functional integral over the gauge fields and this functional integral can only be formulated for sections of a trivial bundle.

2. Gravitational theories

A. To incorporate the effects of background gravitational fields, we need to introduce some formalism for treating spacetime manifolds with a non-trivial intrinsic curvature. We begin with a d-dimensional manifold M with metric $g_{\mu\nu}$, in terms of which the distance ds between two infinitesimally nearby points x^μ and $x^\mu + dx^\mu$ is given by

$$ds^2 = g_{\mu\nu}\, dx^\mu \otimes dx^\nu . \tag{2.1}$$

The metric may always be decomposed into vielbeins $e^a{}_\mu(x)$ satisfying $g_{\mu\nu} = \eta_{ab}\, e^a{}_\mu e^b{}_\nu$, where η_{ab} is a flat metric of the appropriate signature. We use greek letters (μ, ν, ...) to denote coordinate frame indices and latin letters (a, b, ...) to denote orthonormal tangent frame indices. In terms of the vielbein 1-forms

$$e^a = e^a{}_\mu\, dx^\mu ,$$

(2.1) becomes $ds^2 = \eta_{ab}\, e^a \otimes e^b$. Since it also follows that the $e^a{}_\mu$'s satisfy $\eta^{ab} = g^{\mu\nu} e^a{}_\mu e^b{}_\nu$, we can define the inverse vielbein

$$e_a{}^\mu(x) = \eta_{ab}\, g^{\mu\nu}(x) e^b{}_\nu(x)$$

which then satisfies $e_a{}^\mu e^b{}_\mu = \delta^b_a$, $e_a{}^\mu e^a{}_\nu = \delta^\mu_\nu$. The vielbein and its inverse are used to change tensor quantities referred to coordinate frames to tangent frames, and vice-versa.

The orthonormal tangent frames specified by the $e^a{}_\mu$'s are defined only up to local frame rotations: if e^a $(a{=}1, d)$ form an orthonormal basis, then so does $L^a{}_b(x)\, e^b$, where $L^a{}_b$ satisfies $\eta_{cd}\, L^c{}_a L^d{}_b = \eta_{ab}$. Introduction of vielbeins thus automatically involves the group of local frame rotations, also called local Lorentz transformations, generated by the $L^a{}_b(x)$'s. The vielbein formalism is only absolutely required for theories with fields of half-integral spin, since only these fields cannot be represented as ordinary tensor fields $T^{\alpha_1 \cdots \alpha_r}$ ($Gl(d, R)$ has no spinor representations) and must instead be introduced as spinor representations of the orthonormal frame rotation group.

To write down a general coordinate- and local Lorentz- invariant lagrangian incorporating derivatives, we shall need to introduce a so-called spin connection ω_μ to define a spin covariant derivative $D_\mu = \partial_\mu + \omega_\mu$. The spin connection can be defined by requiring that the covariant derivative acting on tensors referred to orthonormal frames transform as a general coordinate covector

$$D_\mu \to \frac{\partial x^\nu}{\partial x'^\mu}\, D_\nu$$

under general coordinate transformations

$$x^\mu \to x'^\mu(x^\nu), \tag{2.2}$$

and that it transform covariantly

$$D_\mu \to L^{-1} D_\mu L$$

under the local frame rotations

$$e^a{}_\mu \to \left(L^{-1}(x)\right)^a{}_b\, e^b{}_\mu(x). \tag{2.3}$$

Recall that in an ordinary gauge theory, we define a covariant derivative $D_\mu = \partial_\mu + A_\mu$ with $A_\mu \to g^{-1}(A_\mu + \partial_\mu)g$ under gauge transformations so that $D_\mu \to g^{-1} D_\mu g$ (this insures that $D_\mu \psi \to g^{-1} D_\mu \psi$ when $\psi \to g^{-1}\psi$). Hence we need a spin connection ω_μ, defined in terms of the vielbein, to transform as $\omega_\mu \to L^{-1}(\omega_\mu + \partial_\mu)L$ under (2.3). It is easily verified that

$$\omega^a{}_{b,\mu} \equiv e^a{}_\nu\, \nabla_\mu\, e_b{}^\nu = e^a{}_\nu \partial_\mu e_b{}^\nu + e^a{}_\nu \Gamma^\nu_{\mu\sigma} e_b{}^\sigma \tag{2.4}$$

has the correct transformation properties and as well satisfies $\omega_\mu \to \omega_\nu \frac{\partial x^\nu}{\partial x'^\mu}$ under (2.2). This last property is insured by the introduction of the usual covariant derivative $\nabla_\mu = \partial_\mu + \Gamma_\mu$ acting on coordinate frame tensors. It is defined in terms of the standard Christoffel connection

$$\Gamma^\nu_{\mu\sigma} = \tfrac{1}{2} g^{\nu\lambda}(\partial_\mu g_{\lambda\sigma} + \partial_\sigma g_{\lambda\mu} - \partial_\lambda g_{\mu\sigma}).$$

The spin connection 1-form is defined in terms of (2.4) as

$$\omega^a{}_b = \omega^a{}_{b,\mu}\, dx^\mu.$$

Lowering an index with the flat metric to give $\omega_{ab} = \eta_{ac}\omega^c{}_b$, we find from (2.4) that $\omega_{ab} = -\omega_{ba}$ and ω_{ab} is hence matrix-valued in the Lie algebra of $SO(d)$.

The Riemann curvature tensor is defined in this formalism as a 2-form

$$R^a{}_b = \tfrac{1}{2} R^a{}_{bcd}\, e^c \wedge e^d = d\omega^a{}_b + \omega^a{}_c \wedge \omega^c{}_b, \tag{2.5}$$

analogous to the field strength tensor F in (1.20). In coordinate frames, the Riemann tensor is given in terms of the Christoffel connection 1-form $\Gamma^\alpha{}_\beta = \Gamma^\alpha_{\mu\beta}\, dx^\mu$ by

$$R^\alpha{}_\beta = \tfrac{1}{2} R^\alpha{}_{\beta\mu\nu}\, dx^\mu \wedge dx^\nu = d\Gamma^\alpha{}_\beta + \Gamma^\alpha{}_\delta \wedge \Gamma^\delta{}_\beta,$$

summarizing the familiar component form $R^{\alpha}{}_{\beta\mu\nu} = \partial_\mu \Gamma^{\alpha}_{\nu\beta} - \partial_\nu \Gamma^{\alpha}_{\mu\beta} + \Gamma^{\alpha}_{\mu\delta}\,\Gamma^{\delta}_{\nu\beta} - \Gamma^{\alpha}_{\nu\delta}\,\Gamma^{\delta}_{\mu\beta}$. The components of the Riemann tensor in the two frames are related in the obvious way according to

$$R^{a}{}_{bcd} = R^{\alpha}{}_{\beta\gamma\delta}\, e^{a}{}_{\alpha}\, e_b{}^{\beta}\, e_c{}^{\gamma}\, e_d{}^{\delta}.$$

The Bianchi identity is derived in this formalism by taking the exterior derivative of (2.5) just as for the gauge case (1.21): $dR = d^2\omega + d\omega\,\omega - \omega\, d\omega = R\omega - \omega R$, so that

$$DR \equiv dR + [\omega, R] = 0. \tag{2.6}$$

Similarly, in coordinate frames we have $dR = d^2\Gamma + d\Gamma\,\Gamma - \Gamma\, d\Gamma = R\Gamma - \Gamma R$, so that

$$\nabla R \equiv dR + [\Gamma, R] = 0. \tag{2.7}$$

To derive the contracted form of the Bianchi identity, we first write the component form of (2.7) as $\nabla_\mu R^{\alpha}{}_{\beta\nu\sigma} + \nabla_\nu R^{\alpha}{}_{\beta\sigma\mu} + \nabla_\sigma R^{\alpha}{}_{\beta\mu\nu} = 0$, and then contract with δ^{ν}_{α} and $g^{\beta\sigma}$ to give the familiar

$$\nabla^{\alpha}(\mathbf{R}_{\alpha\mu} - \tfrac{1}{2} g_{\alpha\mu}\mathbf{R}) = 0.$$

(We have used boldface **R**'s here to denote the Ricci tensor $\mathbf{R}_{\mu\nu} = R^{\sigma}{}_{\mu\sigma\nu}$ and the Ricci scalar $\mathbf{R} = g^{\mu\nu}\mathbf{R}_{\mu\nu}$. Elsewhere R with or without its matrix indices explicitly written always denotes the Riemann curvature 2-form.)

Finally, the Hodge dual is generalized to a d-dimensional curved space by taking the analog of (1.23) in orthonormal frames

$$*e^{a_1} \wedge \ldots \wedge e^{a_p} = \frac{1}{(d-p)!}\, \hat{\epsilon}^{a_1 \ldots a_p}{}_{a_{p+1} \ldots a_d}\, e^{a_{p+1}} \wedge \ldots \wedge e^{a_d}, \tag{2.8}$$

where $\hat{\epsilon}_{1\ldots d} = 1$ and with all other non-vanishing components given by total antisymmetry. This induces the coordinate frame definition

$$*dx^{\mu_1} \wedge \ldots \wedge dx^{\mu_p} = \frac{1}{(d-p)!}\, \epsilon^{\mu_1 \ldots \mu_p}{}_{\mu_{p+1} \ldots \mu_d}\, dx^{\mu_{p+1}} \wedge \ldots \wedge dx^{\mu_d},$$

where $\epsilon_{1\ldots d} = \sqrt{g}$ (note also that $\epsilon^{1\ldots d} = 1/\sqrt{g}$ in euclidean space; in a minkowski signature metric we would instead have $-1/\sqrt{g}$). ϵ is of course just $\hat{\epsilon}$ referred to coordinate frames via $e^{a}{}_{\mu}$, but we have chosen to use the additional hat notation in this case to avoid confusion. The volume form is now simply $e^1 \wedge \ldots \wedge e^d =$

$\sqrt{g}\, dx^1 \wedge \ldots \wedge dx^d$ and integrals of d-forms $\alpha = a(x)\, e^1 \wedge \ldots \wedge e^d$ over (coordinate patches of) a d-dimensional manifold are defined by

$$\int \alpha = \int d^d x \sqrt{g}\; a(x).$$

We should point out that the parallelism between gauge and gravitational formalisms in formulæ such as $F = dA + A^2$ and $R = d\omega + \omega^2$ (or $R = d\Gamma + \Gamma^2$) holds only for background gauge or gravitational fields. The dynamics of the gravitational field involves a kinetic term proportional to the Ricci scalar $\mathbf{R} = R^{\mu}{}_{\alpha\mu\beta}\, g^{\alpha\beta} = R^{a}{}_{b\mu\nu}\, e_a{}^{\mu} e^{b\nu}$. There is no analog of this term for the gauge field lagrangian, since there is no natural object which relates the gauge indices i, j to the spacetime indices μ, ν of the gauge field strength tensor $(F_{\mu\nu})^i{}_j = F^a_{\mu\nu}\,(\lambda^a)^i{}_j$. Instead we have a gauge field kinetic term proportional to $\mathrm{tr}\, F_{\mu\nu} F^{\mu\nu}$.

Having introduced the relevant geometrical quantities, we are now in a position to write down the lagrangian

$$\mathcal{L} = \overline{\psi} \not{D} \psi = \overline{\psi}\, e_a{}^{\mu} \gamma^a (\partial_\mu + A_\mu + \tfrac{1}{2}\omega_{cd,\mu}\, \sigma^{cd})\, \psi \tag{2.9}$$

for a Dirac fermion coupled to an external gauge field in curved spacetime. The gamma matrices γ^a satisfy $\{\gamma^a, \gamma^b\} = 2\eta^{ab}$, and $\sigma^{cd} = \frac{1}{4}[\gamma^c, \gamma^d]$. The spinors ψ^i_α have (typically suppressed) indices acted upon by spinor matrices $(\sigma^{cd})_\beta{}^\alpha$ and the gauge matrices $(\lambda^a)^j{}_i$. The transformation properties of A and ω insure that the lagrangian (2.9) is invariant under the gauge transformations $\psi \to g^{-1}\psi$ and the local Lorentz transformations $\psi \to (L^{-1})_{ab}\, \sigma^{ab} \psi$.

Zero modes of the Dirac operator

$$\not{D} = e_a{}^{\mu} \gamma^a (\partial_\mu + A_\mu + \omega_\mu) \tag{2.10}$$

can be divided into negative and positive chirality according to their eigenvalue with respect to $\gamma_5 = i^n \prod_{a=1}^{2n} \gamma^a$. Generalizing (1.24), the index theorem[11] for (2.10) then gives the difference between the number of positive and negative chirality zero modes as

$$\mathrm{ind} \not{D} = \int_M [\hat{A}(M)\, \mathrm{ch}(F)]_{\mathrm{vol}}. \tag{2.11}$$

The 'vol' subscript on the integrand above is to indicate that the relevant contribution from the expansion of the product of the Dirac genus $\hat{A}(M)$ and the Chern character ch(F) comes from the term proportional to the volume $2n$-form on a $2n$-dimensional manifold M. The Chern character remains as previously defined

$$\begin{aligned}\mathrm{ch}(F) &\equiv \mathrm{tr}\, e^{\frac{i}{2\pi}F} \\ &= r + \frac{i}{2\pi}\mathrm{tr}\, F + \frac{i^2}{2(2\pi)^2}\mathrm{tr}\, F^2 + \ldots + \frac{i^n}{n!\,(2\pi)^n}\mathrm{tr}\, F^n + \ldots,\end{aligned} \tag{2.12}$$

where r is the dimension of the representation of the gauge group. To define $\hat{A}(M)$, we first note that because of its antisymmetry, we can always write the Riemann curvature 2-form $R^a{}_b$ in the skew diagonal form

$$\frac{R_{ab}}{2\pi} = \begin{pmatrix} & x_1 & & & \\ -x_1 & & & & \\ & & \ddots & & \\ & & & & x_n \\ & & & -x_n & \end{pmatrix},$$

where the x_a's ($a = 1, n$) denote the (2-form) skew eigenvalues. The Dirac genus $\hat{A}(M)$ is then written

$$\begin{aligned}\hat{A}(M) &\equiv \prod_a \frac{x_a/2}{\sinh(x_a/2)} \\ &= 1 + \frac{1}{(4\pi)^2}\frac{1}{12}\mathrm{tr}\, R^2 + \frac{1}{(4\pi)^4}\left[\frac{1}{288}(\mathrm{tr}\, R^2)^2 + \frac{1}{360}\mathrm{tr}\, R^4\right] \\ &\quad + \frac{1}{(4\pi)^6}\left[\frac{1}{10368}(\mathrm{tr}\, R^2)^3 + \frac{1}{4320}\mathrm{tr}\, R^2\,\mathrm{tr}\, R^4 + \frac{1}{5670}\mathrm{tr}\, R^6\right] + \ldots,\end{aligned} \tag{2.13}$$

where the formal defining expression on the first line may be straightforwardly expanded to arbitrary order. By antisymmetry of R_{ab}, only the trace of even powers of the Riemann curvature is non-vanishing and since R is a 2-form, (2.13) alone gives a contribution only in $4k$-dimensions. By arguments identical to those leading to (1.8) or (1.25), (2.13) can also be used to give the anomalous divergence of the axial current in the presence of a background gravitational field. For a single Dirac fermion in four dimensions, the result is

$$\partial_\mu j_5^\mu = -2\frac{1}{192\pi^2}\mathrm{tr}\, {*R^2} = -2\frac{1}{768\pi^2}R^\mu{}_{\nu\alpha\beta}R^\nu{}_{\mu\gamma\delta}\,\epsilon^{\alpha\beta\gamma\delta}. \tag{2.14}$$

A derivation of the generalized result (2.11) along the lines of the Fujikawa derivation given earlier here for (1.24) (and intended for physicists) may be found in

[16]. From (2.12) and (2.13), it is easy to pick out the appropriate contribution to the index theorem (2.11) in various dimensions. We give here the results up to dimension $d = 8$:

$$d=2 \qquad \mathrm{ind}\, \not{D} = \frac{i}{2\pi}\int_M \mathrm{tr}\, F$$

$$d=4 \qquad \mathrm{ind}\, \not{D} = \frac{1}{(2\pi)^2}\int_M \left(\frac{i^2}{2}\mathrm{tr}\, F^2 + \frac{r}{48}\mathrm{tr}\, R^2\right)$$

$$d=6 \qquad \mathrm{ind}\, \not{D} = \frac{1}{(2\pi)^3}\int_M \left(\frac{i^3}{6}\mathrm{tr}\, F^3 + \frac{i}{48}\mathrm{tr}\, F\, \mathrm{tr}\, R^2\right)$$

$$d=8 \qquad \mathrm{ind}\, \not{D} = \frac{1}{(2\pi)^4}\int_M \left(\frac{i^4}{24}\mathrm{tr}\, F^4 + \frac{i^2}{96}\mathrm{tr}\, F^2 \mathrm{tr}\, R^2 + \frac{r}{4608}(\mathrm{tr}\, R^2)^2 + \frac{r}{5760}\mathrm{tr}\, R^4\right).$$

In general, we see that there are contributions from terms involving both the gauge field strength F and curvature R in dimensions higher than 4 (and consequently mixed contibutions to the axial anomaly in these dimensions). We note, however, that for a sphere S^{2n} of radius ℓ, the Riemann curvature comes out as $R_{ab} = \frac{1}{\ell^2} e^a \wedge e^b$. It follows that $\mathrm{tr}\, R^{2m} = 0$ for any m and thus that $\hat{A}(S^{2n}) = 1$. This is why we were able to ignore curvature contributions to the index theorems in our earlier discussion of the gauge anomaly when the spacetime manifold was taken to be a sphere.

Another operator whose index is of interest here is the spin-$\frac{3}{2}$ operator $D_{3/2}$ which acts on a Rarita-Schwinger vector-valued spinor ψ_a of definite chirality. With respect to the Dirac operator (2.10), $D_{3/2}$ has an additional piece from the spin connection $\omega_a{}^b$ acting as well directly on the vector index. This modifies the index theorem (2.11) with the additional multiplicative contribution[11][16]

$$\mathrm{tr}\, e^{\frac{i}{2\pi}R} = 2\sum_a \cosh x_a,$$

exactly analogous to the Chern character contribution already present. The field ψ_a is not pure spin-$\frac{3}{2}$, however, so its quantization requires adding extra spin-$\frac{1}{2}$ ghost fields to remove the unphysical degrees of freedom. The constraint $k^a\psi_a = 0$ and invariance under $\psi_a \to \psi_a + k_a\chi$ (χ any spin-$\frac{1}{2}$ field and $k_\mu = k_a e^a{}_\mu$ the momentum) remove two spin-$\frac{1}{2}$ degrees of freedom of the same chirality as ψ_a, and the constraint $\gamma^a\psi_a = 0$ removes one spin-$\frac{1}{2}$ degree of freedom of opposite chirality from ψ_a. The net result of this is that the correct index density is given by subtracting off the contribution $\hat{A}(M)\,\mathrm{ch}(F)$ from a single spin-$\frac{1}{2}$ field. The index theorem for the spin-$\frac{3}{2}$ operator is thus

$$\mathrm{ind}\, D_{3/2} = \int_M \left[\hat{A}(M)\,(\mathrm{tr}\, e^{\frac{i}{2\pi}R} - 1)\,\mathrm{ch}(F)\right]_{\mathrm{vol}}. \tag{2.15}$$

Because of the dimension dependence $\mathrm{tr}\, e^{iR/2\pi} - 1 = \mathrm{tr}\,[e^{iR/2\pi} - \mathbf{1}] + (2n-1)$ (where $\mathbf{1}$ is the $2n\times 2n$ unit matrix), we give here the expression

$$\begin{aligned}\hat{A}(M)\,\mathrm{tr}\,(e^{\frac{i}{2\pi}R} - \mathbf{1})\\ = -\frac{1}{(4\pi)^2}2\,\mathrm{tr}\,R^2 + \frac{1}{(4\pi)^4}\left[-\frac{1}{6}(\mathrm{tr}\,R^2)^2 + \frac{2}{3}\mathrm{tr}\,R^4\right]\\ + \frac{1}{(4\pi)^6}\left[-\frac{1}{144}(\mathrm{tr}\,R^2)^3 + \frac{1}{20}\mathrm{tr}\,R^2\,\mathrm{tr}\,R^4 - \frac{4}{45}\mathrm{tr}\,R^6\right] + \ldots\end{aligned} \tag{2.16}$$

The appropriate contribution to (2.15) can then be obtained by combining terms from (2.13) and (2.16).

B. We now consider anomalies in general coordinate and local Lorentz invariance which may occur in quantum theories based on the curved space lagrangian (2.9). Omitting external gauge fields for the moment, the effective action $\Gamma_{\mathrm{eff}}[e,\omega]$ is defined in terms of the functional integral

$$e^{-\Gamma_{\mathrm{eff}}[e,\omega]} = \int_{\psi,\overline{\psi}} e^{-\int d^{2n}x\sqrt{g}\,\overline{\psi}\,\not{D}\frac{1}{2}(1+\gamma_5)\,\psi}, \tag{2.17}$$

where $\not{D}$ is as in (2.10) with $A=0$. In what follows, we shall frequently define the spacetime integral $\int \equiv \int d^{2n}x\sqrt{g}$, for convenience absorbing the volume element $d^{2n}x\sqrt{g}$ ($\sqrt{g} \equiv \sqrt{|\det g_{\mu\nu}|} = |\det e^a{}_\mu|$).

The classical action in the exponent on the right hand side above is invariant under general coordinate transformations (2.2) and local Lorentz transformations (2.3). Let us examine the consequences for the quantum effective action to share these invariances. Under an infinitesimal local Lorentz transformation $L(x) = 1 + \alpha(x)$, we have $\delta^L_\alpha e^a{}_\mu = -\alpha^a{}_b e^b{}_\mu$, so the variation of Γ_{eff} is $\delta^L_\alpha \Gamma_{\mathrm{eff}} = -\int d^{2n}x\,\alpha^a{}_b e^b{}_\mu\,(\delta\Gamma_{\mathrm{eff}}/\delta e^a{}_\mu)$. But $e_{b\mu}\,\delta\Gamma_{\mathrm{eff}}/\delta e^a{}_\mu = \sqrt{g}\,T_{ab} = e_{a\mu}e_{b\nu}\sqrt{g}\,T^{\mu\nu}$, where $T^{\mu\nu} = \langle \frac{1}{2}\overline{\psi}(\gamma^\mu \overleftrightarrow{D}^\nu_+ + \gamma^\nu \overleftrightarrow{D}^\mu_+)\psi\rangle$ is the expectation value of the energy-momentum tensor in the presence of background curvature, so we find

$$\delta^L_\alpha \Gamma_{\mathrm{eff}} = -\int \alpha^{ab}T_{ab}. \tag{2.18}$$

Since α^{ab} is antisymmetric, it follows that the non-invariance of the effective action under local Lorentz transformations is equivalent to asymmetry of T_{ab}, or equivalently to asymmetry of $T^{\mu\nu}$. It is important to preserve local Lorentz symmetry in order that the unphysical degrees of freedom associated with the introduction of the vielbein play no role, just as ordinary gauge invariance allows us to choose a unitary gauge in which unphysical degrees of freedom are absent.

Under the infinitesimal coordinate transformation, $x^\mu \to x^\mu - \xi^\mu$, the variation in the vielbein is $\delta_\xi^C e^a{}_\mu = e^a{}_\nu \nabla_\mu \xi^\nu + \xi^\nu \nabla_\nu e^a{}_\mu$. The variation of Γ_{eff} is $\delta_\xi^C \Gamma_{\text{eff}} = \int d^{2n}x\, (e^a{}_\nu \nabla_\mu \xi^\nu + \xi^\nu \nabla_\nu e^a{}_\mu)\delta\Gamma_{\text{eff}}/\delta e^a{}_\mu = -\int \xi^\nu (\nabla_\mu T_\nu{}^\mu + \omega_{ab,\nu} T^{ab})$. If Lorentz symmetry is preserved so that T^{ab} is symmetric, then the second term vanishes due to antisymmetry of $\omega_{ab,\nu}$ and we have then that

$$\delta_\xi^C \Gamma_{\text{eff}} = -\int \xi_\nu \nabla_\mu T^{\mu\nu}. \tag{2.19}$$

This identification of a possible non-invariance of the effective action under infinitesimal local coordinate transformations with an anomalous non-conservation of the induced energy-momentum tensor is precisely analogous to the identification (1.13) of non-invariance of $\Gamma_r[A]$ under local gauge transformations with a non-conservation of the induced gauge current. Our interest in theories without gravitational anomalies is thus connected to our preference for theories which conserve energy and momentum.

Reasoning analogous to that in the gauge case suffices to show that the real part of Γ_{eff} is always generally covariant and locally Lorentz invariant. The imaginary part of Γ_{eff}, however, may suffer an anomalous variation[9] which can again be determined either by perturbation theory[9], Fujikawa's method[9], or from a two higher dimensional index theorem[17][2]. Here we shall sketch the results from the index theorem point of view. Instead of a two-parameter family of gauge fields bounded by a one-parameter family of gauge-transformed configurations as used before, we now need to consider a two-parameter family of metrics on our spacetime manifold. Replacing the one-parameter family of gauge transformations, we take a one-parameter family either of general coordinate transformations or local Lorentz transformations. For general coordinate transformations

$$x^\mu \to x'^\mu(\theta, x^\nu), \tag{2.20}$$

we find that the Christoffel connection $\Gamma^\mu{}_\nu = \Gamma^\mu_{\alpha\nu}\, dx^\alpha$ and the curvature $R = d\Gamma + \Gamma^2$ in coordinate frames transform under the passive action of the diffeomorphism group as

$$\begin{aligned} \Gamma(x) &\to \Gamma'(x') = \Lambda^{-1}(\Gamma + d)\Lambda|_x \\ R(x) &\to R'(x') = \Lambda^{-1} R \Lambda|_x \end{aligned} \tag{2.21}$$

$((\Lambda^{-1})^\mu{}_\nu = \partial x'^\mu(\theta)/\partial x^\nu)$, i.e. as $Gl(2n)$ gauge connection and curvature. For a one-parameter family of local Lorentz transformations $L(\theta, x)$, on the other hand, we find that the spin connection $\omega^a{}_b = \omega^a{}_{b,\mu}\, dx^\mu$ and the curvature $R = d\omega + \omega^2$ in orthonormal tangent frames transform as

$$\begin{aligned} \omega &\to L^{-1}(\omega + d)L \\ R &\to L^{-1}RL. \end{aligned} \tag{2.22}$$

The analog of our argument for the pure gauge case is now straightforward. There we used a reference gauge field A to define a $(2n+2)$-dimensional gauge field which reduced to $g^{-1}(A+d)g + g^{-1}d_\theta g$ on the boundary between patches. We then used an index theorem for the winding number of the phase of the functional integral to give its total change, equal to $2\pi i$ times the winding number, in terms of an integral over the $(2n+2)$ dimensional index density tr $\mathcal{F}^{n+1}$. The Chern-Simons form was then used to transform this to an integral over the boundary between patches, giving implicitly a local formula for the winding number density and hence a local formula for the anomalous change in the phase. It follows here as well that a local winding number density is given by the invariant polynomials (2.13) and (2.16) for the appropriate $(2n+2)$-dimensional index theorem.

To express our results, we shall need to work with invariant polynomials $P(R^{n+1})$, homogeneous of degree $n+1$ in R, and invariant under $R \to L^{-1}RL$ (or $R \to \Lambda^{-1}R\Lambda$). Such polynomials can in general be written as products of traces of R to various even powers, as in (2.13) and (2.16). Absorbing all relevant normalization factors directly into the P's, we have from (2.11) and (2.15) that the required polynomials for the cases of spin-$\frac{1}{2}$ and spin-$\frac{3}{2}$ fields are

$$P_{1/2}(R^{n+1}) = 2\pi i[\hat{A}(M^{2n})]_{2n+2} \tag{2.23a}$$

$$P_{3/2}(R^{n+1}) = 2\pi i\Big[\hat{A}(M^{2n})\,(\mathrm{tr}\, e^{\frac{i}{2\pi}R} - 1)\Big]_{2n+2}. \tag{2.23b}$$

The subscript $2n+2$ is meant to indicate that for a $2n$-dimensional theory, we begin with the polynomial form in R appropriate to a $(2n+2)$-dimensional manifold. The a, b indices on $R^a{}_b$ still only act on the $2n$-dimensional space, so one must be careful, for example, to substitute tr $e^{iR/2\pi} = 2n + \mathrm{tr}\,[e^{iR/2\pi} - \mathbf{1}]$ in (2.23b). Finally, the factor of $2\pi i$ in (2.23a, b) converts the winding number to a phase, as in the step from (1.39) to (1.40). A discussion along these lines of another field with a potential gravitational anomaly[9], the self-dual antisymmetric tensor field, may be found in [2].

To manipulate the Chern-Simons forms associated with the invariant polynomials (2.23) we introduce a bit more formalism. Let $P(\Omega_1, \Omega_2, \ldots, \Omega_m)$ denote an arbitrary invariant polynomial in matrix-valued forms Ω_i, symmetrized in its matrix part (i.e. with appropriate sign change for pulling any odd degree form through another as in (1.30)). If some of the arguments are equal, say $\Omega_3 = \Omega_4 = \ldots = \Omega_m = R$, then we use the abbreviated notation $P(\Omega_1, \Omega_2, R, \ldots, R) = P(\Omega_1, \Omega_2, R^{n-2})$. Due to the antisymmetry of R_{ab}, an invariant polynomial $P(R^m)$ of degree m in R vanishes unless m is even, say $m = 2k + 2$. For this reason, the invariant polynomials (2.23) lead to purely gravitational anomalies only in $4k + 2$-dimensions.

The Bianchi identity (2.6),(2.7) insures that $\operatorname{tr} R^m$ is closed, just as in the gauge case: $d\operatorname{tr} R^m = m \operatorname{tr} DR\, R^{m-1} = 0$. By the chain rule, products of traces are consequently closed as well, so we can again realize Poincaré's lemma locally for the invariant polynomials (2.23). In writing down the form for the gravitational anomaly, we are left with the choice of considering the invariant polynomials either as functions of the $Gl(2n)$ or $SO(2n)$ curvature, i.e. we have $P\big((d\Gamma + \Gamma^2)^{n+1}\big) = P\big((d\omega + \omega^2)^{n+1}\big)$ where the two curvatures are related by $R^\alpha{}_\beta = e_a{}^\alpha R^a{}_b e^b{}_\beta$. The two choices give Chern-Simons forms $Q_{2n+1}(\Gamma)$ and $Q_{2n+1}(\omega)$, both having $dQ_{2n+1} = P(R^{n+1})$, and enter respectively into our analysis of anomalies under local Lorentz and general coordinate transformations (that they both stem from the same invariant polynomial already indicates that there is really only one anomaly, resulting from a single higher-dimensional topological obstruction). The generalizations of the Chern-Simons form (1.27), having $dQ_{2n+1}(\omega) = dQ_{2n+1}(\Gamma) = P(R^{n+1})$, are then

$$
\begin{aligned}
Q_{2n+1}(\omega) &= (n+1) \int_0^1 \delta t\, P(\omega, R_t^n), \quad R_t = t d\omega + t^2 \omega^2 \\
Q_{2n+1}(\Gamma) &= (n+1) \int_0^1 \delta t\, P(\Gamma, R_t^n), \quad R_t = t d\Gamma + t^2 \Gamma^2 .
\end{aligned}
\tag{2.24}
$$

Under the infinitesimal form of (2.21), $x^\mu \to x'^\mu = x^\mu - \xi^\mu(x)$, we have $\Lambda = \partial x / \partial x' = 1 + v$ where $v^\alpha{}_\beta = \partial \xi^\alpha / \partial x^\beta$, so (2.21) becomes

$$
\begin{aligned}
\Gamma &\to \Gamma + dv + [\Gamma, v] = \Gamma + \nabla v \\
R &\to R - [v, R].
\end{aligned}
$$

Under the infinitesimal form of (2.22), $L = 1 + \alpha$, we have

$$
\begin{aligned}
\omega &\to \omega + d\alpha + [\omega, \alpha] = \omega + D\alpha \\
R &\to R - [\alpha, R].
\end{aligned}
$$

Both these infinitesimal variations above are formally identical to the infinitesimal gauge variation (1.22*b*). This means we need to define

$$\begin{aligned} Q^1_{2n}(\alpha,\omega) &= n(n+1)\int_0^1 \delta t\, P\big(\alpha, d(\omega, R_t^{n-1})\big) \\ Q^1_{2n}(v,\Gamma) &= n(n+1)\int_0^1 \delta t\, P\big(v, d(\Gamma, R_t^{n-1})\big), \end{aligned} \tag{2.25}$$

analogous to the form $Q^1_{2n}(v,A)$ which appeared in the gauge case. The Q^1_{2n}'s defined above satisfy $\delta Q_{2n+1} = dQ^1_{2n}$ for either infinitesimal general coordinate or local Lorentz transformations, and also satisfy the analog of (1.31*b*).

Reasoning parallel to that which led to (1.40) then tells us that the form of the gravitational anomaly predicted by the index theorem takes either the form

$$\delta^L_\alpha \Gamma_{\text{eff}}[e,\omega] = -\int \alpha^{ab}\, T_{ab} = -\int Q^1_{2n}(\alpha,\omega) \tag{2.26a}$$

or the form

$$\delta^C_\xi \Gamma_{\text{eff}}[e,\omega] = -\int \xi_\mu\, \nabla_\nu T^{\mu\nu} = -\int Q^1_{2n}(v,\Gamma), \tag{2.26b}$$

where Q^1_{2n} is defined by substituting one of the invariant polynomials (2.23*a*, *b*) for the spin-$\frac{1}{2}$ and spin-$\frac{3}{2}$ anomalies, respectively, into (2.25). We notice that the anomaly, described in (2.26*a*) as a violation of local Lorentz invariance, takes a form closely resembling that of an ordinary $SO(4k+2)$ gauge theory, giving it rather naturally the sort of topological interpretation we have already discussed for the gauge case (we have, incidentally, been writing the gauge and gravitational anomalies here in their so-called consistent form, satisfying the Wess-Zumino consistency conditions. There is another form of the anomaly, called the covariant form, which is related to the divergence of gauge currents or energy-momentum tensors differing from the ones used here by the addition of local contact terms[18][2]. The physical content of the anomaly, including the conditions for its removal, is independent of the form in which it is expressed, and we shall continue here to work exclusively in terms of the consistent form).

To verify the equivalence between the two forms of the anomaly (2.26*a*, *b*), we shall now construct a local counterterm which when added to the action can shift the anomaly from one form to the other. The explicit form of the anomaly arrived at via a perturbative calculation will depend on the regularization procedure employed. In the perturbative calculations of [9], a gauge in which $e_{a\mu}$ is symmetric was fixed. This insures symmetry of $\sqrt{g}\, T_{ab} = e_{b\mu}\, \delta\Gamma_{\text{eff}}/\delta e^a{}_\mu$ and hence by (2.18) there are no anomalies in frame rotations. The freedom in choosing the vielbein thus allows studying anomalies exclusively in general coordinate transformations (If one prefers to consider anomalies in local Lorentz transformations as the more natural, either in analogy with $SO(4k+2)$ gauge theories or from a topological standpoint, then in this language it is simply the compensating local Lorentz transformation ($\mathbf{e} \to L^{-1}\mathbf{e}\Lambda$) required to maintain the symmetric gauge condition which causes the anomalies to appear in general coordinate transformations). This assures us that the fermion functional integral can always be used to construct a functional $\Gamma_{\text{eff}}[\Gamma, g]$ (which, due to its Lorentz invariance, depends on the vielbein only through $g_{\mu\nu}$) having only an anomaly in the form (2.26b). To construct a counterterm to shift this anomaly to the local Lorentz form (2.26a), we use a $Gl(2n)$ matrix notation for the vielbein, $\mathbf{e} = e^a{}_\mu$, and consider an interpolation $\mathbf{e}_t$ with $\mathbf{e}_{t=0} = 1$, $\mathbf{e}_{t=1} = \mathbf{e}(x)$. Then integrating up the anomalous variation gives the local counterterm[18][2]

$$S = \int_0^1 \delta t \int Q^1_{2n}(\mathbf{e}^{-1}\partial_t \mathbf{e}, \omega^{\mathbf{e}_t}), \tag{2.27}$$

where $\omega^{\mathbf{e}_t} \equiv \mathbf{e}_t^{-1}(\omega + d)\mathbf{e}_t$. By (2.26), S can also be written

$$S = \Gamma_{\text{eff}}[\omega, \eta] - \Gamma_{\text{eff}}[\omega^{\mathbf{e}}, \mathbf{e}^T \eta \mathbf{e}] = \Gamma_{\text{eff}}[\omega, \eta] - \Gamma_{\text{eff}}[\Gamma, g],$$

where we have noted that the defining relation (2.3) for the spin connection can be rewritten as $\Gamma = \mathbf{e}^{-1}(\omega + d)\mathbf{e}$, giving the the Christoffel connection Γ as a $Gl(2n)$ 'gauge transformed' spin connection ω. Written in the final form above, it is clear that S obeys

$$\delta^C_\xi S = \int Q^1_{2n}(v, \Gamma) \qquad \delta^L_\alpha S = -\int Q^1_{2n}(\alpha, \omega), \tag{2.28}$$

so that $\pm S$ can be used to shift anomalies from one form to the other. The existence of S makes it clear that the cancellation conditions for the gravitational anomaly are independent of whether it is considered to be in local Lorentz or general coordinate transformations.

The treatment of combined gauge and gravitational anomalies, i.e. the anomalous divergences in the local gauge currents or energy momentum tensor in the combined presence of background gauge and gravitational fields, proceeds identically. We now simply use the index density for the appropriate $(2n+2)$-dimensional index theorem including the factor of $\text{ch}(F)$ for the gauge fields. The invariant polynomial $P(R, F)$ is then determined by isolating the $(2n+2)$-form contribution from (2.11) or (2.15). From it we then easily determine a Chern-Simons form Q_{2n+1} satisfying $dQ_{2n+1} = P$, and from that the appropriate Q^1_{2n}. This procedure automatically gives all contributions to the anomalies in gauge and gravitational currents in $2n$-dimensions corresponding to diagrams having both gauge fields and gravitons on external lines.

C. The formalism for treating higher dimensional gauge and gravitational anomalies thus far described here has found its most topical use recently in the superstring anomaly cancellation mechanism discovered by Green and Schwarz. Superstring theories (for reviews and references to the original literature, see [19]) are candidate models for unified theories of all interactions. The superstring approach turns out to be somewhat more constrained than that of ordinary field theories, for example only allowing a consistent formulation in 10 dimensions. The low energy behavior of the massless sector of superstring theories, however, is well approximated by an ordinary field theory which should consequently be free of anomalies if the original string theory is truly consistent in its realization of general coordinate and local gauge invariance. From the standpoint of their low energy field theoretic approximations, it did not seem that any chiral ten-dimensional superstring theories stood any likelihood of being free of gauge and gravitational anomalies. Using their string theory intuition, however, Green and Schwarz discovered[20] the surprising result that the anomalies can be made to cancel for a very restricted choice of gauge group. Our presentation of these results here follows the more recent treatment in [21].

The structure of a superstring theory at low energies is described by $N{=}1$ supergravity coupled to $N{=}1$ super-Yang-Mills theory in ten dimensions with some gauge group G. We shall take the gauge field $A = A^a_\mu \lambda^a dx^\mu$ in the adjoint representation of G and denote traces over forms involving the field strength $F = dA + A^2$ by 'Tr' as a reminder that they are now taken over the adjoint representation of G. The spin connection $\omega = \omega_\mu dx^\mu$ is a $10{\times}10$ antisymmetric matrix and we continue to denote traces over forms involving the curvature $R = d\omega + \omega^2$ by 'tr' since they remain over the fundamental representation of the Lorentz algebra $SO(9,1)$. The chiral fields that contribute to anomalies in hexagon (and higher) one-loop diagrams are a positive chirality Majorana-Weyl gravitino (contributing $\hat{A}\left(\mathrm{tr}\left[e^{iR/2\pi} - \mathbf{1}\right] + 9\right)$ to the anomaly polynomial $P(R^6)$) and a negative chirality Majorana-Weyl spinor from the supergravity multiplet (contributing $-\hat{A}$), as well as $r{=}\dim G$ positive chirality Majorana-Weyl spinors from the Yang-Mills supermultiplet (contributing $\hat{A}\,\mathrm{ch}(F)$). From (2.13) and (2.16), we learn that the invariant polynomial $P(R, F)$ characterizing the Yang-Mills, gravitational, and mixed anomalies due to these loops is equal to a factor $2\pi i \frac{4}{3(4\pi)^6}$ times the 12-form

$$I_{12} = -\frac{1}{15}\mathrm{Tr}\, F^6 + \frac{1}{24}\mathrm{Tr}\, F^4\, \mathrm{tr}\, R^2 - \mathrm{Tr}\, F^2 \left(\frac{1}{192}(\mathrm{tr}\, R^2)^2 + \frac{1}{240}\mathrm{tr}\, R^4\right)$$
$$+ \left(\frac{1}{32} + \frac{r-496}{13824}\right)(\mathrm{tr}\, R^2)^3 + \left(\frac{1}{8} + \frac{r-496}{5760}\right)\mathrm{tr}\, R^2\, \mathrm{tr}\, R^4 + \frac{r-496}{7560}\mathrm{tr}\, R^6 \tag{2.29}$$

(the purely gravitational contribution independent of F, for example, results from (2.16) plus $r+8$ times (2.13)).

In general, no local counterterm can be constructed in terms of the available fields to compensate the leading terms in the gauge and gravitational contributions, $\mathrm{Tr}\, F^6$ and $\mathrm{tr}\, R^6$, so they must somehow be eliminated directly. To remove the term proportional to $\mathrm{tr}\, R^6$, we need to choose a gauge group G with 496 generators. To remove the leading gauge piece, we need that $\mathrm{Tr}\, F^6$ not be an independent sixth-order Casimir invariant of G. Both of these properties are satisfied by $E_8 \times E_8$ and $SO(32)$. For a single E_8 we have

$$\mathrm{Tr}\, F^4 = \frac{1}{100}(\mathrm{Tr}\, F^2)^2 \qquad \mathrm{Tr}\, F^6 = \frac{1}{7200}(\mathrm{Tr}\, F^2)^3, \tag{2.30}$$

as is easily derived by means of a maximal embedding of $SO(16)$. For $SO(32)$, using the property that the adjoint representation is the antisymmetric combination of two fundamentals, we find

$$\begin{gathered}\mathrm{Tr}\, F^2 = 30\, \mathrm{tr}\, F^2 \qquad \mathrm{Tr}\, F^4 = 24\, \mathrm{tr}\, F^4 + 3(\mathrm{tr}\, F^2)^2 \\ \mathrm{Tr}\, F^6 = 15\, \mathrm{tr}\, F^2\, \mathrm{tr}\, F^4,\end{gathered} \tag{2.31}$$

where tr denotes trace in the fundamental representation. It follows from (2.30) and (2.31) that the adjoint representations of both $E_8 \times E_8$ and $SO(32)$ satisfy the condition

$$\mathrm{Tr}\, F^6 = \frac{1}{48}\mathrm{Tr}\, F^2\, \mathrm{Tr}\, F^4 - \frac{1}{14400}(\mathrm{Tr}\, F^2)^3, \tag{2.32}$$

(where for $E_8 \times E_8$, we have used $F = F_1 \oplus F_2$ so that $\mathrm{Tr}\, F^m = \mathrm{Tr}\, F_1^m + \mathrm{Tr}\, F_2^m$). Substituting (2.32) into the first term of (2.29) together with $r = 496$ gives (2.29) the factorized form

$$I_{12} = (\mathrm{tr}\, R^2 - \frac{1}{30}\mathrm{Tr}\, F^2)\, X_8 \tag{2.33a}$$

where

$$X_8 = \frac{1}{24}\mathrm{Tr}\, F^4 - \frac{1}{7200}(\mathrm{Tr}\, F^2)^2 - \frac{1}{240}\mathrm{Tr}\, F^2\, \mathrm{tr}\, R^2 + \frac{1}{32}(\mathrm{tr}\, R^2)^2 + \frac{1}{8}\mathrm{tr}\, R^4. \tag{2.33b}$$

When the form I_{12} so factorizes, it is possible to construct a local counterterm which cancels the anomalies by making use of the two-index antisymmetric tensor field $B_{\mu\nu}$ present in the N=1, $d = 10$ supergravity multiplet. We will choose here to write the gravitational anomaly as a violation of local Lorentz invariance. In terms of the 2-form $B = B_{\mu\nu}\, dx^\mu \wedge dx^\nu$, we construct the 3-form field strength

$$H = dB + Q_{3L} - \frac{1}{30} Q_{3Y}, \tag{2.34}$$

where the Chern-Simons forms $Q_{3Y} = \mathrm{Tr}\,(AF - \frac{1}{3}A^3)$ and $Q_{3L} = \mathrm{tr}\,(\omega R - \frac{1}{3}\omega^3)$ satisfy

$$dQ_{3Y} = \mathrm{Tr}\, F^2 \qquad dQ_{3L} = \mathrm{tr}\, R^2 .$$

The kinetic term in the lagrangian for $B_{\mu\nu}$ is then proportional to $H_{\mu\nu\lambda}H^{\mu\nu\lambda}$. So that H be invariant under local gauge and local Lorentz transformations, B is defined to have the infinitesimal gauge and Lorentz transformation rules

$$\delta^Y_v B = \frac{1}{30} Q^1_{2Y} \qquad \delta^L_\alpha B = -Q^1_{2L}$$

where the 2-forms $Q^1_{2Y} = \mathrm{Tr}\,(v\, dA)$ and $Q^1_{2L} = \mathrm{tr}\,(\alpha\, d\omega)$ satisfy

$$\delta^Y_v Q_{3Y} = dQ^1_{2Y} \qquad \delta^L_\alpha Q_{3L} = dQ^1_{2L} .$$

In analogy with Q_{2n+1} and Q^1_{2n}, we now introduce forms X_7 and X^1_6 satisfying

$$\begin{aligned} X_8 &= dX_7 \\ \delta X_7 &= dX^1_6 . \end{aligned} \tag{2.35}$$

Since $d\delta X_7 = \delta d X_7 = \delta X_8 = 0$, there are always solutions to these equations for arbitrary invariant forms X_8. These solutions can usually be determined by inspection, with any ambiguities associated with the arbitrary addition of an exact form resovable by a well-defined prescription for imposing Bose symmetry. For example, to determine the anomaly associated with I_{12}, we first need to write $I_{12} = dQ_{11}$, but to any Q_{11} we can always add $d\big((Q_{3L} - \frac{1}{30}Q_{3Y})X_7\big)$ with an arbitrary coefficient. Symmetry in the 6 R's or F's of I_{12} (corresponding to symmetry in the 6 external graviton or gauge lines of a hexagon diagram), however, dictates that we write

$$Q_{11} = \frac{1}{3}(Q_{3L} - \frac{1}{30}Q_{3Y})X_8 + \frac{2}{3}(\mathrm{tr}\, R^2 - \frac{1}{30}\mathrm{Tr}\, F^2)X_7$$

since using, for example, the symmetrized trace prescription of (1.30),(2.24), there are twice as many places to make an insertion of an ω or an A in the four R's or F's of X_8 as in the two R's or F's of $(\operatorname{tr} R^2 - \frac{1}{30}\operatorname{Tr} F^2)$.

The anomaly form Q^1_{10} associated with I_{12}, satisfying $\delta Q_{11} = dQ^1_{10}$, then follows easily. Restoring the normalization factor mentioned before (2.29), the resulting anomaly takes the form

$$\begin{aligned} \mathcal{a} = \delta\Gamma_{\text{eff}} = \delta^Y_v \Gamma_{\text{eff}} + \delta^L_\alpha \Gamma_{\text{eff}} &= \int Q^1_{10}(v) + \int Q^1_{10}(\alpha) \\ &= \frac{2i}{9(4\pi)^5} \int \left[(Q^1_{2L} - \frac{1}{30} Q^1_{2Y}) X_8 + 2(\operatorname{tr} R^2 - \frac{1}{30}\operatorname{Tr} F^2) X^1_6 \right]. \end{aligned} \tag{2.36}$$

But now the local counterterm

$$C = \frac{2i}{9(4\pi)^5} \int \left[3B\, X_8 - 2(Q_{3L} - \frac{1}{30} Q_{3Y}) X_7 \right] \tag{2.37}$$

(unique up to terms which are gauge invariant) satisfies

$$\mathcal{a} + \delta C = 0,$$

cancelling both the gauge and gravitational anomalies. The existence of this mechanism and its natural realization in the context of string theories are compelling evidence that string theories embody many magical and profound properties. Much effort is currently being invested to determine whether these properties are as well realized in nature.

D. We conclude our discussion of gauge and gravitational anomalies with a final illustration of the formalism. The mathematical relation between the gauge anomaly in $2n$-dimensions and the index density in $(2n+2)$-dimensions also has a realization in the physical context of fermion zero modes on strings, as pointed out in [22]. We shall see that anomalies in the effective field theory on the string due to chiral zero modes are compensated by quantum number flows in the higher dimensional space in which the string is embedded. What we mean by a string here is substantially different from the preceding subsection. A string now designates an object with two transverse spatial directions. In four dimensions this means that the string is a (1+1)-dimensional object (a one-dimensional object which also propagates in the time direction), whose internal dynamics is described by a 2-dimensional field theory. In $(2n+2)$-dimensions, this makes a string a $((2n-1)+1)$-dimensional object whose internal dynamics is described by a $2n$-dimensional field theory.

We choose to consider a scalar field $\Phi = \Phi_1 + i\Phi_2$ having a non-zero vacuum expectation value with magnitude μ and coupled for definiteness to Dirac fermions in $(2n+2)$-dimensions via

$$\mathcal{L} = \overline{\psi}(\Phi_1 + i\Gamma_5\Phi_2)\psi, \tag{2.38}$$

where $\Gamma_5 = \prod_1^{2n+2}\Gamma^a$ and the Γ's are the Γ-matrices in $(2n+2)$-dimensions. We shall take our string to lie spatially along the $x^2 \dots x^{2n}$ hyperplane so that the internal coordinates are $\{x_{\text{int}}\} = x^1, \dots, x^{2n}$. For the external coordinates $\{x_{\text{ext}}\} = x^{2n+1}, x^{2n+2}$, we shall frequently use cylindrical coordinates $x^{2n+1} = \rho\cos\varphi$, $x^{2n+2} = \rho\sin\varphi$. A string configuration with winding number m is then given by $\Phi = f(\rho)e^{im\varphi}$, where $f(\rho)$ approaches zero at the origin and μ at infinity.

We wish to show that the effective theory of the lowest energy modes is a $2n$-dimensional theory of m chiral fermions on the string. To exhibit the low energy modes, it is convenient to use a basis in which the first $2n$ Γ-matrices take the form $\Gamma^a = 1 \otimes \gamma^a$, $a = 1, \dots, 2n$, where γ^a are the γ-matrices in $2n$-dimensions, and the last two Γ-matrices are $\Gamma^{2n+1} = \sigma_2 \otimes \gamma_5$, $\Gamma^{2n+2} = \sigma_1 \otimes \gamma_5$. We then write $\sum_{a=1}^{2n}\Gamma^a D_a = 1 \otimes \not{D}_{\text{int}}$ and $\sum_{a=2n+1}^{2n+2}\Gamma^a D_a = \not{D}_{\text{ext}} \otimes \gamma_5$, where int and ext refer to internal and external with respect to the string. The internal and external chirality operators, $\Gamma_{\text{int}} \equiv i^n \prod_{a=1}^{2n}\Gamma^a = 1 \otimes \gamma_5$ and $\Gamma_{\text{ext}} \equiv i\Gamma^{2n+1}\Gamma^{2n+2} = \sigma_3 \otimes \gamma_5$, satisfy $\Gamma_{\text{int}}\Gamma_{\text{ext}} = \Gamma_5$, correlating the internal and external chiralities for fixed Γ_5.

Let us now consider the 2-dimensional problem

$$\mathcal{D}\chi \equiv (\not{D}_{\text{ext}} + (\Phi_1 + i\sigma_3\Phi_2))\chi = 0, \tag{2.39}$$

where χ is a Dirac fermion in 2-dimensions and σ_3 is the chirality $i\gamma^1_{\text{ext}}\gamma^2_{\text{ext}} = i\sigma_2\sigma_1$. The existence of fermion zero modes in the presence of a scalar field with a string structure is guaranteed by an index theorem analyzed in [23]. The result is that the index of the 2-dimensional operator $\mathcal{D}$, defined to be the number of zero modes of $\mathcal{D} = \not{D} + (\Phi_1 + i\sigma_3\Phi_2)$ minus the number of zero modes of $\mathcal{D}^\dagger = -\not{D} + (\Phi_1 - i\sigma_3\Phi_2)$, is equal to the winding number m of the complex scalar field $\Phi(x)$.

We now proceed in a manner similar to the approach taken in Kaluza-Klein theories, in which the massless fermions left over in the field theory of the 'external' uncompactified dimensions result from the zero modes of the 'internal' Dirac operator on the compactified dimensions (here, however, the role of 'internal' and 'external' is reversed since we shall associate massless fermions in the effective $2n$-dimensional internal field theory on the string with the 2-dimensional zero modes external to the string).

The $(2n+2)$-dimensional operator acting on Dirac fermions

$$\not{D} + (\Phi_1 + i\Gamma_5\Phi_2) = 1 \otimes \not{D}_{\rm int} + \left(\not{D}_{\rm ext} \otimes \gamma_5 + (\Phi_1 + i\sigma_3 \otimes \gamma_5\, \Phi_2)\right) \qquad (2.40)$$

then reduces to $1 \otimes \not{D}_{\rm int} + \mathcal{D} \otimes \mathbf{1}$ acting on wave functions of the form $\psi = \chi\eta_+$ and to $1 \otimes \not{D}_{\rm int} + \mathcal{D}^\dagger \otimes \mathbf{1}$ acting on wave functions of the form $\psi = \chi\eta_-$. Here $\eta_\pm$ are positive and negative chirality Weyl spinors in $2n$-dimensions. The only modes without a large mass coming from the second term of (2.40) are thus either of the form $\chi(x_{\rm ext})\,\eta_+(x_{\rm int})$ where χ is a zero mode of $\mathcal{D}$, or of the form $\chi'(x_{\rm ext})\,\eta_-(x_{\rm int})$ where χ' is a zero mode of $\mathcal{D}^\dagger$. But in the presence of a string of winding number m we have $\operatorname{ind}\mathcal{D} = m$, so we learn that the lowest energy modes are described by a $2n$-dimensional theory comprised of a net number m of chiral fermions η_+ (or $m\,\eta_-$'s if $m < 0$).

The original theory of a Dirac fermion in $(2n+2)$-dimensions coupled to an axionic string and as well having a gauge charge is overall anomaly-free and thus has local conservation of gauge charge and energy-momentum. The string zero modes, on the other hand, are summarized by $2n$-dimensional chiral fermion fields. Since the zero modes have a definite chirality, according to (1.29) and (2.26*b*) their coupling to gauge fields or gravity is subject to the anomalies

$$\begin{aligned} \operatorname{tr} v\, D_\mu j_0^\mu &= m\, Q_{2n}^1(v) \\ \xi_\nu D_\mu T_0^{\mu\nu} &= m\, Q_{2n}^1(v_\xi) \quad \left((v_\xi)^\alpha{}_\beta = \partial_\beta \xi^\alpha\right), \end{aligned} \qquad (2.41)$$

indicating that the charge and energy-momentum carried by the string zero modes are not conserved (we have chosen to express the gravitational anomaly as a violation of general coordinate invariance). The anomaly polynomials Q_{2n}^1 here satisfy $\delta_v Q_{2n+1} = dQ_{2n}^1(v)$ and $\delta_\xi Q_{2n+1} = dQ_{2n}^1(v_\xi)$, where the Chern-Simons form Q_{2n+1} is determined from the $(2n+2)$-dimensional index density according to $2\pi i[\hat{A}\operatorname{ch}(F)]_{2n+2} = dQ_{2n+1}$ (this normalization of Q_{2n}^1 agrees with (2.25) for the gravitational anomaly but we have for convenience absorbed the factor of $N = \frac{i^{n+2}}{(2\pi)^n(n+1)!}$ with respect to (1.29*b*) for the gauge Q_{2n}^1). For $(2n+2=4)$-dimensional spacetime, for example, the zero modes on the string correspond to fermions traveling at the speed of light in a single direction along the string determined by their chirality, and we have

$$Q_2^1(v) = -\frac{i}{4\pi}\operatorname{tr}(v\, dA) \quad Q_2^1(v_\xi) = \frac{i}{96\pi}\operatorname{tr}(v_\xi\, d\Gamma).$$

In order for the full theory to have locally conserved quantities, the string/fermion system must be able to compensate the anomalous non-conservation (2.41) on the string with a net flow of charge from the outside world. This flow cannot be mediated by straightforward emission or absorption of fermions because the scalar field expectation value μ gives the fermions a mass (which can be arbitrarily large) off the string. Equivalently, the anomalies above tell us that the part of the overall fermion determinant coming from the string zero modes has the anomalous variations under gauge or general transformations

$$\delta_v \Gamma^0_{\text{eff}} = -m \int Q^1_{2n}(v) \qquad \delta_\xi \Gamma^0_{\text{eff}} = -m \int Q^1_{2n}(v_\xi) \tag{2.42}$$

which must be cancelled by something in the rest of the fermion determinant (coming from massive degrees of freedom living off the string) to restore gauge and coordinate invariance.

We now describe how the string can exchange charges with the outside world. Away from the core of the string we can write $\Phi(x) = \mu\, e^{i\theta(x)}$, where $\theta(x)$ is the dimensionless axion field. The θ dependence may then be removed from the interaction term $\mu\,\overline{\psi} e^{i\theta\Gamma_5}\psi$ by a chiral rotation $\psi \to e^{-i\theta\Gamma_5/2}\psi$, $\overline{\psi} \to \overline{\psi} e^{-i\theta\Gamma_5/2}$. According to (1.4), (1.7), and (2.11), however, this anomalous change of basis induces an effective interaction term

$$\Gamma_{\text{eff}}[\theta] = -i \int \theta [\hat{A}\, \text{ch}(F)]_{2n+2}. \tag{2.43}$$

In 4-dimensions, for example, this results in the effective interaction of axions with gauge and gravitational fields

$$\Gamma_{\text{eff}}[\theta] = \frac{i}{8\pi^2} \int \theta \,\text{tr}\, F^2 - \frac{i}{192\pi^2} \int \theta \,\text{tr}\, R^2.$$

(2.43) is ill-defined in the presence of a string since θ is then multivalued. It only makes sense if it is integrated by parts using $2\pi i [\hat{A}\, \text{ch}(F)]_{2n+2} = dQ_{2n+1}$ to yield

$$\Gamma_{\text{eff}}[\theta] = \frac{1}{2\pi} \int d\theta\, Q_{2n+1} \tag{2.44}$$

(retaining the Chern-Simons normalization described after (2.41)).

It is now straightforward to see that $\Gamma_{\text{eff}}[\theta]$, once written sensibly as in (2.44), has an anomalous variation which exactly compensates the anomalous contribution of the zero mode effective action under gauge or general coordinate transformations. For either type of infinitesimal transformation, we have

$$\delta \Gamma_{\text{eff}}[\theta] = \frac{1}{2\pi} \int d\theta\, \delta Q_{2n+1} = \frac{1}{2\pi} \int d\theta\, dQ^1_{2n} = \frac{1}{2\pi} \int d^2\theta\, Q^1_{2n}. \tag{2.45}$$

$d^2\theta$ is non-vanishing in the above expression because $d\theta$ (despite its appearance) is not globally exact due to the multivaluedness of θ. In fact, for $\theta = m\varphi$ (or any other configuration in which θ wraps m times around the origin in x^{2n+1}, x^{2n+2} space) $d^2\theta$ should be interpreted as $2\pi m\, \delta^{(2)}(x_{\rm ext})\, dx^{2n+1} \wedge dx^{2n+2}$. This is because any surface integral of $d^2\theta$ can be converted to a boundary integral of $d\theta$, which in turn gives $2\pi m$ if and only if the boundary encircles the origin (as can easily be verified by playing with simple 2-dimensional integrals). The appearance of $d^2\theta$ is in any event an artifact of our approximation of the string as a point singularity in the external two dimensions. (2.45) then reduces to the $2n$-dimensional integral

$$\delta\Gamma_{\rm eff}[\theta] = m \int Q^1_{2n}, \tag{2.46}$$

exactly compensating the anomalous variations (2.42).

To gain a little more intuition for this compensation mechanism, we consider the current

$$J = \frac{\delta}{\delta A}\Gamma_{\rm eff}[\theta] \tag{2.47}$$

(the current is a matrix-valued $(2n+1)$-form $J = *J_\mu dx^\mu$ with dual components equal to the usual current J_μ, and its divergence satisfies $DJ \equiv dJ + [A, J] = *(D_\mu J^\mu)$). J is the current induced in the axion field vacuum by the presence of the background fields. It has a ρ component if there are gauge electric and magnetic fields filling out the internal directions of the string. In $(2n+2=4)$-dimensions, for example, a J_ρ is induced by an electric field along the (spatially 1-dimensional) string. From the infinitesimal gauge transformation $\delta_v\Gamma_{\rm eff}[\theta] = \int {\rm tr}\,[Dv\, J] = -\int {\rm tr}\,[v\, DJ]$ together with (2.45), we can pick out the divergence of the current

$${\rm tr}\,[v\, DJ] = -m\, Q^1_{2n}(v)\; \delta^{(2)}(x_{\rm ext})\; dx^{2n+1} \wedge dx^{2n+2}. \tag{2.48}$$

We see that the current is conserved everywhere except on the string itself where the net inflow of charge is exactly what is needed to account for the charge appearing on the string via the anomaly (2.41) in the string zero mode sector.

Similarly we can define the energy-momentum tensor induced by the background fields

$$\frac{1}{2}T^{\mu\nu} = \frac{\delta}{\delta g_{\mu\nu}}\Gamma_{\rm eff}[\theta], \tag{2.49}$$

(here a $(2n+2)$-form) which describes the inflow of energy momentum from the region outside the string in the presence of background gauge and gravitational fields. From the infinitesimal coordinate transformation $\delta_\xi \Gamma_{\text{eff}}[\theta] = -\int \xi_\nu D_\mu T^{\mu\nu}$ together with (2.45), we find

$$\xi_\nu D_\mu T^{\mu\nu} = -m\, Q^1_{2n}(v_\xi)\, \delta^{(2)}(x_{\text{ext}})\; dx^{2n+1} \wedge dx^{2n+2}, \tag{2.50}$$

exactly compensating the anomalous variation (2.41) in the energy-momentum carried by the string due to the chiral zero modes, and elsewhere conserved.

It is interesting to see that a theory with anomalies can be given a consistent interpretation by imbedding it in a higher dimensional spacetime, and having its intrinsic lack of charge conservation compensated by a flow of quantum numbers from the outside.

3. Sigma-models

We shall now consider how the formalism and ideas thus far introduced may be applied to the anomalies which occur[24] in non-linear σ-models. These are theories with scalar fields $\varphi(x)$ living on an arbitrary manifold M coupled to fermion fields $\psi(x)$. The scalar fields $\varphi(x)$ are simply maps from spacetime (which we continue to take to be S^{2n}) into M. The fermion fields, on the other hand, are introduced geometrically as sections of some N-dimensional complex vector bundle $\mathcal{B}$ over M.

In an effective lagrangian approach, one considers a theory which incorporates a group G of chiral symmetries (linearly realized in some underlying theory) spontaneously broken to a subgroup H below some scale. The low energy theory then describes goldstone boson fields parametrized by the homogeneous space G/H (barring accidental degeneracies) with G non-linearly realized as a symmetry group. The standard low energy theorems follow from the Ward identities, manifesting the original presence of G symmetry in the low energy world[25]. When chiral fermions are coupled to the G/H goldstone boson fields, however, there exists the possibility that the naive Ward identities may be broken by anomalies similar to those appearing in the gauge case. If such anomalies cannot be removed, then there would be no consistent way of coupling such fermion fields so as to preserve the standard G/H current algebra predictions. Our aim in what follows is to detail when such anomalies may appear and cause problems.

A. We first develop some aspects of the theory of coset spaces relevant to our discussion. The systematic application of general G/H spaces to low energy effective lagrangians and current algebra was developed in the classic papers of Callan, Coleman, Wess, and Zumino (CCWZ)[26]. We shall rewrite these results here in a way which makes their geometrical interpretation more evident. This approach has recently been very popular in the context of Kaluza-Klein theories and we refer the reader to the literature[27][28] for further details.

Given a Lie group G and a subgroup H⊂G, let H_i be the generators of H, X_a the broken generators of G, and T_A the complete set of generators of G: $T = (H, X)$ (all of which are taken to be anti-hermitian). We will assume that the Lie algebra of G admits a reductive splitting, which means that the group commutation relations take the form

$$[H_i, H_j] = f_{ij}{}^k H_k \tag{3.1a}$$

$$[H_i, X_a] = f_{ia}{}^b X_b \tag{3.1b}$$

$$[X_a, X_b] = f_{ab}{}^i H_i + f_{ab}{}^c X_c. \tag{3.1c}$$

(3.1a) simply expresses the fact that H is a subgroup of G and (3.1b) is the condition for the splitting (H_i, X_a) to be reductive (which is possible, in particular, whenever H is compact). A homogeneous space G/H is called symmetric[28] if it is possible to arrange things so that $f_{ab}{}^c = 0$. Simple examples of this are given by the Grassmannian spaces $U(p+q)/U(p)\times U(q)$.

In the local formulation of G/H spaces we start by choosing a canonical coset representative $\ell(\varphi^\alpha) \in$ G, where φ^α are coordinates which parametrize the left cosets in a given coordinate patch, and $\alpha = 1, N_B$ (N_B = number of broken generators). The map $\ell(\varphi)$: G/H $\rightarrow$ G is thus a local section of the principal H-bundle G $\rightarrow$ G/H. The left action of G on $\ell(\varphi)$ is given by

$$g\,\ell(\varphi) = \ell(\varphi')\,h(\varphi, g), \tag{3.2}$$

where $h(\varphi, g) \in$ H is a compensating H transformation depending on φ and g which brings us back to our canonical choice for coset representative in the new coset specified by φ'. In terms of the intrinsic coordinates φ^α, the G-action generates a non-linear G-realization $\varphi^\alpha \rightarrow \varphi'^\alpha(\varphi, g)$. Under infinitesimal transformations, $g \approx 1 + \epsilon^A T_A$, we have $\varphi^\alpha \rightarrow \varphi^\alpha + \epsilon^A k_A^\alpha(\varphi)$, where the k_A's are by definition the Killing vectors on G/H associated with the infinitesimal symmetry action of G acting by left translation. Under this transformation, we may also parametrize the compensating $h(\varphi, g)$ infinitesimally as $h(\varphi, g) \approx 1 + \epsilon^A \Omega^i{}_A(\varphi) H_i$, where $\Omega^i{}_A H_i \equiv \Omega_A$ is known as the 'H-compensator'[27]. The infinitesimal form of (3.2) then reads

$$k_A^\alpha(\varphi)\frac{\partial}{\partial\varphi^\alpha}\ell(\varphi) = T_A \ell(\varphi) - \ell(\varphi)\,\Omega_A(\varphi). \tag{3.3}$$

The vielbein and connection follow from the CCWZ prescription. We decompose the 1-form $\ell^{-1}d\ell = \left(\ell^{-1}(\varphi)\,\frac{\partial}{\partial\varphi^\alpha}\ell(\varphi)\right)d\varphi^\alpha$ into its pieces along H and X

$$\ell^{-1}d\ell = \left(\omega^i{}_\alpha(\varphi)H_i + e^a{}_\alpha(\varphi)X_a\right)d\varphi^\alpha = \omega^i H_i + e^a X_a. \tag{3.4}$$

$e^a{}_\alpha(\varphi)$ represents the vielbein in G/H, and $\omega^i{}_\alpha(\varphi)$ is an H-connection (called the canonical connection[28]) which plays a very important role in defining the coupling of matter fields to the goldstone boson manifold. (Our index conventions henceforth are $a, b, \ldots$ = flat coset indices; $\alpha, \beta, \ldots$ = curved coset indices; $i, j, \ldots$ = H indices; $A, B, \ldots$ = G indices; and $\mu, \nu, \ldots$ = spacetime indices.) Explicit forms for k_A^α and Ω^i_A can be given in terms of ω and e defined above, and the matrix elements $D_A{}^B(g)$ of the adjoint representation of G defined by

$$g^{-1}T_A\,g = D_A{}^B(g)\,T_B. \tag{3.5}$$

These are derived by combining (3.3) and (3.4) to give

$$k^\alpha_A \frac{\partial}{\partial\varphi^\alpha}\ell = \ell(D_A{}^B(\ell)\,T_B - \Omega_A)$$
$$= k^\alpha_A \ell\,(e^a{}_\alpha X_a + \omega^i{}_\alpha H_i),$$

from which it follows that $k^\alpha_A\, e^a{}_\alpha = D_A{}^a(\ell)$ and $k^\alpha_A\,\omega^i{}_\alpha = D_A{}^i - \Omega^i_A$. In terms of the inverse vielbein $e_a{}^\alpha$, we thus find the explicit forms

$$\begin{aligned} k^\alpha_A(\varphi) &= D_A{}^a(\ell(\varphi))\, e_a{}^\alpha(\varphi), \\ \Omega^i_A(\varphi) &= D_A{}^i(\ell(\varphi)) - D_A{}^a(\ell(\varphi))\, e_a{}^\alpha(\varphi)\,\omega^i{}_\alpha(\varphi). \end{aligned} \tag{3.6}$$

Since we will be interested in determining whether the G-action on G/H is anomalous, it is important to work out how $e^a{}_\alpha$ and $\omega^i{}_\alpha$ transform under isometries, i.e. under the group of motions generated by the Killing vectors $k^\alpha_A \frac{\partial}{\partial\varphi^\alpha}$. Geometrically, this is equivalent to computing the Lie derivatives of e and ω. From (3.2), we see that a finite transformation $\ell \to g\ell h^{-1}$ gives

$$\ell^{-1}d\ell \to h\ell^{-1}g^{-1}\,d(g\ell h^{-1}) = h(\ell^{-1}d\ell)h^{-1} + h\,dh^{-1}.$$

In terms of the decomposition (3.4), this implies

$$\omega \to h\omega h^{-1} + h\,dh^{-1} \qquad e \to heh^{-1}. \tag{3.7}$$

Infinitesimally, we have $h \approx 1 + \epsilon^A\Omega_A(\varphi)$, with $\Omega_A = \Omega^i_A H_i$ the H-compensator, so the infinitesimal form of (3.7) is

$$\pounds_{k_A}\omega^i = -d\Omega^i_A - f_{jk}{}^i\omega^j\,\Omega^k_A = -(d\Omega_A + [\omega,\Omega_A])^i \quad (\omega = \omega^i{}_\alpha\, H_i\, d\varphi^\alpha) \tag{3.8a}$$

$$\pounds_{k_A} e^a = -f_{bj}{}^a\, e^b\,\Omega^j_A = [\Omega_A, e]^a \quad (e = e^a{}_\alpha\, X_a\, d\varphi^\alpha). \tag{3.8b}$$

(3.8a) means that the canonical connection transforms like an H-gauge connection under global G-transformations; (3.8b) means that the vielbein e transforms according to the same gauge transformation like a field in the adjoint representation of H. The vielbein and connection are thus not G-invariant, but only G-invariant up to H-gauge transformations.

(3.8) provides the basic properties used in [26] to incorporate matter fields into a phenomenological theory with underlying symmetry group G and unbroken subgroup H. One need only choose an action which classically is H-gauge invariant in order that the transformations induced by the Killing vectors on G/H remain symmetries. The transformation properties of $e_\alpha = e^a{}_\alpha(\varphi)X_a$ under G first of all provide a natural G-invariant metric on G/H:

$$g_{\alpha\beta}(\varphi) = \mathrm{tr}[e_\alpha e^\dagger_\beta], \tag{3.9}$$

in terms of which the kinetic term for the goldstone bosons may be written

$$\mathcal{L}_0 = \frac{f_\pi}{2} g_{\alpha\beta}\,\partial_\mu\varphi^\alpha\,\partial^\mu\varphi^\beta. \tag{3.10}$$

The fermionic matter fields are chosen to transform according to some representation of H, reducible or irreducible. Let $\mathcal{H}_i$ denote the generators of H in this representation. Then the lagrangian

$$\begin{gathered}\mathcal{L}_m = \overline{\psi} i\gamma^\mu(\partial_\mu + \omega_\mu)P_+\,\psi,\\ \partial_\mu = \frac{\partial}{\partial x^\mu}, \quad \omega_\mu = \omega^i{}_\alpha\,\partial_\mu\varphi^\alpha\,\mathcal{H}_i, \quad P_+ = \frac{(1+\gamma_5)}{2}\end{gathered} \tag{3.11}$$

is invariant under the global G-transformations

$$\ell(\varphi) \to g\,\ell(\varphi)\,h^{-1}(\varphi, g) \tag{3.12a}$$

$$\omega \to h(\omega + d)h^{-1} \tag{3.12b}$$

$$\psi \to h\psi. \tag{3.12c}$$

In (3.12*b*, *c*), $h(\varphi, g)$ is the compensating H transformation in the representation $\mathcal{H}$ (we represent it for convenience with the same abstract symbol as used for elements of H itself; no confusion should arise).

The classically conserved G-currents follow easily from (3.10)–(3.12):

$$\begin{aligned}J^\mu_A &= (k^\alpha_A\frac{\delta}{\delta\partial_\mu\varphi^\alpha} + \Omega_A\psi\frac{\delta}{\delta\partial_\mu\psi})(\mathcal{L}_0 + \mathcal{L}_m)\\ &= f_\pi\, g_{\alpha\beta}\, k^\alpha_A(\varphi)\,\partial^\mu\varphi^\beta + i\overline{\psi}\gamma^\mu\mathcal{H}_i P_+\,\psi\;(\omega^i{}_\alpha k^\alpha_A(\varphi) + \Omega^i_A(\varphi))\\ &= f_\pi\,\mathrm{tr}[X_a e^\dagger_\beta]\,D_A{}^a(\ell)\,\partial^\mu\varphi^\beta + i\overline{\psi}\gamma^\mu\mathcal{H}_i P_+\,\psi\;D_A{}^i(\ell).\end{aligned} \tag{3.13}$$

We refer to the consequences of the Ward identities associated with these G-currents as current algebra predictions.

This completes our discussion of the isometries associated with the left action of G on G/H. If the normalizer N(H) of H in G is larger than H, then there are also isometries associated with the right action of N(H)/H on G/H. We have also only been considering global G-symmetries generated by a spacetime independent $g \approx 1 + \epsilon$. We mention that it is straightforward to transcribe the results of [26] and promote this to a local G-symmetry, effectively gauging the isometry group G with a gauge field $A(x) \in \mathcal{L}(\mathrm{G})$ (a lie-algebra valued 1-form on spacetime). Further details on each of these issues may be found in [3].

B. We are now in a position to consider the anomalous change in the σ-model effective action, defined as follows: we take our spacetime manifold to be S^{2n} and begin by considering a topologically trivial background bosonic configuration $\ell(\varphi)$, i.e. with the map $\ell: S^{2n} \to \mathrm{G/H}$ in the trivial homotopy class (removal of this restriction on the image of $\varphi(x)$ is discussed in [3]). As described in the previous section, we couple a set of Weyl fermions transforming according to the representation $\mathcal{H}$ of H with lagrangian (3.11). Performing the functional integral over the fermion fields gives

$$e^{-\Gamma_{\text{eff}}[\omega]} = \int_{\psi,\overline{\psi}} e^{-\int d^{2n}x\ \mathcal{L}_m(\omega(\varphi),\overline{\psi},\psi)}\ . \tag{3.14}$$

We wish to study the transformation properties of the fermion effective action $\Gamma_{\text{eff}}[\omega]$ under G-isometries. If the fermions transform according to a complex representation of H, the theory does not admit a gauge invariant regulator (e.g. a Pauli-Villars regulator) to cut off the ultraviolet divergences in the integral over the fermion fields. Given the form of the coupling of the fermions to the H-connection (3.11), we see that the calculation is precisely analogous to that performed in the case (1.1) for ordinary gauge theories. Specifically, we see from (3.8a) that the canonical connection ω_μ transforms like an H-gauge field under infinitesimal G-transformations

$$\omega \to \omega - d(\epsilon^A \Omega_A) - [\omega, \epsilon^A \Omega_A] = \omega - D(\epsilon^A \Omega_A).$$

It follows immediately that the change of the effective action $\Gamma_{\text{eff}}[\omega]$ under an infinitesimal G-transformation is simply given by the standard formulæ for gauge anomalies as described in sec. 1, with $\epsilon^A \Omega_A$ playing the role of the infinitesimal gauge transformation parameter v. We have then for the change in the fermion effective action

$$\delta_\epsilon \Gamma_{\text{eff}}[\omega] = N \int_{S^{2n}} Q^1_{2n}(\epsilon^A \Omega_A, \omega), \tag{3.15}$$

where both Ω_A and ω are matrix-valued in the matter representation $\mathcal{H}$ of the Lie algebra of H. The anomalous change (3.15) results from the local change in the fibre basis (3.8) induced by the G-action, because the symmetry under such local frame rotations is spoiled by a quantum anomaly. This anomaly then has the effect of breaking the invariance of the quantum effective action $\Gamma_{\text{eff}}[\varphi]$ under the full isometry group G. This does not necessarily make the theory non-sensical since we are not trying to couple the G-currents to dynamical gauge fields. It does, on the other hand, mean that we are not realizing the desired theory with a dynamically broken G symmetry. The fields φ, for example, may gain mass at the quantum level or not satisfy the usual scattering theorems because the full G Ward identities no longer act to insure their status as goldstone bosons.

From (3.15), it is clear that the theory is free from anomalies in G-transformations if the matter fields belong to an anomaly-free representation of H (in $2n = 4$ dimensions, for example, if we take G/H $= SO(N+1)/SO(N) = S^N$ with $N \neq 6$, then the fermion representations are unconstrained since all representations of $SO(N)$, $N \neq 6$, are anomaly free). This condition, while suggested by a naive formulation of the G/H σ-model in terms of maps into G with lagrange-multiplier H-gauge fields added to eliminate the extra degrees of freedom, is not obviously necessary, however, since it may be possible to construct a φ-dependent counterterm whose anomalous variation under G-transformations exactly cancels the fermion-induced anomaly. Adding such a counterterm to the original bosonic action will then lead to a combined bosonic plus fermionic theory free of anomalies (it is always legitimate, of course, to add any such term as long as it is local in the physical pion fields φ^α). We shall indeed find that certain anomalous representations of H do allow the construction of this counterterm and thus can be allowed in G/H current algebra without spoiling G-invariance. We note that the counterterm may significantly affect low energy predictions based on G/H current algebra, just as inclusion of the usual Wess-Zumino term in an $SU(N)_L \times SU(N)_R/SU(N)_V$ phenomenological lagrangian can affect the phenomenolgy of the pseudoscalar octet[29]. We also point out that (3.15) will automatically vanish for arbitrary fermion content when the dimension of the target space G/H is less than the dimension $2n$ of the spacetime manifold. This is because (3.15) can also be viewed as in integral over the image of S^{2n} in G/H and any $2n$-form such as Q^1_{2n} pushed forward to G/H necessarily vanishes on a manifold of dimension less than $2n$ (for $2n = 4$ dimensions, for example, this will happen if we take G/H $= SU(2)/U(1) = S^2$).

Now we must consider possible counterterms which may be added to the effective action to compensate the anomalous variation (3.15) under G-transformations. We start by denoting the generators of G in some representation $\mathcal{T}$ of $\mathcal{L}(\mathrm{G})$ as $\mathcal{T}_A$, and write $\tilde{\omega} = \omega^i \mathcal{T}_i$, $\tilde{\Omega}_A = \Omega^i_A \mathcal{T}_i$, etc. Within a given patch of G/H, it is always possible to find an interpolation $\ell_t(\varphi)$ such that $\ell_t = 1$ at $t = 0$, and $\ell_t = \ell(\varphi)$ at

$t = 1$. In terms of $\tilde{\omega}^{\ell_t^{-1}} = \ell_t(\tilde{\omega} + d)\ell_t^{-1}$, we may then construct the quantity

$$-N \int_0^1 \delta t \int_{S^{2n}} Q^1_{2n}(\ell_t d_t \ell_t^{-1}, \tilde{\omega}^{\ell_t^{-1}}) = \Gamma_{\text{eff}}[\tilde{\omega}^{\ell^{-1}}] - \Gamma_{\text{eff}}[\tilde{\omega}]. \tag{3.16}$$

This is essentially the Wess-Zumino-Witten lagrangian[29] for the goldstone boson fields $\ell(\varphi)$ coupled to the gauge field $\tilde{\omega}$. Although perhaps not obvious in the above form, (3.16) is actually independent (up to an additive $2\pi i\cdot$(integer) for interpolations ℓ_t in different topological classes) of the specific choice of interpolation ℓ_t and has only a local dependence on $\tilde{\omega}$ and ℓ. This is because the integrand $Q^1_{2n}(\ell_t d_t \ell_t^{-1}, \tilde{\omega}^{\ell_t^{-1}}) = Q_{2n+1}(\tilde{\omega}^{\ell_t^{-1}} + \ell_t d_t \ell_t^{-1})$ is a closed form, satisfying $(d + d_t)Q_{2n+1}(\tilde{\omega}^{\ell_t^{-1}} + \ell_t d_t \ell_t^{-1}) = \text{tr}(\tilde{R}^{\ell_t^{-1}})^{n+1} = \text{tr}\tilde{R}^{n+1} = 0$ (since it is a $2n+2$-form and the curvature 2-form $\tilde{R} = d\tilde{\omega} + \tilde{\omega}^2$ pulled back to spacetime has only form components in the $2n$ spacetime directions).

Under a G-transformation we have $\tilde{\omega}^{\ell^{-1}} \to (\tilde{\omega}^{h^{-1}})^{h\ell^{-1}g^{-1}} = (\tilde{\omega}^{\ell^{-1}})^{g^{-1}}$, so a rigid (space-time independent) G-transformation leaves the first term on the right hand side of (3.16) invariant (global gauge transformations, as mentioned, produce no anomalous variation; the case of local G-transformations is discussed further in [3]). From (1.29a), it follows that the infinitesimal variation of the second term on the right hand side of (3.16) is

$$N \int_{S^{2n}} Q^1_{2n}(\epsilon^A \tilde{\Omega}_A, \tilde{\omega}). \tag{3.17}$$

Now in general the anomaly Q^1_{2n} for a given representation $\mathcal{T}$ is proportional, according to (1.29b), to the $(n+1)$-fold symmetrized trace

$$d_{A_1 A_2 \ldots A_{n+1}}(\mathcal{T}) = \text{tr}\,\mathcal{T}_{\{A_1}\mathcal{T}_{A_2}\ldots\mathcal{T}_{A_{n+1}\}}. \tag{3.18}$$

Cancellation of the anomalous variation (3.15) by the first term in (3.17) then requires that the representation $\mathcal{T}$ satisfy

$$d_{i_1 i_2 \ldots i_{n+1}}(\mathcal{T}) = d_{i_1 i_2 \ldots i_{n+1}}(\mathcal{H}), \tag{3.19}$$

where the left hand side above is just (3.18) restricted to the generators $\mathcal{T}_i$ of the subgroup $H \subset G$. Adding the counterterm (3.16) to Γ_{eff} thus cancels the anomalous variation (3.15), resulting in a theory with the desired G-invariance.

It should be clear that unless the representation $\mathcal{H}$ satisfies the condition (3.19) for some $\mathcal{T}$, then there can be no local counterterm which eliminates the anomalous variation (3.15). This is because the choice of the representation $\mathcal{T}$ exhausts the freedom in constructing, out of the available fields, terms like (3.16) with the proper form to compensate (3.15). But adding (3.16) for arbitrary $\mathcal{T}$ to the action gives a new action with anomalous variation equal to that of $\Gamma_{\text{eff}}[\omega]-\Gamma_{\text{eff}}[\tilde{\omega}]$. This quantity is simply the effective action for an ordinary H-gauge theory with fermion content $\mathcal{H} \oplus \overline{\mathcal{T}}|_{\text{H}}$ and is anomalous unless (3.19) is satisfied. If (3.19) is not satisfied, then, we see that the anomalous variation (3.15) cannot be cancelled by a local counterterm for the same reason that the anomaly cannot be so cancelled for an ordinary non-abelian gauge theory with an anomalous fermion content.

Finally, we should point out why the existence of an appropriate anomaly cancelling counterterm should have also been clear a priori in the case when the fermions transform in H-representations which fit together to form a full G-representation. This is because we could then have written an alternative form for the fermionic lagrangian $\mathcal{L}_m = \overline{\psi}' i\gamma^\mu \partial_\mu P_+ \psi'$ in terms of fermions ψ' defined to transform under global G-transformations as $\psi' \to g\psi'$. This lagrangian is automatically invariant under global G-transformations and since no local frame rotations ever enter, we see that whenever the fermions happen to have a well-defined transformation property under all of G, the quantum effective action admits a realization which is manifestly anomaly-free. The form of the lagrangian (3.11) in terms of the fermions ψ can then be recovered by the change of fermionic basis $\psi' = \ell\psi$. This change of basis is anomalous, however, and induces (à la Fujikawa) identically the counterterm (3.16) required for an anomaly-free effective action when the theory is written in terms of the ψ's (as reminiscent of the case discussed in [30]).

The condition, in any event, on the allowable representations $\mathcal{H}$ of H for the fermion fields is then simply the condition (3.19) requiring that any H-anomalies exactly match those of some representation $\mathcal{T}$ of G when restricted to H⊂G. As an example, consider the Grassmannian spaces G/H $= U(p+q)/U(p)\times U(q)$. The representation $(p,1) \oplus (1,q)$ of $U(p)\times U(q)$ is an allowable representation for the fermions as long as the $U(p)$ and $U(q)$ charges are set equal. The anomalies then match those of the representation $(p+q)$ of $U(p+q)$, allowing the construction of the necessary counterterm (3.16) and hence resulting in a well-defined fermion effective action. It is not necessary, however, that the representations of H fit together to form a complete representation of G, as exemplified by the case of $CP^N = SU(N+1)/SU(N)\times U(1)$. The representation (N) of $SU(N)$ is then allowed (without an additional $SU(N)$ singlet) as long as it is given the specific $U(1)$ charge induced by the canonical embedding $SU(N)\times U(1) \subset SU(N+1)$ so that its anomalies then match those of the representation $(N+1)$ of $SU(N+1)$.

There is a close relation between the anomaly matching condition (3.19) given here and that given by 't Hooft[10]. The G/H σ-model may be considered as the effective low energy theory describing the massless modes which result from the dynamical breakdown of a higher energy theory with an underlying flavor G-symmetry. The fermions ψ transforming under the unbroken group H here are then those fermions which remain in the low energy theory, presumably kept massless by some unbroken chiral symmetry. In the absence of any fermions remaining massless, a flavor G-anomaly in multi-current greens functions of the underlying theory could be reproduced in the low energy theory only by goldstone boson poles. The underlying fermions would then have to have been anomaly-free under the subgroup H, since unbroken currents do not produce goldstone bosons from the vacuum and hence can't give any contribution to the low energy G-anomaly to match that of the underlying G-symmetry. If there are remaining massless fermions, on the other hand, then they may transform anomalously under H as long as their anomalies match those of the underlying H$\subset$G. We have thus discovered that the existence of a well-defined fermion effective action and hence the counterterm (3.16) is ***required*** when the condition (3.19) is satisfied, simply because retaining only the massless modes in an effective lagrangian is a legitimate approximation to a well-defined underlying theory. The non-trivial result here is that massless chiral fermions may be consistently coupled to a G/H σ-model only when they ***could*** have resulted from such a dynamical breakdown. The only exception to this occurs when the group manifold G is non-compact. Then there can be no underlying theory with a linearly realized G symmetry because the kinetic term would not be positive definite. But σ-models with G/H $= SO(n,m)/SO(n) \times SO(m)$ or G/H $= SU(n,m)/SU(n) \times SU(m) \times U(1)$, for example, can be well defined with a linearly realized (compact) H symmetry as long as the H representations are again chosen to have an anomaly which matches that of some representation of G when restricted to the H generators.

One context in which σ-models of the form discussed here has arisen recently is in the context of supersymmetric preon model building (see [24] for references). In this case, however, the fermion representation content is fixed by supersymmetry to lie in the holomorphic tangent bundle of G/H with no further freedom in choosing fermions to cancel anomalies. Various models excluded on the basis of their anomalous content are mentioned in [24] and [3].

As discussed further in [3], it turns out that addition of the appropriate local counterterm (3.16) required for a locally well-defined action insures that the total effective action is equally well-defined globally with respect to the symmetry transformation (3.12). In the language of bundles, this absence of a global anomaly can be expressed as the statement that the complex exponential (3.14) is well-defined as a section of a line bundle with trivial transition functions. This is important to maintain in order to allow a functional integral over the scalar fields φ which are also to be treated ultimately as dynamical quantum fields.

C. We shall now generalize these considerations to anomalies which may occur when fermions are coupled to a σ-model defined on an arbitrary riemannian manifold (i.e. not necessarily a coset space). We shall introduce and analyze anomalies which may occur in general coordinate reparametrizations and local frame rotations in these more general theories and explain how our results are related to those of the previous section.

The most general coupling of fermions to a σ-model defined on an arbitrary riemannian manifold M is described by the classical lagrangian

$$\mathcal{L} = \frac{f_\pi}{2}\, g_{\alpha\beta}\, \partial_\mu\varphi^\alpha\, \partial^\mu\varphi^\beta + \overline{\psi} i\gamma^\mu(\partial_\mu + \omega_\mu)P_+\,\psi, \tag{3.20}$$

generalizing (3.10) and (3.11). $g_{\alpha\beta} = \eta_{ab}\, e^a{}_\alpha\, e^b{}_\beta$ is here the metric on the manifold M, $e^a{}_\alpha$ the vielbein, and $\omega_\mu = \partial_\mu\varphi^\alpha\,\omega_\alpha$ is the spin conection $\omega^a{}_{b,\alpha}$ on M pulled back to spacetime via the map $\varphi(x)$. In general, the spin connection on a manifold is valued in the lie algebra of the holonomy group H of the manifold[28]. We thus introduce the fermions in (3.20) in some representation of H with matrix generators λ^{ab} so that $\omega_\alpha = \omega_{ab,\alpha}\,\lambda^{ab}$ (orthonormal frame indices $a, b, \ldots$ are raised and lowered using the flat metric η_{ab} having the same signature as $g_{\alpha\beta}$). The lagrangian (3.20) is then invariant under local frame rotations defined by

$$e^a{}_\alpha \to L^a{}_b\, e^b{}_\beta, \qquad \omega_\mu \to L(\omega_\mu + \partial_\mu)L^{-1}, \tag{3.21}$$

with $L(\varphi) \in$ H. Under coordinate reparametrizations

$$\varphi \to \varphi'(\varphi), \tag{3.22}$$

on the other hand, we have $g_{\alpha\beta}(\varphi) \to g_{\alpha\beta}(\varphi')$, $\omega_\alpha(\varphi) \to \omega_\alpha(\varphi')$, and (3.20) is not generally invariant. Only when the change of coordinates $\varphi \to \varphi'$ corresponds to an isometry of the metric $g_{\alpha\beta}$ on M (so that $g_{\gamma\delta}(\varphi')\frac{\partial\varphi'^\gamma}{\partial\varphi^\alpha}\frac{\partial\varphi'^\delta}{\partial\varphi^\beta} = g_{\alpha\beta}(\varphi)$, and similarly for $\omega_\beta(\varphi')$ up to a local frame rotation) is (3.20) classically invariant.

The quantum theory associated with (3.20) is defined formally by

$$e^{-\Gamma_{\text{eff}}[\varphi, e, \omega]} = \int_{\psi,\overline{\psi}} e^{-\int d^{2n}x\,\mathcal{L}}. \tag{3.23}$$

In order for (3.23) to have geometrical significance as a quantum theory which depends only on the intrinsic properties of M, we need to demand at least that

$$\Gamma_{\text{eff}}[\varphi', e', \omega'] = \Gamma_{\text{eff}}[\varphi, e, \omega] \tag{3.24}$$

with $$e'^a{}_\alpha(\varphi') = e^a{}_\beta(\varphi)\frac{\partial\varphi^\beta}{\partial\varphi'^\alpha} \quad \text{and} \quad \omega'_\alpha(\varphi') = \omega_\beta(\varphi)\frac{\partial\varphi^\beta}{\partial\varphi'^\alpha},$$

and that

$$\Gamma_{\text{eff}}[\varphi, e'', \omega''] = \Gamma_{\text{eff}}[\varphi, e, \omega] \tag{3.25}$$

with $$e'' = Le \quad \text{and} \quad \omega'' = L(\omega + d)L^{-1}.$$

(3.24) expresses the coordinate independence of the effective action, i.e. that any ultimately calculated Greens functions should not have any residual dependence on the initial (and arbitrary) choice of coordinate system. (3.25) similarly expresses the independence of the effective action with respect to the choice of vielbein, i.e. that Greens functions should not have any residual dependence on the initial (and arbitrary) choice of the section of the frame bundle chosen for introducing the fermions. The two properties together thus result in an effective action for the theory which depends only on the metric and connection chosen for M. Invariance under (3.24) is automatic in the vielbein formalism because the fermions $\psi(x)$ are taken to be invariant under the transformation. Invariance under (3.25), on the other hand, is not immediate because of a potential anomaly in local frame rotations. Indeed, under an infinitesimal local frame rotation, $L^a{}_b(\varphi) \approx 1+\alpha^a{}_b(\varphi)$, the connection transforms by (3.21) just like a gauge field, $\omega \to \omega - (d\alpha + [\omega, \alpha])$, so we have for the change in the action in (3.23)

$$\delta_\alpha \Gamma_{\text{eff}}[\varphi, e, \omega] = N \int_{S^{2n}} Q^1_{2n}(\alpha, \omega), \tag{3.26}$$

where both α and ω are matrix-valued in the Lie algebra of the holonomy group H. We see that the anomaly in local frame rotations, which we call a holonomy anomaly, takes the form of an ordinary non-abelian gauge anomaly. For example, in $2n = 4$ spacetime dimensions, the anomaly automatically vanishes when the connection is $SO(m)$ valued, because the symmetrized trace of three $SO(m)$ generators always vanishes. When M is a Kähler manifold with an $SU(m)$ valued connection, on the other hand, anomalies may arise in any number of spacetime dimensions. When M is flat, of course, we can work in coordinates in which $\omega = 0$ so there is no anomaly regardless of the fermion content.

As we have pointed out, when the metric on M admits isometries there are associated symmetries of (3.23) at the classical level. The conditions for their preservation at the quantum level were precisely the subject of our discussion of isometry anomalies earlier. The major subtlety there was that the proper realization of the symmetry frequently required the addition of appropriate local counterterms to the action. We must now ask the same question concerning the anomalous variation (3.26) under *arbitrary* frame rotations and determine when there might be a counterterm which exactly compensates it.

The generalization of the construction of (3.16) is straightforward. Generalizing the natural embedding of G/H in G employed there, we now need to embed the space M in a higher dimensional space whose connection $\tilde{\omega}$ allows coupling to fermion representations with anomalies which match those of the connection ω on M. If M bounds a one higher dimensional manifold B in the higher dimensional space, $M = \partial B$, we can construct the analog of (3.16):

$$-N \int_B Q_{2n+1}(\tilde{\omega}). \tag{3.27}$$

This is insensitive to deformations of B because the integrand is closed in the higher dimensional sense, just as discussed after (3.16). (3.27) thus qualifies as a local counterterm whose anomalous variation cancels that of (3.26). Since this construction may depend on the detailed nature of the manifold M, we shall not belabor it in much generality. As a specific example, however, we may consider the sphere S^N bounding an $(N+1)$-ball B^{N+1} in R^{N+1}. Then the connection on B^{N+1} written in spherical coordinates (so that it is $SO(N+1)$ valued) has the requisite properties for the construction of (3.27). We thus recover the result that the $SO(N)$ anomalies must match those of some $SO(N+1)$ representation, the same result obtained by regarding S^N as a coset space $SO(N+1)/SO(N)$ (although we no longer need be constrained to the symmetric S^N metric). This will be a general property of manifolds M which happen to be G/H spaces: since (3.27) will automatically cancel any isometry anomalies and we have already argued that this is *only* possible if the anomaly of the H representation matches that of some representation of the isometry group G, we do not expect to find any less restrictive conditions for the construction of (3.27). Finally, we should add that our consideration of counterterms has relied only upon the bosonic fields intrinsic to the σ-model. Much more latitude in construction of possible counterterms would be available if the theory had additional dynamical fields such as gauge fields or antisymmetric tensor fields (in particular, when the holonomy group can be embedded in an external gauge group, the anomaly can be cancelled by giving the gauge field an expectation value equal to the spin connection[31]).

For completeness, we mention that depending on the fermion content, it is frequently possible to define a σ-model model without introducing a vielbein at all. For example, when the manifold M has $SO(m)$ holonomy and the fermions are taken in the vector representation of $SO(m)$, the fermionic lagrangian may be written as $\mathcal{L}_m = g_{\alpha\gamma}\overline{\psi}^{\alpha} i\gamma^{\mu}(\delta^{\gamma}_{\beta}\,\partial_{\mu} + \partial_{\mu}\varphi^{\delta}\,\Gamma^{\gamma}_{\delta\beta})P_{+}\,\psi^{\beta}$, i.e. purely in terms of the Christoffel symbol. The fermions ψ^{α} now transform as $\psi^{\alpha} \to \frac{\partial\varphi'^{\alpha}}{\partial\varphi^{\beta}}\psi^{\beta}$ under (3.22). More generally, whenever the fermion content is such that it may be introduced directly in coordinate frames, we may write the lagrangian

$$\mathcal{L} = \frac{f_{\pi}}{2}\, g_{\alpha\beta}\,\partial_{\mu}\varphi^{\alpha}\,\partial^{\mu}\varphi^{\beta} + \overline{\psi} i\gamma^{\mu}(\partial_{\mu} + \Gamma_{\mu})P_{+}\,\psi, \tag{3.20$'$}$$

where Γ_{μ} represents the Christoffel symbol on M valued in the appropriate fermion representation $(\partial_{\mu}\varphi^{\delta}\,\Gamma^{\alpha}_{\delta\beta})(\lambda^{\beta}{}_{\alpha})^{\rho_1\cdots}_{\sigma_1\cdots}$ pulled back to spacetime via the map $\varphi(x)$. Since we no longer have frame transformations (3.22) to contend with, we need only ask whether the effective action defined by

$$e^{-\Gamma_{\rm eff}[\varphi, g, \Gamma]} = \int_{\psi,\overline{\psi}} e^{-\int d^{2n}x\ \mathcal{L}} \tag{3.23$'$}$$

now satisfies

$$\Gamma_{\rm eff}[\varphi', g', \Gamma'] = \Gamma_{\rm eff}[\varphi, g, \Gamma] \tag{3.24$'$}$$

with $\quad g'_{\alpha\beta}(\varphi') = g_{\delta\gamma}(\varphi)\dfrac{\partial\varphi^{\delta}}{\partial\varphi'^{\alpha}}\dfrac{\partial\varphi^{\gamma}}{\partial\varphi'^{\beta}} \quad$ and $\quad \Gamma'(\varphi') = \Lambda(\Gamma + d)\Lambda^{-1}|_{\varphi},$

where $\Lambda = \frac{\partial\varphi'^{\alpha}}{\partial\varphi^{\beta}}\lambda^{\beta}{}_{\alpha}$. Under an infinitesimal coordinate transformation, $\varphi'^{\alpha} \approx \varphi^{\alpha} + \xi^{\alpha}(\varphi)$, the Christoffel connection evidently transforms like a $Gl(n)$ gauge field (for a review see [2]), $\Gamma \to \Gamma - (dv + [\Gamma, v])$, where v is constructed from $v^{\alpha}{}_{\beta} = \partial\xi^{\alpha}/\partial\varphi^{\beta}$ times the appropriate matrix generator $(\lambda^{\beta}{}_{\alpha})^{\rho_1\cdots}_{\sigma_1\cdots}$. We can thus once again carry over our results from the non-abelian gauge anomaly to find the anomalous variation

$$\delta_{\xi}\,\Gamma_{\rm eff}[\varphi, g, \Gamma] = N\int_{S^{2n}} Q^{1}_{2n}(v, \Gamma) \tag{3.28}$$

under infinitesimal coordinate reparametrizations. But were we now to introduce a vielbein $e^{a}{}_{\alpha}$, as is always our perogative, we have already seen (eq. (2.27)) that there exists a local counterterm whose anomalous variation under coordinate reparametrizations exactly compensates (3.28) and whose variation under local frame rotations (3.21) takes the form (3.26) (it is simply the anomaly functional induced by the (anomalous) change of fermionic basis from coordinate frames to orthonormal tangent frames, in which the vielbein plays the role of a $Gl(m)$ gauge transformation interpolating between the Christoffel connection Γ and the spin connection ω). This counterterm may thus be employed to shift the anomalies from coordinate reparametrizations to local frame rotations, and our preceding discussion then carries over in full. Just as in the gravitational case, there is a close relation between realizing coordinate reparametrization invariance and local frame rotation invariance, and the conditions for realizing one are sufficient for realizing both.

In closing, we comment upon the difference between the gravitational anomalies considered earlier and the coordinate reparametrization and frame rotation anomalies currently under consideration. In the gravitational case, invariance under coordinate reparametrizations is a true symmetry of the theory which leads to a conserved current, the energy-momentum tensor. Loss of coordinate reparametrization invariance then leads to a loss of conservation of energy-momentum. Invariance under local frame rotations, on the other hand, is an accessory symmetry associated with the appearance of the vielbein, required to introduce fermions transforming as local Lorentz spinors. Removing this extra freedom in gravitational theories is necessary to prevent extra degrees of freedom associated with the introduction of the vielbein from propagating. This may be seen, as mentioned earlier, by imposing a symmetric gauge fixing condition on the vielbein, and then the anomaly under general coordinate transformations is generated by the compensating local lorentz transformation required to preserve the gauge-fixing condition. Loss of invariance under local frame rotations then leads to a violation of unitarity because the antisymmetric degrees of freedom of the vielbein become propagating quantum fields. For the models considered in this section, the situation is somewhat different. Loss of invariance under coordinate reparametrizations or local frame rotations does not mean that the theory is necessarily sick, only that it is probably not the sort of theory we desire. It would lead not to a theory with purely geometrical significance, depending only on the metric and connection on M, but to a theory with an extraneous dependence on the specific choice of coordinates or local frames used to define it (just as the isometry anomalies of the previous section result not necessarily in an inconsistent theory, but in a theory whose quantum effective action, not properly realizing the desired G-symmetry, has a smaller group of symmetries than the manifold itself). Given such a theory in one coordinate system, we could always express the same theory in another coordinate system by adding the appropriate induced anomaly functional, but we would always find an implicit dependence on our initial choice of coordinates. Our better geometrical instincts would then have us cast such theories aside.

References

[1] L. Alvarez-Gaumé and P. Ginsparg, Nucl. Phys. B243 (1984) 449.

[2] L. Alvarez-Gaumé and P. Ginsparg, Ann. Phys. 161 (1985) 423.

[3] L. Alvarez-Gaumé and P. Ginsparg, Geometry Anomalies, Harvard preprint HUTP-85/A015, to appear in Nucl. Phys. B.

[4] S. L. Adler, Phys. Rev. 177 (1969) 2426;
J. Bell and R. Jackiw, Nuovo Cimento 60A (1969) 47.

[5] G. 't Hooft, Phys. Rev. Lett. 37 (1976) 8, Phys. Rev. D14 (1976) 3432.

[6] R. Delbourgo and A. Salam, Phys. Lett. B40 (1972) 381;
T. Eguchi and P. Freund, Phys. Rev. Lett. 37 (1976) 1251.

[7] D. J. Gross and R. Jackiw, Phys. Rev. D6 (1972) 477.

[8] C. Bouchiat, J. Iliopoulos, and Ph. Meyer, Phys. Lett. 38B (1972) 519;
H. Georgi and S. Glashow, Phys. Rev. D6 (1972) 429.

[9] L. Alvarez-Gaumé and E. Witten, Nucl. Phys. B234 (1983) 269.

[10] G. 't Hooft, "Naturalness, Chiral Symmetry, and Spontaneous Chiral Symmetry Breaking," in *Recent developments in gauge theories,* ed. G. 't Hooft et al. (Plenum, New York, 1980).

[11] M. F. Atiyah and I. M. Singer, Ann. Math. 87 (1968) 485, 546, Ann. Math. 93 (1971) 119, 139;
M. F. Atiyah and G. B. Segal, Ann. Math. 87 (1968) 531.

[12] K. Fujikawa, Phys. Rev. Lett. 42 (1979) 1195, 44 (1980) 1733, Phys. Rev. D21 (1980) 2848, D22 (1980) 1499(E), D23 (1981) 2262.

[13] S. Iyanaga and Y. Kawada, eds., *Encyclopedic Dictionary of Mathematics,* MIT Press (Cambridge, Mass, 1980), p. 1417.

[14] P. H. Frampton and T. W. Kephart, Phys. Rev D28 (1983) 1010.

[15] M. Atiyah and I. Singer, Proc. Nat. Acad. Sci. USA, 81 (1984) 2597;
B. Zumino, "Chiral Anomalies and Differential Geometry," in *Relativity, Groups and Topology II,* eds. B. S. DeWitt and R. Stora (North-Holland, Amsterdam, 1984);
R. Stora, "Algebraic Structure and Topological Origin of Anomalies," in *Progress in Gauge Field Theory,* ed. G. 't Hooft et al. (Plenum, New York, 1984).

[16] L. Alvarez-Gaumé, Comm. Math. Phys. 90 (1983) 161, J. Phys. A16 (1983) 4177;
D. Friedan and P. Windey, Nucl. Phys. B235 (1984) 395.

[17] O. Alvarez, I. M. Singer, and B. Zumino, Comm. Math. Phys. 96 (1984) 409.

[18] W. Bardeen and B. Zumino, Nucl. Phys. B244 (1984) 421.

[19] J. H. Schwarz, Phys. Rept. 89 (1982) 223;
M. B. Green and D. J. Gross, eds., Proceedings of the Unified String Workshop (Santa Barbara, 1985).

[20] M. B. Green and J. H. Schwarz, Phys. Lett. 149B (1984) 117.

[21] M. B. Green, J. H. Schwarz, and P. C. West, Nucl. Phys. B254 (1984) 327.

[22] C. G. Callan and J. A. Harvey, Nucl. Phys. B250 (1985) 427.

[23] E. Weinberg, Phys. Rev. D24 (1981) 2669.

[24] G. Moore and P. Nelson, Phys. Rev. Lett. 53 (1984) 1519, Comm. Math. Phys. 100 (1985) 83.

[25] S. Weinberg, "Dynamic and Algebraic Symmetries," in *Lectures on Elementary Particles and Quantum Field Theory*, eds. S. Deser, M. Grisaru, H. Pendleton (MIT Press, Cambridge, 1970).

[26] S. Coleman, J. Wess, and B. Zumino, Phys. Rev. 177 (1969) 2239;
C. Callan, S. Coleman, J. Wess, and B. Zumino, Phys. Rev. 177 (1969) 2247.

[27] A. Salam and J. Strathdee, Ann. Phys. 141 (1982) 316;
P. van Nieuwenhuizen, "An introduction to simple supergravity and the Kaluza-Klein program," in *Relativity, Groups and Topology II*, eds. B. S. DeWitt and R. Stora (North-Holland, Amsterdam, 1984).

[28] S. Helgason, *Differential Geometry, Lie Groups, and Symmetric Spaces* (Academic Press, N. Y., 1978);
A. Lichnerowicz, *General Theory of Connections and the Holonomy Group* (Noordhoff, Holland, 1976).

[29] J. Wess and B. Zumino, Phys. Lett. 37B (1971) 95;
E. Witten, Nucl. Phys. B223 (1983) 422.

[30] A. Manohar and G. Moore, Nucl. Phys. B243 (1984) 55.

[31] P. Candelas, G. Horowitz, A. Strominger, and E. Witten, Nucl. Phys. B258 (1985) 46.

QUANTUM GRAVITY AND BLACK HOLES

G. 't Hooft

Institute for Theoretical Physics
Princetonplein 5, P.O. Box 80.006
3508 TA Utrecht, The Netherlands

Abstract

After a sketchy resumé of the General Theory of Relativity the problem of quantizing this theory is addressed. The perturbative problem seems to be hardly different from gauge theory models in particle physics, but then some fundamental divergences and instabilities are pointed out. It is argued that "black holes" should be more closely studied in attempting to solve the problems that have arisen. The author then presents a discussion of his views on black holes that has also appeared elsewhere.

1. Curved space-time

There are many excellent treatizes on the subject of General Relativity[1)] and the beginning student of this subject is highly advised to consult one or several of those. Only for pedagogical reasons and for the sake of establishing notations we will give here a short resumé of this beautiful theory, in as far as it is needed for our further considerations. A reader who is familiar with the principles of general relativity may decide to skip sections 1 and 2.

For representing points in space-time one uses four coordinates x^μ ($\mu = 0,1,2,3$). Contrary to the case in more conventional physics one will not be able in general to distinguish rectangular from curved coordinates, so any (continuous) representation of points by four numbers is acceptable.

For a pair of neighbouring points x^μ and x'^μ with $x'^\mu = x^\mu + dx^\mu$ (dx^μ being infinitesimal) one defines a distance ds by

$$ds^2 = g_{\mu\nu}\, dx^\mu\, dx^\nu \tag{1.1}$$

where summation over repeated indices is assumed, and $g_{\mu\nu}$ is a symmetric 4×4 matrix that may depend on x^μ:

$$g_{\mu\nu} = g_{\mu\nu}(x) \ . \tag{1.2}$$

In order to make contact with special relativity one assumes that the matrix $g_{\mu\nu}$ has one negative and three positive eigenvalues. Usually (but not always) we will have

$$g_{oo} < 0 \tag{1.3}$$

and identify x^o with time t.

For integrations over space and time an infinitesimal volume element is needed, besides the notion of distance. For this one chooses

$$dV = \sqrt{-g}\, dx^0\, dx^1\, dx^2\, dx^3 , \tag{1.4}$$

where g is the determinant:

$$g = \det(g_{\mu\nu}) . \tag{1.5}$$

In flat space:

$$g_{\mu\nu} = \begin{pmatrix} -1 & & & \emptyset \\ & 1 & & \\ & & 1 & \\ \emptyset & & & 1 \end{pmatrix} ; \quad g = -1 . \tag{1.6}$$

We now need the notion of a scalar field $\varphi(x)$. It might represent any measurable quantity at x but we often use it to describe a spinless particle. The Lagrangian for a free spinless particle in any "curved" space-time may be written as

$$\mathcal{L} = - \sqrt{-g} \left\{ g^{\mu\nu} \partial_\mu \phi^* \partial_\nu \phi + m^2 \phi^* \phi \right\} ; \tag{1.7}$$

where

$$g^{\mu\nu} = (g_{\mu\nu})^{-1} . \tag{1.8}$$

The Euler-Lagrange equation for this field is obtained from the extremal principle

$$\delta \int \mathcal{L}\, d^4x = 0 , \tag{1.9}$$

which should hold for any infinitesimal variation $\varphi(x) \rightarrow \varphi(x) + \delta\varphi(x)$. Consequently, the Klein-Gordon equation is now written as

$$\partial_\mu \left[g^{\mu\nu} \sqrt{-g}\, \partial_\nu \phi \right] = m^2 \sqrt{-g}\, \phi . \tag{1.10}$$

86

The fundamental theme underlying these equations is invariance or covariance under coordinate transformations of the form

$$x^{\mu} \Rightarrow u^{\mu}(x) \quad ; \quad x^{\mu} = x^{\mu}(u) \ , \tag{1.11}$$

where the mapping $x \leftrightarrow u$ may be any differentiable function. In the new coordinates the scalar field is $\widetilde{\phi}(u)$, with

$$\widetilde{\phi}(u) = \phi(x(u)) \ . \tag{1.12}$$

The gradient of a scalar field is a vector field. Its transformation properties under (1.11) can be read off from

$$\phi(x+dx) = \phi(x) + \phi_{\mu} dx^{\mu} \ , \tag{1.13}$$

$$\widetilde{\phi}(u+du) = \widetilde{\phi}(u) + \widetilde{\phi}_{\mu} du^{\mu} \ , \tag{1.14}$$

$$dx^{\mu} = \frac{\partial x^{\mu}}{\partial u^{\nu}} du^{\nu} \ , \tag{1.15}$$

so that

$$\widetilde{\phi}_{\mu}(u) = \frac{\partial x^{\lambda}}{\partial u^{\mu}} \phi_{\lambda}(x(u)) \ . \tag{1.16}$$

A "co-vector field" is now any quantity $A_{\mu}(x)$ that transforms this way,

$$\widetilde{A}_{\mu}(u) = \frac{\partial x^{\lambda}}{\partial u^{\mu}} A_{\lambda}(x(u)) \ . \tag{1.17}$$

A "contra-vector field" $B^{\mu}(x)$ is defined to transform in such a way that for any vector field $A_{\mu}(x)$, the quantity $B^{\mu}(x)A_{\mu}(x)$ transforms as a scalar. Since the inverse of the matrix $\partial x^{\lambda}/\partial u^{\mu}$ is $\partial u^{\mu}/\partial x^{\lambda}$ (under the usual definitions of partial integration), we find that B^{μ} must transform as

$$\tilde{B}^{\mu}(u) = \frac{\partial u^{\mu}}{\partial x^{\lambda}} B^{\lambda}(x(u)) \ . \tag{1.18}$$

Note that co-vectors and contra-vectors are distinguished by placing the index either below or above the field symbol.

Tensor fields $A_{\mu\nu}$ are defined to transform just as products of vector fields:

$$\tilde{A}_{\mu\nu}(u) = \frac{\partial x^{\lambda}}{\partial u^{\mu}} \frac{\partial x^{\kappa}}{\partial u^{\nu}} A_{\lambda\kappa}(x(u)) \ , \tag{1.19}$$

and similarly for fields $B^{\mu\nu}$ and mixed tensors $F_{\mu\nu\ldots}^{\ \ \alpha\beta\ldots}$.

Now a gradient of a scalar field transforms as a vector. A gradient $\partial_{\mu}A_{\nu}$ of a vector field A_{ν} is *not* a proper tensor. Of course, *products* $A_{\mu}A_{\nu}$ are. In order to obtain correction terms to a gradient such that it does transform as a tensor one introduces the "connection field" $\Gamma^{\alpha}_{\ \beta\gamma}(x)$. In spite of the special choice of upper and lower indices, Γ does not transform as a tensor. Instead, we require

$$\tilde{\Gamma}^{\alpha}_{\ \beta\gamma}(u) = \frac{\partial u^{\alpha}}{\partial x^{\mu}} \left[\frac{\partial x^{\lambda}}{\partial u^{\beta}} \frac{\partial x^{\kappa}}{\partial u^{\gamma}} \Gamma^{\mu}_{\ \lambda\kappa}(x) + \frac{\partial^2 x^{\mu}}{\partial u^{\beta} \partial u^{\gamma}} \right] . \tag{1.20}$$

One may convince oneself that (1.20) is consistent with repeated transformations, and that now

$$D_{\mu}A_{\nu} \underset{\mathrm{def}}{=} \partial_{\mu}A_{\nu} - \Gamma^{\alpha}_{\ \mu\nu}A_{\alpha} \ , \tag{1.21}$$

does transform as a true tensor.

Similarly we have the true tensor

$$D_{\mu}B^{\nu} \underset{\mathrm{def}}{=} \partial_{\mu}B^{\nu} + \Gamma^{\nu}_{\ \mu\alpha}B^{\alpha} \ , \tag{1.22}$$

for a contravector field B^μ. The symbol D_μ is called "covariant derivative". So far, the quantity $\Gamma^\alpha_{\beta\gamma}$ is an arbitrary field. We usually restrict ourselves to the case that it is symmetric in the lower indices β,γ.

Now in spaces with a metric tensor $g_{\mu\nu}$ there is *another* way to define covariant derivatives. Consider any vector field $A_\mu(x)$. How do we produce a tensor $D_\mu A_\nu(x)$?

1. Consider first one point $x = x_{(1)}$.
2. Choose special coordinates u at x close to $x_{(1)}$, such that

$$ds^2 = du^{1^2} + du^{2^2} + du^{3^2} - du^{0^2} + \mathcal{O}\left(du^2\ (x-x_{(1)})^2\right) , \tag{1.23}$$

   or

$$\begin{aligned} \tilde{g}_{\mu\nu}(u(x_{(1)})) &= \eta_{\mu\nu} \\ \partial_\alpha \tilde{g}_{\mu\nu}(u(x_{(1)})) &= 0 . \end{aligned} \tag{1.24}$$

3. In these coordinates we define that, at $u = u(x_{(1)})$, $D_\mu A_\nu(u) = \partial_\mu A_\nu(u)$.
4. Transform to any other coordinate system by the transformation rule of true tensors.
5. Do the same for all other points x.

One can convince oneself that this procedure is uniqe. With this definition one always gets

$$D_\mu g_{\alpha\beta} = 0 \quad ; \quad D_\mu g^{\alpha\beta} = 0 \tag{1.25}$$

($g^{\alpha\beta}$ is defined to be the inverse of the matrix $g_{\alpha\beta}$). Now the previous definition with a Γ field would give

$$D_\mu g_{\alpha\beta} = \partial_\mu g_{\alpha\beta} - \Gamma^\lambda_{\mu\alpha} g_{\lambda\beta} - \Gamma^\kappa_{\mu\beta} g_{\alpha\kappa} . \tag{1.26}$$

This vanishes if we choose

$$\Gamma^{\alpha}{}_{\beta\gamma} = \tfrac{1}{2}g^{\alpha\mu}\left(\partial_\beta g_{\mu\gamma} + \partial_\gamma g_{\mu\beta} - \partial_\mu g_{\beta\gamma}\right) . \tag{1.27}$$

Thus, a metric tensor $g_{\mu\nu}$ generates uniquely a symmetric connection field $\Gamma^{\alpha}{}_{\beta\gamma}$.

The Riemann curvature tensor is now defined by looking at the commutator of covariant derivatives:

$$D_\mu D_\nu A_\alpha - D_\nu D_\mu A_\alpha = - R^{\lambda}{}_{\alpha\mu\nu} A_\lambda , \tag{1.28}$$

for all fields A_α. Writing it out in terms of the connection field we get

$$R^{\alpha}{}_{\beta\gamma\delta} = \partial_\gamma \Gamma^{\alpha}{}_{\beta\delta} - \partial_\delta \Gamma^{\alpha}{}_{\beta\gamma} + \Gamma^{\alpha}{}_{\gamma\sigma}\Gamma^{\sigma}{}_{\beta\delta} - \Gamma^{\alpha}{}_{\delta\sigma}\Gamma^{\sigma}{}_{\beta\gamma} . \tag{1.29}$$

We find the following properties:

1. $R^{\alpha}{}_{\beta\gamma\delta}$ is a tensor. This is obvious in eq. (1.28) but not if (1.29) is used as a definition.

2. $$R^{\alpha}{}_{\beta\gamma\delta} = - R^{\alpha}{}_{\beta\delta\gamma} \tag{1.30}$$

3. $$R^{\alpha}{}_{\beta\gamma\delta} + R^{\alpha}{}_{\gamma\delta\beta} + R^{\alpha}{}_{\delta\beta\gamma} = 0 \tag{1.31}$$

4. $$D_\mu R^{\alpha}{}_{\beta\gamma\delta} + D_\gamma R^{\alpha}{}_{\beta\delta\mu} + D_\delta R^{\alpha}{}_{\beta\mu\gamma} = 0 \tag{1.32}$$

5. For a contravector field B^μ we have

$$D_\mu D_\nu B^\alpha - D_\nu D_\mu B^\alpha = R^{\alpha}{}_{\lambda\mu\nu} B^\lambda \tag{1.33}$$

6. If $R^{\alpha}{}_{\beta\gamma\delta}(x)$ is zero in some region then space-time is flat there: there exist coordinates x^μ such that

$$g_{\mu\nu}(x) = \begin{pmatrix} -1 & & & \emptyset \\ & 1 & & \\ & & 1 & \\ \emptyset & & & 1 \end{pmatrix} . \tag{1.34}$$

The Ricci tensor is

$$R_{\mu\nu} = R^{\alpha}{}_{\mu\alpha\nu} = R_{\nu\mu} \tag{1.35}$$

and the scalar curvature

$$R = g^{\mu\nu}R_{\mu\nu} . \tag{1.36}$$

The Ricci tensor satisfies the Bianchi identity

$$g^{\alpha\beta}D_{\alpha}R_{\beta\nu} - \tfrac{1}{2}D_{\nu}R = 0 . \tag{1.37}$$

Defining

$$G_{\mu\nu} = R_{\mu\nu} - \tfrac{1}{2}Rg_{\mu\nu} \tag{1.38}$$

we have

$$g^{\alpha\beta}D_{\alpha}G_{\beta\nu} = 0 . \tag{1.39}$$

2. Einstein's equation

Usually, matter is introduced in the theory by writing down the stress-energy-momentum tensor $T_{\mu\nu}$. We take

$$T_{oo} = \rho = \text{energy (mass-)density}. \tag{2.1}$$

$$-T_{oi} = -T_{io} = P_i = \text{energy flow} = \text{momentum density}. \tag{2.2}$$

$$T_{ij} = t_{ij} = \text{stress tensor}. \tag{2.3}$$

$$\tfrac{1}{3}t_{ii} = p = \text{pressure}. \tag{2.4}$$

The stress tensor t_{ij} describes the force $d\vec{F}$ on a surface element $d\vec{S}$:

$$d\vec{F} = t.d\vec{S} \ . \tag{2.5}$$

We have conservation of energy and momentum:

$$\partial_\mu T_{\mu\nu} = 0 \ . \tag{2.6}$$

If we ignore the effect of a gravitational field on the energy-momentum then we also have

$$g^{\alpha\beta}D_\alpha T_{\beta\nu} = 0 \ . \tag{2.7}$$

The idea is now that (2.7) describes matter only, whereas (2.6) would include the energy and momentum of the gravitational field.

Einstein first guessed

$$R_{\mu\nu} \propto T_{\mu\nu} \; ? \; , \qquad (2.8)$$

but it turned out that then (2.7), in combination with (1.37) would give

$$D_\mu R = 0 \quad ; \quad D_\mu T_{\lambda\lambda} = 0 \; ? \qquad (2.9)$$

which cannot be true. He then found the correct equation:

$$G_{\mu\nu} = R_{\mu\nu} - \tfrac{1}{2} R g_{\mu\nu} = 8\pi G T_{\mu\nu} \qquad (2.10)$$

where G is the gravitational constant:

$$G = 6.672 \; . \; 10^{-8} \; \mathrm{cm}^3 \; \mathrm{g}^{-1} \; \mathrm{sec}^{-2} \; . \qquad (2.11)$$

This is the only equation of this sort that is consistent with both (1.39) and (2.7). Eq. (2.10) can be cast into a Lagrange form. Consider

$$\mathcal{L}_1 = \sqrt{-g}\, R \; . \qquad (2.12)$$

And consider all variations

$$g_{\mu\nu} \to g_{\mu\nu} + \delta g_{\mu\nu} \; . \qquad (2.13)$$

One finds that

$$\delta \int \sqrt{-g}\, R\, d^4x = - \int \sqrt{-g}\, G^{\mu\nu} \, \delta g_{\mu\nu} \, d^4x \; . \qquad (2.14)$$

This vanishes if

$$G_{\mu\nu} = 0 \qquad (2.15)$$

which is Einstein's equation in the absence of matter.

If we consider the Lagrangian

$$\mathcal{L} = \sqrt{-g}\left(\frac{R}{16\pi G} + \mathcal{L}^{\text{matter}}\right) \qquad (2\ .16)$$

with for instance

$$\mathcal{L}^{\text{matter}} = -\ g^{\mu\nu}\ \partial_\mu \phi^* \partial_\nu \phi - m^2 \phi^* \varphi \qquad (2.17)$$

then one finds, for variations $g_{\mu\nu} \to g_{\mu\nu} + \delta g_{\mu\nu}$,

$$\delta\left(\int \sqrt{-g}\ \mathcal{L}^{\text{matter}}\ d^4x \right) = \tfrac{1}{2} \int T^{\mu\nu}\ \sqrt{-g}\ \delta g_{\mu\nu}\ . \qquad (2.18)$$

So that

$$\delta \int \mathcal{L}\ d^4x = \int \sqrt{-g}\ \delta g_{\mu\nu} \left(-\ \frac{G^{\mu\nu}}{16\pi G} + \tfrac{1}{2} T^{\mu\nu} \right) d^4x\ . \qquad (2.19)$$

Requiring now that $\delta \int \mathcal{L}\ d^4x$ vanish for all choices of $\delta g_{\mu\nu}$ corresponds to imposing Einstein's equation (2.10).

Having a Lagrangian we can now apply "standard" Hamilton-Lagrange theory to this system, and consider the quantization of gravity.

3. The temporal gauge

General coordinate transformations, eqs. (1.11) - (1.19), must be seen as gauge transformations. In some sense, quantum gravity is a gauge field theory. Like in a gauge theory, we must choose a gauge-fixing procedure. Writing

$$u^\mu = x^\mu + \eta^\mu \tag{3.1}$$

with η^μ infinitesimal, we find that a gauge transformation for the metric tensor can be expressed as

$$g_{\mu\nu} \rightarrow g_{\mu\nu} + D_\mu\eta_\nu + D_\nu\eta_\mu \ , \tag{3.2}$$

to be compared with

$$A_\mu \rightarrow A_\mu + D_\mu\Lambda \tag{3.3}$$

in a vector gauge theory. Because of this invariance, the physical Hilbert space is "smaller" than what would be suggested by the number of field variables. Conceptually, the best way to characterize the Hilbert space is by choosing the equivalent of the so-called "temporal gauge",

$$A_o = 0 \tag{3.4}$$

in vector theories, namely

$$g_{oo} = 1 \quad ; \quad g_{oi} = 0 \qquad (i=1,2,3) \ . \tag{3.5}$$

In this gauge we have

$$\Gamma^o{}_{oo} = \Gamma^o{}_{oi} = \Gamma^i{}_{oo} = 0 \ . \tag{3.6}$$

Now $\Gamma^i{}_{oo}$ can be identified with the leading components of the gravitational field. Apparently we have coordinates such that it vanishes: we have freely falling coordinates.

As long as we work in perturbation theory, (3.5) can always be imposed. The first part can be realized by choosing an η in (3.2) with a certain condition on $\partial_o \eta_o$. This will essentially fix η_o. The other conditions determine

$$\partial_o \eta_i + \partial_i \eta_o \tag{3.7}$$

and since η_o was already fixed, these conditions determine η_i. But clearly, there are free integration constants.

This means that, after having realized (3.5) by some choice of η, we will still have a subclass of coordinate transformations that will leave (3.5) unaffected. Since now we also have (3.6), the infinitesimal η that generate this class of residual gauge transformations satisfy

$$\partial_o \eta_o = 0 \qquad \rightarrow \qquad \eta_o(\vec{x},t) = \eta_o(\vec{x}) \tag{3.8}$$

and

$$D_o \eta_i + D_i \eta_o = 0 \tag{3.9}$$

which implies

$$\partial_o \eta^i + g^{ij} \partial_j \eta_o = 0 \ . \tag{3.10}$$

So that

$$\eta^i(\vec{x},t) = \eta^i(\vec{x}) - \int g^{ij} \partial_j \eta_o(\vec{x}) dt \ . \tag{3.11}$$

The $\eta_o(\vec{x})$ and $\eta^i(\vec{x})$ generate local but time-independent symmetry transformations. Due to Noether's theorem they must be associated with conservation laws. These are found by subjecting the Lagrangian in which

(3.5) was substituted, to infinitesimal gauge transformations not satisfying (3.8) - (3.11). The only terms that survive are those due to the fact that δg_{oo} and δg_{oi} were omitted, since otherwise $\delta \int \mathcal{L}\, d^4x$ would vanish identically. So we can also look at the terms generated by δg_{oo} and δg_{oi} only. They give

$$- \delta \int \mathcal{L}\, d^4x = \int \sqrt{-g}\, d^4x \left(T^{oo}-2G^{oo}\right) \partial_o \eta_o$$

$$+ \int \sqrt{-g}\, d^4x \left(T^o{}_k-2G^o{}_k\right)\left(\partial_o \eta^k + g^{kj} \partial_j \eta_o\right) = 0 \ , \qquad (3.12)$$

where we have put $16\pi G = 1$ (this convention may seem to be an obvious choice, but differs from another obvious choice by a factor 16π. Apparently "large" numbers such as 16π can grow easily in quantum gravity).

From (3.12) we find the following conservation laws:

$$\partial_o\left(\sqrt{-g}\left(T^o{}_k-2G^o{}_k\right)\right) = 0 \ , \qquad (3.13)$$

and

$$\partial_o\left(\sqrt{-g}\ T^{oo}-2G^{oo}\right) + \partial_k\left(\sqrt{-g}\ T^{ok}-2G^{ok}\right)\Big) = 0 \ . \qquad (3.14)$$

The oo and oi components of Einstein's equations follow, if we put these "conserved currents" equal to zero. The other (ij) components follow from the Euler-Lagrange equations directly.

Let us compare this situation with an ordinary symmetry such as translation invariance. The associated conservation law is the conservation of momentum. Apparently we must require that the momentum itself is zero: the wave function $|\psi\rangle$ is invariant under translations. In quantum gravity, the wave function $|\psi\rangle$ must be invariant under the transformations generated by the η satisfying (3.8) - (3.11).

The importance of this is the following. Remember that the temporal gauge corresponds to freely falling coordinates. This would be quite cumbersome for describing e.g. the planetary system, or a black hole, because such coordinates would become very chaotic or even singular after a small lapse of time. Fortunately at any moment t we can choose to perform a gauge rotation of the subclass that leaves the gauge condition (3.5) intact. The wave function $|\psi>$ will not change. Thus at every t we can decide to "unwind" the coordinates.

A problem on the other hand is that η_o generates time translations. Apparently the Hamiltonian in this representation should vanish also. But this is not so. The correct attitude is to consider only those η that vanish at spatial infinity. Then $|\psi>$ needs not be invariant under time translations that also affect time at infinity. In short: time is defined by clocks located at spatial infinity. It is crucial to observe that apparently the definition of time involves the boundary conditions at spatial infinity. Without explicitly fixing the boundary conditions at spatial infinity there would not exist such a thing as a quantum-mechanical Hamiltonian.

4. Perturbation expansion and the Wick rotation

To set up a perturbative formalism for quantum gravity is deceptively simple[2]. We define

$$g_{\mu\nu} = \eta_{\mu\nu} + \lambda_{\mu\nu} \tag{4.1}$$

with $\eta_{\mu\nu} = \mathrm{diag}\,(-1,1,1,1)$ and $\lambda_{\mu\nu}$ is infinitesimal. Expanded up to second order in λ the Lagrangian is

$$\mathcal{L} = -\tfrac{1}{4}\left(\partial_\mu \lambda_{\alpha\beta}\right)^2 + \tfrac{1}{8}\left(\partial_\mu \lambda_{\alpha\alpha}\right)^2 + \tfrac{1}{2}\Lambda_\mu^2 - \tfrac{1}{2}\lambda_{\mu\nu}T_{\mu\nu} + \mathcal{L}^{\mathrm{matter}} \tag{4.2}$$

where

$$\Lambda_\mu = \partial_\alpha \lambda_{\alpha\mu} - \tfrac{1}{2}\partial_\mu \lambda_{\alpha\alpha} \ . \tag{4.3}$$

In the temporal gauge we require

$$\lambda_{o\mu} = 0 \ , \tag{4.4}$$

but of course other gauge choices, such as requiring Λ_μ to vanish, are also possible.

Let us Fourier transform in the three space coordinates and take as our dynamical variables $\lambda_{ij}(\vec{k},t)$. In $\vec{k}$ space the quadratic part of the Lagrangian diagonalizes, so that we can concentrate on one particular $\vec{k}$. Let us rotate $\vec{k}$ such that

$$\vec{k} = (0,0,k_3) \ . \tag{4.5}$$

Let us choose the beginning of the alphabet for indices that take the values 1,2 only (transverse indices), and define the traceless, transverse part $\tilde{\lambda}_{ab}$ of λ_{ij}:

$$\lambda_{ab} = \tilde{\lambda}_{ab} + \lambda\delta_{ab} \qquad (a,b=1,2) \tag{4.6}$$

$$\tilde{\lambda}_{aa} = 0 \quad ; \quad \lambda = \tfrac{1}{2}\lambda_{aa} \; . \tag{4.7}$$

We find that the Lagrangian then splits in three parts

$$\mathcal{L} = \mathcal{L}_1 + \mathcal{L}_2 + \mathcal{L}_3 \; , \tag{4.8}$$

with

$$\mathcal{L}_1 = -\tfrac{1}{4}k_3^2\,\tilde{\lambda}_{ab}\tilde{\lambda}_{ab} + \tfrac{1}{4}\dot{\tilde{\lambda}}_{ab}\dot{\tilde{\lambda}}_{ab} - \tfrac{1}{2}\tilde{T}_{ab}\tilde{\lambda}_{ab} \; , \tag{4.9}$$

where $\tilde{T}_{ab}$ is the transverse, traceless part of T_{ij};

$$\mathcal{L}_2 = \tfrac{1}{2}(\dot{\lambda}_{a3})^2 - \dot{\lambda}_{a3}\frac{i}{k_3}T_{oa} \; ; \tag{4.10}$$

$$\mathcal{L}_3 = \tfrac{1}{8}k_3^2\lambda^2 - \tfrac{1}{8}\dot{\lambda}^2 - \tfrac{1}{4}T_{oa}\lambda - \dot{\lambda}\dot{\lambda}_{33} - \tfrac{1}{2}\dot{\lambda}_{33}\dot{T}_{oo}/k_3^2 \; . \tag{4.11}$$

We make use of energy-momentum conservation:

$$ik\,T_{3\mu} = \dot{T}_{o\mu} \; . \tag{4.12}$$

Now $\mathcal{L}_1$ looks just like any Lagrangian for a driven harmonic oscillator and indeed $\tilde{\lambda}_{ab}$ which has two independent degrees of freedom, is an ordinary dynamical variable. It describes gravitons.

$\mathcal{L}_2$ implies $\partial_o\Pi_a = 0$, with, classically:

$$\Pi_a = \dot{\lambda}_{a3} - \frac{i}{k_3}T_{oa} \; . \tag{4.13}$$

As discussed in the previous section, we must impose $\Pi_a = 0$. Then $\mathcal{L}_2$ generates a Hamiltonian

$$H_2 = \tfrac{1}{2}\left|\Pi_a - \frac{i}{k_3}T_{oa}\right|^2 = \frac{1}{2k_3^2}T_{oa}^2 \; . \tag{4.14}$$

In $\mathcal{L}_3$ something new happens. The $\dot{\lambda}_{33}$ occurs only linearly. It acts as a Lagrange multiplier. This ensures:

$$\frac{d}{dt}\left(\lambda+T_{oo}/k_3^2\right) = 0 \ . \qquad (4.15)$$

Again, we require more restrictively:

$$\lambda = - T_{oo}/k_3^2 \qquad (4.16)$$

so that $\mathcal{L}_3$ does not contain any independent dynamical variables. It contributes to the Hamiltonian

$$H_3 = - \frac{1}{8k_3^2} T_{oo}^2 + \frac{1}{8k_3^2} T_{o3}^2 - \frac{1}{4k_3^2} T_{aa} T_{oo} \ , \qquad (4.17)$$

of which the first part generates Newton's law.

We went through this exercise in perturbation theory in order to make a point. It is often useful in quantum field theory to perform the so-called Wick-rotation: $k_4 = ik_o$; $x_4 = it$. The functional integral then becomes

$$\int Dg \exp\left(i \int \mathcal{L}\, d^3x\, dt\right) \rightarrow \int Dg \exp \int \mathcal{L}\, d^4x \ . \qquad (4.18)$$

In conventional quantum field theories the kinetic parts of the Lagrangian are all negative:

$$\mathcal{L} = - \tfrac{1}{2}(\partial_\mu\varphi)^2 - \frac{m^2}{2}\varphi^2 \ \ldots \qquad (4.19)$$

so that the functional integral (4.18) can be well-defined. Now in our case $\dot{\lambda}_{33}$ acted as a Lagrange multiplier, α, securing some constraint, f = 0. In Minkowski space one has

$$\int d\alpha\, e^{i\alpha f} \rightarrow \delta(f) \ . \tag{4.20}$$

After a Wick rotation one deals with Euclidean space, in which usually all fields and coordinates, including f, are real. But in order to obtain $\delta(f)$ as the result of a functional integration, we have to keep the i in (4.20). This means that in Euclidean space all Lagrange multipliers must be imaginary! Therefore, somewhat surprisingly, in the temporal gauge $\lambda_{o\mu} = 0$ (which becomes $\lambda_{4\mu} = 0$ in Euclidean space), we must require the field λ_{33} to be imaginary and not real. Actually, as in all integrals, there is some freedom in choosing the integration contours in the space of field variables. We could write

$$\alpha f = \tfrac{1}{4}(\alpha+f)^2 - \tfrac{1}{4}(\alpha-f)^2 \ . \tag{4.21}$$

The functional integral will be well defined if α is imaginary and f real, but it converges better if α-f is chosen real, and α+f imaginary. So we see from (4.11) that it will be even better to choose

$$\sum_\mu \lambda_{\mu\mu}$$

to be imaginary, and all components of λ orthogonal to that to be real. In fact we found this requirement to hold in all gauge choices, not only the temporal gauge. We can understand this as follows. Write

$$g_{\mu\nu} = e^{\varphi}\, \hat{g}_{\mu\nu} \tag{4.22}$$

with

$$\det(\hat{g}_{\mu\nu}) = 1 \ . \tag{4.23}$$

The Lagrangian then corresponds to

$$\int \sqrt{g}\, R = \int e^{\varphi}\Big(\hat{R} + \tfrac{3}{2}(\partial_\mu\varphi)^2 + \text{matter}\Big) \tag{4.24}$$

$$= \int e^{\varphi}\Big[\tfrac{3}{2}(\partial_\mu\varphi)^2 - \tfrac{1}{12}\Big(\partial_\mu\hat{g}_{\alpha\beta}+\partial_\alpha\hat{g}_{\beta\mu}+\partial_\beta\hat{g}_{\mu\alpha}\Big)^2\Big] , \tag{4.25}$$

where $\hat{R}$ is the curvature generated by $\hat{g}$ only, and $(\partial_\mu\varphi)^2$ stands for

$$\hat{g}^{\mu\nu}\ \partial_\mu\varphi\partial_\nu\varphi , \tag{4.26}$$

etc.

In the perturbative regime we can neglect the exponent in front, and we notice that in order for the functional integral to be properly defined, φ must be imaginary and $\hat{g}$ real.

Naturally, we ask the question what requirement to use in the non-perturbative regime where the exponent e^{φ} cannot be ignored. We write

$$e^{\varphi}(\partial_\mu\varphi)^2 \propto (\partial_\mu e^{\frac{1}{2}\varphi})^2 . \tag{4.27}$$

We see that the functional integral is well-defined if variations $\delta e^{\frac{1}{2}\varphi}$ of $e^{\frac{1}{2}\varphi}$ satisfy

$$|\mathrm{Re}(\delta e^{\frac{1}{2}\varphi})| \leqslant |\mathrm{Im}(\delta e^{\frac{1}{2}\varphi})| \tag{4.28}$$

whereas we must also insist on

$$\mathrm{Re}(e^{\varphi}) \geqslant 0 \tag{4.29}$$

to make the terms with $\hat{g}$ converge.

It is just barely possible to meet these requirements:

$$e^{\varphi} = 1 + ia , \quad a \text{ real} . \tag{4.30}$$

In the presence of scalar and vector fields S and A_μ, the matter-Lagrangian contributes

$$- \tfrac{1}{4}F_{\mu\nu}F_{\mu\nu} - \tfrac{1}{2}e^{\varphi}(\partial_\mu S)^2 - \tfrac{1}{2}e^{2\varphi}\left(m^2S^2 + \frac{\lambda}{12} S^4\right) . \tag{4.31}$$

We see that the kinetic parts are well behaved, but with (4.30), the mass term and scalar coupling terms diverge dangerously. The problem will not show up in perturbation theory, but will force us to choose φ dependent integration contours in S space.

With the choice (4.30) the integration over the conformal factor φ is just barely convergent. We suspect that this rather bad behaviour of the functional integral is related to an intrinsic instability in the theory of quantum gravity: collapse of matter into a black hole.

Another fundamental difficulty in quantum gravity was known for a long time: the theory is not renormalizable[2)]. Renormalization counter-terms with higher derivatives such as $(R_{\alpha\beta\gamma\delta})^2$, etc. would be necessary, which would add up eventually to non-local effects and violate causality. Now this problem seems to occur only at length scales in the order of the Planck scale, and its resolution will probably require drastic changes in the theory such as the "superstring" approach. But gravitational collapse may take place at length scales much larger than the Planck scale. Therefore we might be able to attack this problem using conventional quantum field theory only. This is the philosophy of our present research.

5. Black hole

The philosophy used as a starting point in quantizing gravity is that a Hilbert space exists, with a Hamiltonian, to be constructed for instance in the temporal gauge, where time is defined by clocks located at space-like infinity. Now one consequence of standard general relativity is well-known: if enough matter is accumulated in a small enough volume, then it collapses under the action of the gravitational force into a configuration called a "black hole". The black hole is the only naturally stable end product of this process. However, it is only stable in terms of the unquantized theory. If we set up a Hilbert space according to the above prescriptions, we find that its dimensionality is growing perpetually, even if one restricts oneself to states with total energy within certain bounds[3].

This situation is unsatisfactory for various reasons. In the case of a very large black hole the problem does not look very serious: all regions of space-time visible to the outside observer are regular, so if we "solve" our field theories in those regions, we can predict the observations by using standard coordinate transformations. As first discovered by Hawking, these coordinate transformations produce a non-trivial background of radiation: the Hawking radiation emitted by a black hole[4]. The fact that Hilbert-space is "expanding" is related to the ideal randomness of this thermal radiation: it corresponds to black body radiation with a temperature

$$T = \lambda/8\pi M , \tag{5.1}$$

where λ is probably 1, but other values are not entirely excluded[5]. It has been deduced from thermodynamical arguments that this radiation gives black holes an entropy proportional to the area of the horizon:

$$S = 4\pi M^2/\lambda \tag{5.2}$$

But black-hole solutions could have sizes as small as or smaller than elementary particles. In fact, in some sense, elementary particles are black holes themselves. Is the internal Hilbert space of these particles bounded or infinite? If infinite then all axioms of quantum mechanics may fall apart.

Of course we are unable to prove that the axioms of quantum mechanics hold for black holes, but this author chose for the theory that quantum mechanics is valid, but the rules for general coordinate transformations might have to be adapted.

Of course what we need foremost is a mathematically unique prescription for obtaining the laws of physics for every imaginable system. This "theory" should as much as possible reproduce all known results of ordinary quantum mechanics on the one hand and general relativity on the other. We will be quite content if this "theory" is first formulated in a coordinate-invariant way and then allows us to construct a Hamiltonian suitable to describe anything seen by any observer. But this construction might be dependent on the observer and in particular his "horizon". It could even be that the "probabilities" experienced by one observer are not the same as those of another. All is well if the two "classical limits" are as they should be.

We will now make the assumption that the black hole quantum properties[9)] somehow follow from Lagrange quantum field theory at the same length scale. We are very well aware of the risk that this may be wrong. Still, we like to know how far one can get. Regrettably, the results to be reported in this paper will be extremely modest.

We will start by making a simplification that caused some confusion for some readers of my previous publication: we first concentrate on the steady state black hole: every now and then something falls in and something else comes out. *Nowhere a distinction is made between "primordial" black holes and black holes that have been formed by collapse.* It has been argued that Hawking's derivation in particular holds for collapsed black holes and not necessarily for ones eternally in equilibrium. However if we succeed to describe infalling things in a satisfactory way then one might expect that inclusion of the entire collapse (and the entire evaporation in the end) can naturally be incorporated at a later stage. Our main concern at present will be time scales of order $M \log M$ in Planck units, which is much shorter than the black hole's history. As we will see, understanding in- and outgoing things at this scale will be difficult enough, and indeed Hawking's radiation can very well be understood at this time scale.

In the absence of matter, the metric of a black hole is*

$$ds^2 = -\left(1 - \frac{2M}{r}\right)dt^2 + \left(1 - \frac{2M}{r}\right)^{-1} dr^2 + r^2 d\Omega^2 . \tag{5.3}$$

The Kruskal coordinates u, v are defined by

$$uv = \left(1 - \frac{r}{2M}\right)e^{r/2M} , \tag{5.4}$$

$$v/u = -e^{t/2M} , \tag{5.5}$$

and then we have

$$ds^2 = -\frac{32M^3}{r} e^{-r/2M} du dv + r^2 d\Omega^2 , \tag{5.6}$$

which is now entirely regular at $r > o$. However (5.4) and (5.5) admit two solutions at every (r,t): we have two universes connected by a

-----

* We now use units in which $G = 1$; see the remarks on units in section 3.

"whormhole". The Schwarzschild region, I, is v > 0, u < 0. The other regions are indicated in Fig. 1.

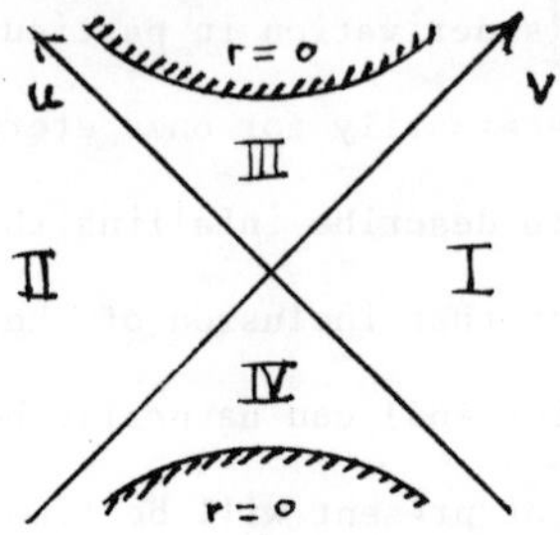

FIGURE I

Now the classical picture of a black hole formed by collapse only shows regions I and III, the others being shielded by the imploding matter which accumulates at the past horizon (the u-axis). Similarly, an evaporating black hole (sometimes called a "white hole") only has regions I and IV. In both cases it is convenient to extend analytically the particle content in regions III or IV towards region II, and a black hole in equilibrium is perhaps best described by the entire system I-II-III-IV.

Are black holes fundamental particles or solitons? It would be tempting to view upon these objects as being the "magnetic monopoles" of gravity theory, extended solutions of localized equations. There are two important differences however. One is that, unlike magnetic monopoles, black holes can be arbitrarily small and their total mass reduces when they shrink. The other is that in the case of monopoles, a Callan-Rubakov procedure of quantization is possible[8]. But when we do quantum field theory near a black hole, we find that an ingoing wave reflects back. In the Schwarzschild coordinates, the speed of light,

$$\frac{dr}{dt} = 1 - \frac{2M}{r} , \tag{5.7}$$

tends to zero at the horizon (the points r = 2M). Wave packets accumulate there and never return: an infinite amount of information can be stored "indefinitely" at this horizon.

6. The brick wall model

It is not inconceivable that gravitational interactions will remove the infinities described in the previous section. If we look at wave packets that arrived at the region $r \simeq 2M$ long ago and those that will leave that region in the late future, then in the Kruskal frame we will see one wave packet squeezed onto the u axis and one onto the v-axis. One wave packet is Lorentz-boosted to tremendous energies with respect to the other. Because of these enormous relative energies one expects non-negligible gravitational interactions between the two. This will happen typically when they both come closer to the horizon than a distance comparable to the Planck length. Apparently then, at some distance h from the horizon the particles are no longer described by free wave packets.

This is how we got motivated to look at the following simplistic model[3)]. It should be regarded as an excercise rather than a theory. Let us assume that all wave functions vanish within some fixed distance h from the horizon:

$$\varphi(x) = 0 \quad \text{if} \quad x \leqslant 2M + h \ , \tag{6.1}$$

where M is the black hole mass. For simplicity we take $\varphi(x)$ to be a scalar wave function for a light ($m \ll 1 \ll M$) spinless particle. Later we will give them a multiplicity Z as a first attempt to mimic more closely the real world.

In view of a freely falling observer, condition (6.1) corresponds to a uniformly accelerated mirror which in fact will create its own energy-momentum tensor due to excitation of the vacuum. As in sect. 1 we stress that this presence of matter and energy may be observer dependent, but above all this model should be seen as an elementary exercise, rather than an attempt to describe physical black holes accurately.

Let the metric of a Schwarzschild black hole be given by

$$ds^2 = -\left(1 - \frac{2M}{r}\right)dt^2 + \left(1 - \frac{2M}{r}\right)^{-1} dr^2 + r^2\, d\Omega^2 \ . \tag{6.2}$$

Furthermore, we need an "infrared cutoff" in the form of a large box with radius L:

$$\varphi(x) = 0 \quad \text{if} \quad x = L \ . \tag{6.3}$$

The quantum numbers are l, l_3 and n, standing for total angular momentum, its z-component and the radial excitations. The energy levels $E(l,l_3,n)$ can then be found from the wave equation

$$\left(1 - \frac{2M}{r}\right)^{-1} E^2\varphi + \frac{1}{r^2}\,\partial_r r(r-2M)\partial_r\varphi - \left(\frac{l(l+1)}{r^2} + m^2\right)\varphi = 0 \ . \tag{6.4}$$

As long as $M \gg 1$ (in Planck units) we can rely on a WKB approximation. Defining a radial wave number $k(r,l,E)$ by

$$k^2 = \frac{r^2}{r(r-2M)}\left(\left(1 - \frac{2M}{r}\right)^{-1} E^2 - r^{-2}l(l+1)-m^2\right) , \tag{6.5}$$

as long as the r.h.s. is non-negative, and $k = 0$ otherwise, the number of radial modes n is given by

$$\pi n = \int_{2M+h}^{L} dr\, k(r,l,E) \ . \tag{6.6}$$

The total number N of wave solutions with energy not exceeding E is then given by

$$\pi N = \int (2l+1)dl\ \pi n \stackrel{\text{def}}{=} g(E)$$

$$= \int_{2M+h}^{L} dr\left(1 - \frac{2M}{r}\right)^{-1} \int (2l+1)dl\ \sqrt{E^2 - \left(1 - \frac{2M}{r}\right)\left(m^2 + \frac{l(l+1)}{r^2}\right)}\ , \tag{6.7}$$

where the l-integration goes over those values of l for which the argument of the square root is positive.

What we have counted in (6.7) is the number of classical eigenmodes of a scalar field in the vicinity of a black hole. We now wish to find the thermodynamic properties of this system such as specific heat etc. Every wave solution may be occupied by any integer number of quanta. Thus we get for the free energy F at some inverse temperature β,

$$e^{-\beta F} = \sum e^{-\beta E} = \prod_{n,l,l_3} \frac{1}{1-e^{-\beta E}}\ , \tag{6.8}$$

or

$$\beta F = \sum_{N} \log\left(1-e^{-\beta E}\right)\ : \tag{6.9}$$

and, using (6.7),

$$\pi\beta F = \int dg(E)\ \log\left(1-e^{-\beta E}\right)$$

$$= -\int_{0}^{\infty} dE\ \frac{\beta g(E)}{e^{\beta E}-1}$$

$$= -\beta \int_{0}^{\infty} dE \int_{2M+h} dr\left(1 - \frac{2M}{r}\right)^{-1} \int \left(2l+1\right)dl$$

$$\times\ \left(e^{\beta E}-1\right)^{-1} \sqrt{E^2 - \left(1 - \frac{2M}{r}\right)\left(m^2 + \frac{l(l+1)}{r^2}\right)}\ . \tag{6.10}$$

Again the integral is taken only over those values for which the square root exists. In the approximation

$$m^2 \ll 2M/\beta^2 h \ , \qquad L \gg 2M \ , \tag{6.11}$$

we find that the main contributions are

$$F \simeq -\frac{2\pi^3}{45h}\left(\frac{2M}{\beta}\right)^4 - \frac{2}{9\pi} L^3 \int_m^\infty \frac{dE(E^2-m^2)^{3/2}}{e^{\beta E}-1} \ . \tag{6.12}$$

The second part is the usual contribution from the vacuum surrounding the system at large distances and is of little relevance here. The first part is an intrinsic contribution from the horizon and it is seen to diverge linearly as $h \to 0$.

The contribution of the horizon to the total energy U and the entropy S are

$$U = \frac{\partial}{\partial\beta}(\beta F) = \frac{2\pi^3}{15h}\left(\frac{2M}{\beta}\right)^4 Z \ , \tag{6.13}$$

$$S = \beta(U-F) = \frac{8\pi^3}{45h} 2M\left(\frac{2M}{\beta}\right)^3 Z \ . \tag{6.14}$$

We added a factor Z denoting the total number of particle types.

Let us now adjust the parameters of our model such that the total entropy is

$$S = 4\lambda^{-1} M^2 \ , \tag{6.15}$$

as in eq. (5.2), and the inverse temperature is

$$\beta = 8\pi\lambda^{-1}M \ . \tag{6.16}$$

This is seen to correspond to

$$h = \frac{Z\lambda^4}{720\pi M} \ . \tag{3.17}$$

Note also that the total energy is

$$U = \frac{3}{8} M , \tag{6.18}$$

independent of Z, and indeed a sizeable fraction of the total mass M of the black hole! We see that it does not make much sense to let h decrease much below the critical value (6.17) because then more than the black hole mass would be concentrated *at our side* of the horizon.

Eq. (6.17) suggests that the distance of the "brick wall" from the horizon depends on M, but this is merely a coordinate artifact. The invariant distance is

$$\int_{r=2M}^{r=2M+h} ds = \int \frac{dr}{\sqrt{1-2M/r}} = 2\sqrt{2}\ Mh = \sqrt{\frac{Z\lambda^4}{90\pi}} \ . \tag{6.19}$$

Thus, the brick wall may be seen as a property of the horizon independent of the size of the black hole.

The conclusion of this section is that not only the infinity of the modes near the horizon should be cut-off, but also the value for the cut-off parameter is determined by nature, and a property of the horizon only. The model described here should be a reasonable description of a black hole aslong as the particles near the horizon are kept at a temperature as given by (6.16) and all chemical potentials are kept close to zero. The reader is invited to investigate further properties of the model such as the average time spent by one particle near the horizon, etc.

The model automatically preserves quantum coherence completely, but it is also unsatisfactory: there might be several conserved quantum numbers, such as baryon number*. What is wrong, clearly, is that we

* One may postpone this difficulty by inserting explicitly baryon number violating interactions near the horizon.

abandoned the principle of invariance under coordinate transformations at the horizon. The question that we should really address is how to keep not only the quantum coherence but also general invariance, while dropping all global conservation laws.

7. The equivalence theorem and Hawking radiation

Consider now the mapping from Schwarzschild coordinates to Kruskal coordinates and back. The equivalence theorem should now relate the Hilbert space as needed by an observer in the wormhole ("Kruskal observer") to the one needed to describe the "physical" world I as experienced by an outside observer ("Schwarzschild observer"). Imagine a limited number of soft particles that can be described by the Kruskal observer using standard physics. With "soft" we mean that the energies of these particles are so small that gravitational effects on the metric can be neglected. We have then a reasonable description of an important part of the Hilbert space for the wormhole observer. The evolution of this system is described by a Hamiltonian

$$H = \int \mathcal{H}(x)dx > 0 \ , \tag{7.1}$$

with one ground state

$$H|0\rangle_k = 0 \tag{7.2}$$

where k stands for Kruskal. Due to curvature this vacuum is not exactly but only approximately conserved. H describes the evolution in the time coordinate $\tau = u+v$.

Now theoutside observer uses t as his time coordinate, and a generator of a boost in t produces

$$\delta v = \frac{v}{2M}\,\delta t \ , \tag{7.3}$$

$$\delta u = -\frac{u}{2M}\,\delta t \ , \tag{7.4}$$

so the generator of this boost is

$$h = \frac{1}{2M} \int dx\, \rho \mathcal{H}(x) \quad ; \quad \rho = v-u\ . \tag{7.5}$$

We split $h = H_I - H_{II}$:

$$H_I = \frac{1}{2M} \int \rho \mathcal{H}(\vec{x}) d\vec{x}\ \theta(\rho) \quad ; \quad H_{II} = \frac{1}{2M} \int |\rho|\ \mathcal{H}(\vec{x}) d\vec{x}\ \theta(-\rho)\ . \tag{7.6}$$

We have

$$[H_I, H_{II}] = 0\ , \tag{7.7}$$

and we can write the eigenstates of H_I and H_{II} as $|n,m\rangle$ with

$$H_I |n,m\rangle = n|n,m\rangle \quad ; \quad H_{II}|n,m\rangle = m|n,m\rangle\ . \tag{7.8}$$

Extensive but straightforward calculations show that the "Kruskal vacuum" $|0\rangle_k$ does not coincide with the "Schwarzschild vacuum $|0,0\rangle$, but instead, we have

$$|0\rangle_k = C \sum_n |n,n\rangle e^{-4\pi M n}\ , \tag{7.9}$$

where C is a normalization factor. Note that we do have

$$h|0\rangle_k = 0\ , \tag{7.10}$$

which is due to Lorentz-invariance of $|0\rangle_k$.

If we consider the equivalence theorem in its usual form and consider all those particles that are trapped into region IV as lost and therefore unobservable then without any doubt the correct prescription for describing the observations of observers in I is to average over the unseen particles. Let $\mathcal{O}$ be an operator built from a field $\phi(\vec{x},t)$ with $\vec{x}$ in region I, then

$$[\mathcal{O}, H_{II}] = 0\ , \tag{7.11}$$

$$\mathcal{O}|n,m\rangle = \sum_k \mathcal{O}_{nk}|k,m\rangle \ , \qquad (7.12)$$

and

$$\langle\mathcal{O}\rangle = {}_k\langle 0|\mathcal{O}|0\rangle_k = C^2 \sum_{n,n'} e^{-4\pi M(n+n')} \langle n',n'|\mathcal{O}|n,n\rangle = C^2 \sum_n e^{-8\pi Mn} \mathcal{O}_{nn} \ . \qquad (7.13)$$

We recognize a Boltzmann factor $e^{-\beta n}$ with $\beta = 8\pi M$, corresponding to a temperature

$$T = 1/8\pi M \ . \qquad (7.14)$$

This is Hawking's result in a nutshell. Black holes radiate and the temperature of their thermal radiation is given by (7.14). The only way in which the horizon entered in this calculation is where it acts as a shutter making part of Hilbert space invisible.

As stated in section 5 this result would imply that black holes are profoundly different from elementary particles: they turn pure quantum mechanical states into mixed, thermal, states. Our only hope for a more complete quantum mechanical picture where black holes also show pure transitions, that in principle allow for some effective Hamiltonian is to reformulate the equivalence principle. Let us assume that the location of the horizon has a more profound effect on the interpretation that one should give to a wave function.

A pair of horizons (the u- and the v-axis in Fig. 1) always separate regions where a boost in t goes in opposite directions with respect to a regular time coordinate such as u+v. As before[5] we speculate that these regions act directly as the spaces of bra states and ket states, respectively. Any "state" as described by a Kruskal observer actually looks like the product of a bra and a ket state to the Schwarzschild observer. More precisely, it looks like an element of his density matrix, ρ:

$$|n,m\rangle \rightarrow |n\rangle\langle m| = \rho \ . \tag{7.15}$$

Just like any density matrix its evolution is given by the commutator with H_I:

$$\frac{d}{dt}\rho_{nm} = -ih|n,m\rangle = -i(n-m)|n,m\rangle = -i[H_I,|n\rangle\langle m|] = -i[H_I,\rho] \ . \tag{7.16}$$

Now the Kruskal vacuum $|0\rangle_k$ corresponds to the density matrix

$$\rho_{nn'} = C|n\rangle e^{-4\pi Mn}\langle n|\delta_{nn'} \ , \tag{7.17}$$

which is a thermal state at temperature

$$T = 1/4\pi M \ , \tag{7.18}$$

twice the usual result. The usual result would require not ρ from eq. (7.15) but $\rho\rho^\dagger$ to be the density matrix, from which of course (7.14) follows.

As long as we consider *stationary black holes with only soft particles* our mapping (7.15) is perfectly acceptable. The Hamiltonian (7.1) may ad libitum be extended to include any kind of interactions including those of curious observers. In the two classical limits we reproduce quantum mechanics and general relativity as required.

The only possible way to settle the question which of the procedures is correct and which of the temperatures (7.14) or (7.18) describe a black hole's radiation spectrum, is to include the effects of "hard" particles. This is also a necessary requirement for understanding the effects of implosion and explosion of black holes. Hard particles are particles whose rest masses may be small, but whose energies are so large that their gravitational effects may not be ignored.

8. Hard particles

The black holes considered in the previous section were only exactly time-translation-invariant if they were covered by a Kruskal vacuum $|0\rangle_k$. This is because translations in t correspond to Lorentz-transformations at the origin of the Kruskal coordinate frame and only a vacuum can be Lorentz-invariant. Naturally, $|0\rangle_k$ corresponds to a Schwarzschild density matrix ρ which is diagonal in the energy-representation.

Any other state will undergo boosts in t as if the Kruskal observer continuously applies Lorentz-transformations to his states, and eventually any "soft" particle will turn into a hard particle. This is why hard particles, particles with enormously large Lorentz γ factors are unavoidable if we want to understand how a system evolves over time scales only slightly larger than $\mathcal{O}(M \log M)$. Hard particles alter their surrounding space-time metric. Some basic features of their effects on space-time are now well-known.

A hard particle in Minkowski space produces a gravitational shock wave[6], sometimes called "impulsive wave", not unlike Cerenkov radiation. Before and behind this shock wave space-time is flat, but the way in which these flat regions are connected at the location of the shock wave produces delta-distributed curvature. Writing

$$\begin{aligned} u &= t-z \\ v &= t+z \end{aligned} \tag{8.1}$$

we find that a particle moving in the positive z direction with momentum p, at $\tilde{y} = 0$, produces a shock wave on the v axis where the two half-spaces are connected after a shift

$$\delta v = -4p \ \ln(\tilde{y}^2) \ . \tag{8.2}$$

Here $\tilde{y}$ is the transverse coordinate.

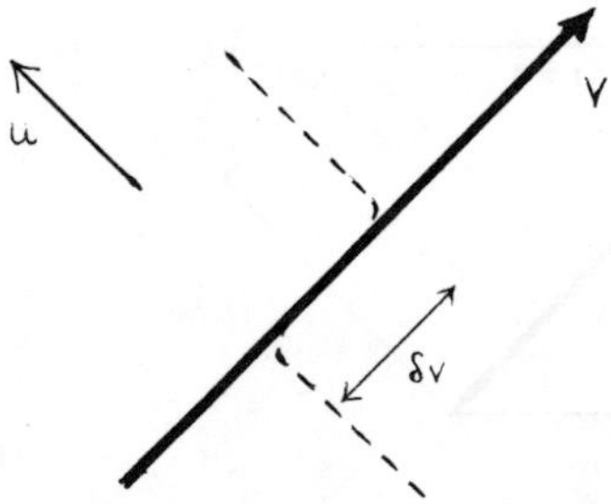

FIGURE 2

A way to picture this is to choose $g_{\mu\nu} = \eta_{\mu\nu}$ everywhere except at u=0, where all geodesics make a jump δv from past to future. See Fig. 2.

For us it is interesting to consider now a hard particle on one of the black hole's horizons. It was found that again a displacement of a form similar to (8.2) solves Einstein's equations. In Kruskal's coordinates u,v a hard particle with momentum p again produces a shift δv, with

$$\delta v(\vec{\Omega}) = pf(\vec{\Omega},\vec{\Omega}') \ , \tag{8.3}$$

where Ω' is the angle where the particle goes through the horizon and p its momentum. f is given by

$$\Delta f - f = -2\pi\kappa \ \delta(\theta) \ , \tag{8.4}$$

where θ is the angle between Ω and Ω'; Δ the angular Laplacian and κ a dimensionless numerical constant. The solution to (3.4),

$$f = \kappa \sum_{\ell} \frac{\ell+\frac{1}{2}}{\ell(\ell+1)+1} P_{\ell}(\cos\theta) \ , \tag{8.5}$$

can be seen to be positive for all θ.

Because of the shift, the causal structure of space-time is slightly changed. The Penrose diagram for a hard particle coming in along the past horizon is given in Fig. 3.

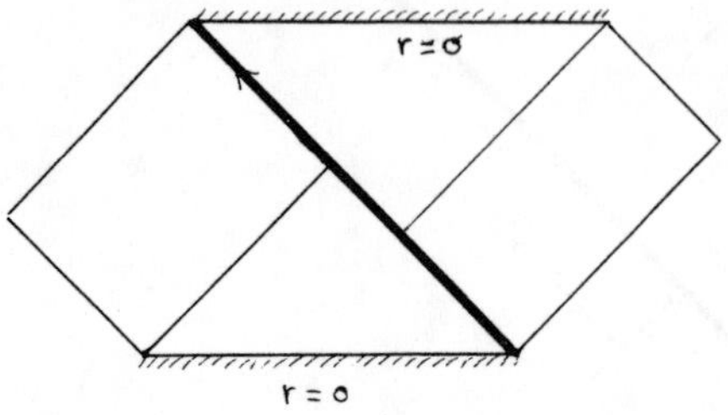

FIGURE 3

In Fig. 3 the geodesics are defined to go straight through the shock wave but enter into a more or less badly curved metric.

When two hard particles meet each other from opposite directions the curvature due to the resulting gravitational radiation is not easy to describe. We do need some description of this situation and therefore we introduced a simplification by imposing spherical symmetry. Hard particles are now replaced by spherically symmetric hard shells of matter entering or leaving the black hole. We guessed correctly that then Einstein's equations are also solved by connecting shifted Schwarzschild solutions with different mass parameters[7)]. The space-time structure of Fig. 4 results.

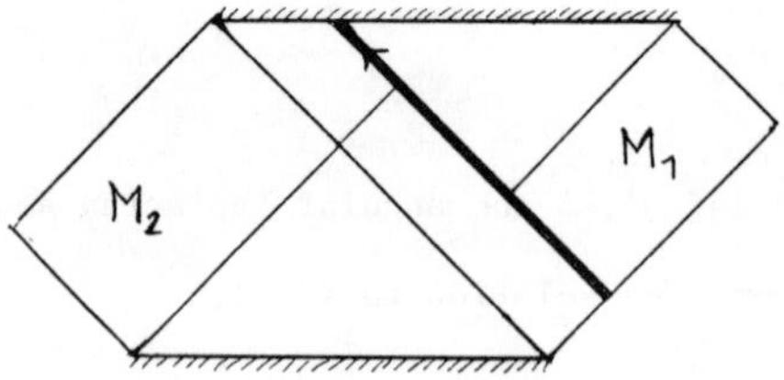

FIGURE 4

In Fig. 4 matter hits the future singularity at some distance from the past-horizon. In that case $M_1 > M_2$, if we require that the energy content of the shell of matter be positive.

This solution allows us now to combine various shells of ingoing and outgoing matter. One gets the Penrose diagram of Fig. 5.

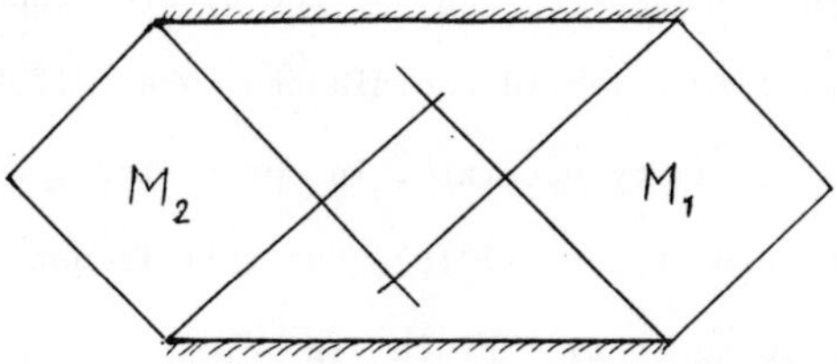

FIGURE 5

The algebra of the allowed amounts of energy in the shells and the resulting mass parameters M_i is fairly complicated.

An interesting limiting case occurs if one of the internal mass parameters tends to zero. If we require all shell-energies to be positive then such a zero mass region must always connect the future- with the past singularity by an r=0 line. This r=0 line is the origin of a polar coordinate representation of a flat space and one easily convinces oneself that then no longer any wormhole exists that connects us with another space. Bra- and ket-space are clearly disconnected and indeed we will argue that such a no-bra-space may perhaps be a way to describe a pure state for the Schwarzschild observer.

9. Purification of black hole states and the effect on the metric

We can now understand qualitatively some aspects of the mapping from Schwarzschild coordinates to Kruskal coordinates and back.

Suppose we have in Kruskal space an eigenstate of the Hamiltonian (7.1), say the lowest, the vacuum. The general coordinate transformation that maps this state onto Schwarzschild coordinates, where (7.5) is the Hamiltonian, produces some density matrix ρ. In our picture, a likely candidate is (7.17), with temperature (7.18), but this is not so crucial for the following. Our mapping is a1 $\leftrightarrow$ b1 in Fig. 6.

Now let us consider another state in Kruskal space, such that in Schwarzschild space most of the ingoing particles disappear. We are searching then for solutions of equations of the form

$$a_s|\psi\rangle = 0 \tag{9.1}$$

where a_s are annihilation operators in Schwarzschild space, whereas before we had

$$a_k|\psi\rangle = 0 \tag{9.2}$$

where a_k are the particle-annihilation operators in Kruskal space. The relation between a_k and a_s is

$$a_s\sqrt{e^{\pi\omega}-e^{-\pi\omega}} = a_k(\omega)e^{\pi\omega/2} + a_k^{\dagger}(-\omega)e^{-\pi\omega/2} \tag{9.3}$$

$$b_s\sqrt{e^{\pi\omega}-e^{-\pi\omega}} = a_k(-\omega)e^{\pi\omega/2} + a_k^{\dagger}(\omega)e^{-\pi\omega/2} \tag{9.4}$$

where ω is the energy eigenvalue of an ingoing or outgoing wave in Schwarzschild space. b_s are annihilation operators in section II of the Schwarzschild world. We may or may not require $b_s|\psi\rangle \to 0$.

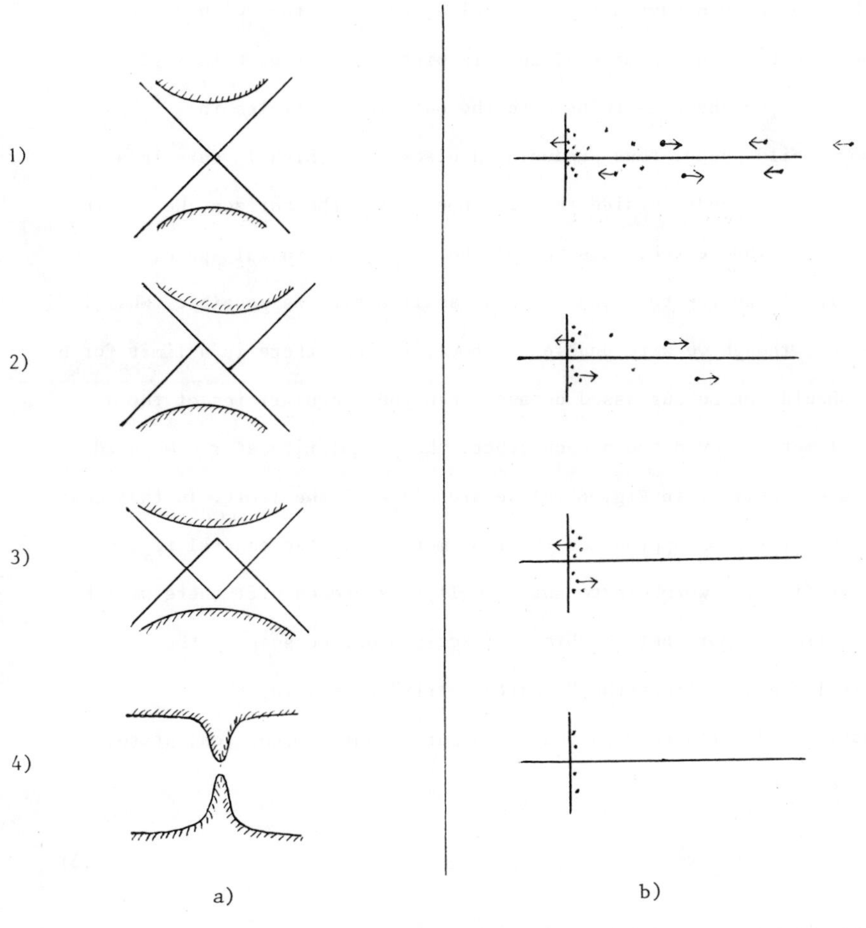

FIGURE 6

Mapping from Kruskal coordinates (a) to Schwarzschild coordinates (b).

Clearly (9.1) and (9.2) cannot simultaneously be satisfied. So, imposing (9.1) instead of (9.2) implies admitting particles in the Kruskal world. If all a_s would annihilate $|\psi>$ then an infinite amount of matter would appear in the Kruskal world. That is more than we can handle. So rather we decide to annihilate all particles further from the horizon than some distance h in Schwarzschild space. The particles in

Kruskal space then have a finite total energy-momentum, with a corresponding effect on the metric. This is pictured in Fig. 6 (a2, b2).

We can do the same thing with the outgoing particles in Schwarzschild space, thus producing a black hole which is pure in a region that extends outside the distance h from the horizon. We see in Fig. 6 (3) what effect this has on the metric in Kruskal space.

Finally we try to reduce h to as small a size as possible. What is expected, though we were unable to prove, is that there is a limit for h that should not be surpassed because then the singularities of the Kruskal metric would touch each other: the singularity at $r \to 0$ would become space-like. In Fig. 6 (4) we are close to the limit. In that case the central region approaches the flat metric. If the central region becomes flat the wormhole to universe II is squeezed off: there will be no coupling anymore between bra- and ket-states. We suspect that precisely in this limit the "density matrix" describing the Schwarzschild world becomes that of a pure quantum mechanical state, that is

$$\rho = |\psi\rangle \langle\psi| \ , \qquad (9.5)$$

so that it has one eigenvalue 1 and the others are zero.

We believe that the scenario sketched in this section can replace the "brick wall model" to obtain a more natural description of a black hole in a pure state.

10. Conclusion

The aim of our investigation was to apply quantum field theory at moderately large distance scales in sufficiently smooth but non-trivial space-times of deduce the quantum properties of black holes. This aim was not achieved. Although the black hole is classically a stationary configuration, it is a run-away solution in the quantum-mechanical sense: the dimensionality of Hilbert space seems to explode indefinitely. On the one hand this is of course a disappointment. A more detailed theory of black holes could suggest a link between them and the elementary constituents of some "unified field theory". On the other hand this situation is extremely interesting, because it strongly indicates that even at large distance scales quantum field theory may have to be amended. Our "dissident" ideas about the density matrix at the horizon might be an example of such a change, but in any case they are not sufficient.

Curiously, some very fundamental notions in quantum mechanics may have to be reconsidered. In General Relativity a Hamiltonian only exists if time is defined by "clocks" at spatial infinity, as we saw in section 3. But which definitions should we use when an observer falls into a black hole? He should take his clocks with him.

It is quite likely that black holes can be described in many different ways, which are essentially equivalent. This is because we can put anything we like in space II without observing any difference in space I. This resembles the situation in a gauge theory where also the dynamical variables can be described in many different (but "gauge equivalent") ways. Perhaps black holes are huge gauge particles.

References

1) R. Adler, M. Bazin, M. Schiffer, "Introduction to General Relativity", McGraw-Hill 1965.

P.A.M. Dirac, "General Relativity", Wiley Interscience 1975.

S. Weinberg, "gravitation and Cosmology: Principles and Applications of the General Theory of Relativity", J. Wiley & Sons.

S.W. Hawking and G.F.R. Ellis, "The large scale Structure of Space-time", Cambridge Univ. Press 1975.

S. Chandrasekhar, "The Mathematical Theory of Black Holes", Clarendon Press, Oxford Univ. Press.

R.M. Wald, "General Relativity", 1984, Univ. of Chicago Press.

2) B.S. DeWitt, "Quantum Theory of Gravity I, II, III", Phys. Rev. 160 (1967) 1113; 162 (1967) 1195, 1239.

G. 't Hooft and M.Veltman, "One loop divergences in the Theory of Gravitation", Ann. Inst. H. Poincaré 20 (1974) 69.

3) G. 't Hooft, "On the quantum structure of a black hole", Nucl. Phys. B256 (1985) 727.

4) S.W. Hawking, "Particle creation by black holes", Comm. Math. Phys. 43 (1975) 199.

J.B. Harkle and S.W. Hawking, "Path integral derivation of black hole radiance", Phys. Rev. D13 (1976) 2188.

W.G. Unruh, "Notes on black-hole evaporation", Phys. Rev. D14 (1976) 870.

5) G. 't Hooft, "Ambiguity of the equivalence principle and Hawking's temperature", J. Geom. Phys. 1 (1984) 45.

6) T. Dray and G. 't Hooft, "The gravitational shock wave of a massless particle", Nucl. Phys. B253 (1985) 173.

7) T. Dray and G.'t Hooft, "The effect of spherical shells of matter on the Schwarzschild black hole", Comm. Math. Phys. 99 (1985) 613.

8) V. Rubakov, "Superheavy magnetic monopoles and the decay of the proton", JETP Lett. 33 (1981) 644; "Adler-Bell-Jackiw Anomaly and Fermion-number Breaking in the Presence of a Magnetic Monopole", Nucl. Phys. B203 (1982) 311.

C.G. Callan, "Disappearing Dyons", Phys. Rev. D25 (1982) 2141; "Dyon-Fermion Dynamics", Phys. Rev. D26 (1982) 2058; "Monopole Catalysis of Baryon Decay", Nucl. Phys. B212 (1983) 391.

RANDOM GEOMETRY, LATTICES AND FIELDS

Claude Itzykson
Service de Physique Théorique,
CEA -Saclay, 91191 Gif-sur-Yvette Cedex, France

I present in these notes a review of the formulation of statistical or field theoretic models on random lattices. This is a subject rejuvenated by Christ, Friedberg and Lee[1], which has attracted attention in the past in the context of condensed matter physics[2,3,4], and other domains. A notable mathematical text book is the one of Santalo[5]. A related subject is the one of discretized gravity initiated by Regge[6]. Numerous recent works scrutinize and apply these ideas particularly in the context of random surfaces[7-13].

The introduction of a random lattice in flat or curved space is in some sense equivalent to the use of an arbitrary coordinate system. It forces us to formulate invariants in a neat geometric fashion (and there remains much to do in this direction). On the other hand it offers a new approach to the restoration of continuous symmetries -if any- like translational, rotational or scale invariance.

In this respect the problems encountered are closely akin to those studied in the context of disordered systems, among them localization.

Our presentation is a summary of published work[14-19]. The main headings are

I. POISSONIAN LATTICES

A- Two Dimensions

I owe a great deal to my collaborators, M. Bander, E. Gardner, B. Derrida and J.M. Drouffe. It is a great pleasure to express them my gratitude. Discussions with A. Billoire, D. Gross, F. David and E. Marinari are gratefully acknowledged.

I. POISSONIAN LATTICES

In a large regular volume Ω, of Euclidean d-dimensional space, N points are chosen independently with uniform probability. With $|\Omega|$ the measure of the volume, we always understand the finite density limit

$$N \mapsto \infty, \qquad |\Omega| \mapsto \infty, \qquad \frac{N}{|\Omega|} \mapsto \rho = a^{-d} \tag{1}$$

with a playing the role of elementary length scale, proportional to the average spacing between points. Units are chosen such that a = 1.

In one dimension, points can be ordered, separated by intervals of lengths $\ell_{i,i+1}$, which in the limit (1) obey a Poisson distribution

$$p(\ell)\,d\ell = e^{-\ell} d\ell \tag{2}$$

Hence the general name Poissonian lattice. For the methods used below see[20,1,17].

A. Two Dimensions

Let us look at the case d=2. The generalization to higher dimension will then be straightforward. The following construction is attributed to Dirichlet and Voronoi.

To each point M_i we assign a closed cell C_i of all those points of the plane which are nearer to M_i than to any other M_j, $j \neq i$. C_i is a convex polygon (the intersection of half planes) and for $j \neq i$ the intersection $C_i \cap C_j$ is either empty or a segment of the bisecting line between M_i and M_j. In this case (M_i, M_j) form a pair of neighbors. Let q_i be the number of neighbors of M_i, or equivalently, the number of sides of the cell C_i (local coordination number). Similarly for (ijk) distinct, if $C_i \cap C_j \cap C_j$ is not empty, it is a vertex of each of the cells C_i, C_j, C_k and we call the triangle an elementary 2-simplex as $(M_i M_j)$ was an elementary 1-simplex. The plane region Ω is now paved by elementary 2-simplices as it is paved by elementary cells C_i. The two sets of objects are dual.

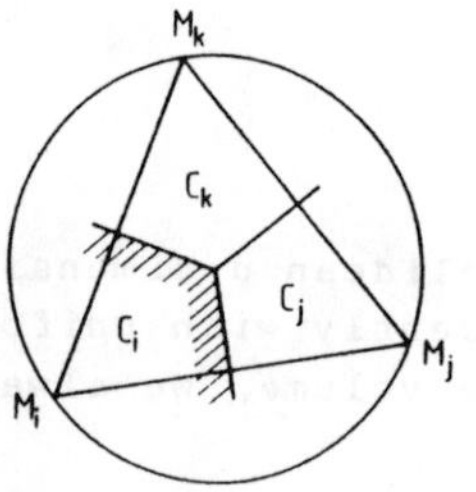

Fig. 1 : The Dirichlet-Voronoi construction in 2 dimensions.

Let

$N_0 = N$ = number of points = number of cells
N_1 = number of links (between neighbors) = number of cell vertices.
N_2 = number of 2-simplices = number of cell vertices.

$$N_0 - N_1 + N_2 = \chi \tag{3}$$

where the Euler characteristics χ (2 for a sphere, 0 for a torus...) is not extensive, i.e. in the infinite volume limit (1) $\chi/N \mapsto 0$. Cutting the plane along the edges of the cells we get N_0 pieces (the cells) with $2N_1 = \sum_i q_i$ edges and

$3N_2$ vertices. Since each cell has as many vertices as edges (Euler's relation in dimension 1) we find

$$2N_1 = 3N_2 = \sum_i q_i$$
$$N_o = \sum_i 1 \qquad (4)$$

If

$$q = \lim_{N\mapsto\infty} \frac{\sum_i q_i}{\sum_i 1} \qquad (5)$$

is the average coordination, then in the limit $N\mapsto\infty$

$$q = 6 \qquad \frac{N_1}{N_o} = 3 \qquad \frac{N_2}{N_o} = 2 \qquad (6)$$

Equivalently if $n_{i,j}$, $i>j$, is the average number of i-simplices incident on the j simplex, we have

$$n_{1,o} = n_{2,o} = 6 \qquad (7)$$

A remarkable property of the two-dimensional case is the existence of a regular triangular lattice where (6) is true locally. The random lattice may thus be described as a deformation of a regular triangular lattice with defects (points at which $q_i \neq 6$). Those points may be thought as carrying a "charge" or "curvature", while globally the system is neutral.

Up to now we have only used topology. Let us turn to metrical properties. We use the notations

ℓ_{ij} = distance between the neighbors (M_iM_j) $\langle\ell_{ij}\rangle = \ell_1$
length of 1-simplex (ij)
ℓ_{ijk} = area of 2-simplex (ijk) $\langle\ell_{ijk}\rangle = \ell_2$

Similarly

σ_i = area of cell C_i $\langle\sigma_i\rangle=\sigma_2$
σ_{ij} = length of edge perpendicular to the link (ij) $\langle\sigma_i\rangle=\sigma_2$

Since the density is unity

$$\sigma_2 = 1 \tag{8}$$

and

$$\ell_2 = 1/2 \tag{9}$$

To compute ℓ_1 and σ_1 one makes the following observation. Let (ijk) be a 2-simplex and O the dual vertex common to the cells C_i, C_j, C_k. The circle Γ of radius R circumscribed to the triangle (ijk) has O as its center. Then there are no lattice points inside Γ. Indeed the distance $|OM_\ell|$ for $\ell\neq\{i,jk\}$ is greater or equal to $|OM_i|$, $|OM_j|$, $|OM_k|$ by definition.

Given a point M_o of the lattice, the probability that M_1 and M_2 lying withing d^2x_1 and d^2x_2 respectively altogether form a 2-simplex is therefore

$$dp = \underbrace{n_{2,o}^{-1}}_{\neq\text{of 2-simplices/point}} \times \underbrace{C_{N-1}^{2}}_{\text{choice of 2 out of N-1 points}} \times \underbrace{\frac{d^2x_1d^2x_2}{\Omega^2}}_{\text{uniform probabilities for } M_1 \text{ and } M_2} \times \underbrace{\frac{(\Omega-\pi R^2)^{N-3}}{\Omega^{N-3}}}_{\text{N-3 other points required to lie outside }\Gamma}$$

and in the infinite limit volume

$$dp = \frac{e^{-\pi R^2}}{12} d^2x_1d^2x_2 \tag{10}$$

Using as parameters the radius R of the circumscribed circle and the polar angles φ_o, φ_1, φ_2 of OM_o, OM_1, OM_2

$$d^2x_1 d^2x_2 = R^3 |\sin(\varphi_1-\varphi_2)+\sin(\varphi_2-\varphi_o)+\sin(\varphi_o-\varphi_1)| dR\, d\varphi_o d\varphi_1 d\varphi_2$$

one can check that $\int dp=1$ as follows

$$\int dp = \frac{1}{12}\int_0^\infty dR\, R^3 e^{-\pi R^2} \times 2\pi \times 2 \int_o^{2\pi} d\varphi_2$$

$$\int_o^{\varphi_2} d\varphi_1 \{\sin(\varphi_2-\varphi_1)-\sin\varphi_2+\sin\varphi_1\} = 2\pi^2 \int_o^\infty dR\, R^3 e^{-\pi R^2} = 1$$

The average separation between two neighbors ℓ_1 is the mean of

$|M_oM_1| = 2R \left|\sin \frac{(\varphi_1-\varphi_o)}{2}\right|$ over dp, i.e.

$$\ell_1 = \frac{1}{12}\int_o^\infty dR\, R^3 e^{-\pi R^2} \times 2\pi \times 2 \int_o^{2\pi} d\varphi_2$$

$$\int_o^{\varphi_2} d\varphi_1\, 2R \sin \frac{1}{2}\varphi_1 \{\sin(\varphi_2-\varphi_1)-\sin\varphi_2+\sin\varphi_1\}$$

$$= \frac{32}{9\pi} = 1.1317684... \qquad (11)$$

This procedure is justified in dimension 2, since each link belongs to two triangles

$$\ell_1 = \frac{1}{N_1}\sum_{\text{links}} \ell_{ij} = \frac{1}{2N_1}\sum_{\text{triangles}} \sum_{\substack{\ell_{ij} \text{ belongs to} \\ \text{a given triangle}}} \ell_{ij} = \frac{3N_2}{2N_1} \times \text{average of a triangle side}$$

and $3N_2/2N_1=1$. In three dimensions, the number of tetrahedra sharing a link is not constant so the average length of a tetrahedron side is not the average of a link of the lattice. But again the procedure is correct for the average area of a triangular face. For lack of a better procedure the quantities referred to as $\ell_{(i)}$ in dimension d are in fact computed as the averages of the corresponding elements of a typical d-simplex.

One verifies that the average area of a triangle is

1/2 as indicated in (9) and its relative variance

$$\frac{\delta \ell_2}{\ell_2} = \left[\frac{\langle \ell_{ijk}^2 \rangle}{\langle \ell_{ijk} \rangle^2} - 1 \right]^{1/2} = \left(\frac{35}{2\pi^2} - 1 \right)^{1/2} = 0.8793 \qquad (12)$$

The average perimeter of the cells has been found by Meijering[20]. Let M be a lattice point and consider for any other lattice point M_k the bisecting line Δ_k of the segment NM_k. The number of such lines at a distance ρ, up to dρ, is the number of points M_k at distance 2ρ up to d2ρ

$$N \frac{8\pi\rho d\rho}{|\Omega|} = 8\pi\rho d\rho$$

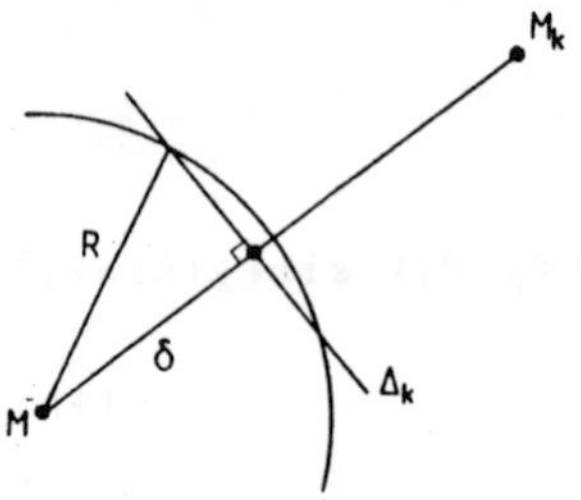

Fig. 2 : Geometric construction involved in Meijering's argument.

The fraction length of such a line Δ_k at a distance between R and R+dR is $(R^2-\rho^2)^{-1/2}$ 2RdR.

The total amount of "would be" cell-boundary length between R and R+dR is obtained by integrating in ρ from zero to R the product of these factors

$$2RdR \int_0^R \frac{8\pi\rho d\rho}{\sqrt{R^2-\rho^2}} = 16\pi \, R^2 dR$$

An element of this length will belong to the boundary of the cell around M if the circle of radius R centered at the length element and passing through M does not contain any interior latice point. This occurs with probability $e^{-\pi R^2}$. Thus the perimeter of the cell boundary at a distance between R and R+dR is given by

$$dP = 16\pi R^2 dR\, e^{-\pi R^2} \quad (13)$$

Integrating over R gives the mean perimeter

$$P_{\bullet} = 16\pi \int_0^{\infty} R^2\, e^{-\pi R^2}\, dR = 4 \quad (14)$$

Since on average a cell has six sides, we find

$$\sigma_1 = \frac{2}{3}$$

Table 1 summarizes the results and compares them with the regular triangular lattice (numbers in brackets) of equal density.

Simplex	Number	Direct lattice average number incident on a point	mean length or area	Dual lattice mean length or area
point	$N_o = N$	$n_{o.o} = 1$		$\sigma_2 = 1$
1-simplex	$N_1 = 3N$	$N_{1.o} = 6$	$\ell_1 = \frac{32}{9\pi}\left[\frac{2^{1/2}}{3^{1/4}}\right]$	$\sigma_1 = \frac{2}{3}\left[\frac{2^{1/2}}{3^{3/4}}\right]$
2-simplex	$N_2 = 2N$	$n_{2.o} = 6$	$\ell_2 = \frac{1}{2}$	

Table I - Averages in two dimensions
(quantities in brackets refer to the triangular lattice)

It may be amusing to report here some findings obtained in collaboration with J.M. Drouffe [16]. Following earlier investigators [21], but using different methods, we have computed several quantities pertaining to the distribution of cells of n sides ($\bar{n}=q=6$) in particular the probability p_n of the occurrence of such a cell up to very large n (of order 50 where $p_n \sim 10^{-73}$!!) using analytical (and untractable) integral representation for p_n. Table II shows that n=6 only occurs in approximately 30% of the cases with 4, 5, 7 and 8 sided faces still in appreciable amounts

n	100 p_n
3	1.13 ± 0.01
4	10.70 ± 0.2
5	25.7 ± 0.77
6	29.4 ± 0.3
7	19.8 ± 0.3
8	9.3 ± 0.8
9	2.88 ± 0.07
10	0.69 ± 0.02
20	$(1.5 \pm 0.8)\ 10^{-11}$
35	$(3.6 \pm 2.7)\ 10^{-38}$
50	$(1.5 \pm 1.5)\ 10^{-73}$

Table II : Probability distribution for n-sided cells

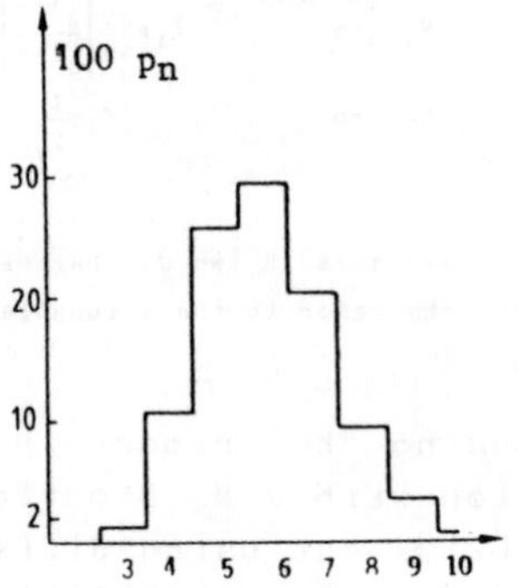

Fig. 3 : Probability distribution for cells with n sides.

We were able to show that there exists two constants A, B such that $\frac{A}{n^{2n}} \le p_n \le \frac{B}{n^n}$. The data seem to favor a behavior corresponding to the lower bound.

B- Three Dimensions

The Voronoi-Dirichlet construction goes through with

$N = N_o$, the number of points, N_1 of links, N_2 of triangles, N_3 of tetrahedra. Euler's relation is

$$0 = N_o - N_1 + N_2 - N_3 \qquad (15)$$

Each triangle is shared by two tetrahedra and each tetrahedron has four triangles

$$2N_2 = 4N_3 \qquad (16)$$

We define as above $n_{k,o}$ as the number of k-simplices incident on a point

$$n_{k,o} = (k+1)\,\frac{N_k}{N} \qquad (17)$$

It follows from topology alone that

$$n_{1,o} = 2 + \frac{1}{2}\,n_{3,o}$$
$$n_{2,o} = \frac{3}{2}\,n_{3,o} \qquad (18)$$

leaving as unknown $n_{3,o}$.

The probability to find a tetrahedron with one vertex M_o at the origin and the three other ones M_i, $i=1,2,3$, at $\vec{x}_i$ within d^3x_i is

$$dp = \frac{1}{3!\,n_{3,o}} \prod_1^3 d^3x_i \; e^{-\frac{4\pi R^3}{3}} \qquad (19)$$

with

$$n_{3,o} = \frac{1}{3!} \int \prod_1^3 d^3x_i \; e^{-\frac{4\pi R^3}{3}} \qquad (20)$$

If O is the center of the tetrahedron, n_o, n_1, n_2, n_3 unit vectors along OM_o, OM_1, OM_2, OM_3 and R^3w the volume of the tetrahedron.

$$w = \frac{1}{3!}\left|\det\begin{bmatrix}1 & 1 & 1 & 1\\ n_o & n_1 & n_2 & n_3\end{bmatrix}\right| \tag{21}$$

then

$$\frac{1}{3!}\, d^3x_1 d^3x_2 d^3x_3 = R^8 dR\; w\, d^2n_o d^2n_1 d^2n_2 d^2n_3 \tag{22}$$

a formula which generalises in d-dimensions as

$$\frac{1}{d!}\, d^dx_1 \ldots d^dx_d = R^{d^2-1}\, dR\; w \prod_o^d d^{d-1} n_i \tag{23}$$

Returning to dimension 3 with $\langle w\rangle$ denoting the rotational average of w,

$$\langle w\rangle = \int \prod_o^3 \frac{d^2\hat{n}_i}{4\pi}\, w \tag{24}$$

we find

$$n_{3,o} = 72\pi \langle w\rangle \tag{25}$$

We shall present below a general formula for $\langle w\rangle$ which restricted to d=3 yields

$$n_{3,o} = \frac{96}{35}\pi^2 = 27.0709\ldots \tag{26}$$

From (18) we deduce $n_{1,o}$ and $n_{2,o}$. These numbers are irrational and do not correspond to any regular lattice. The relative number of tetrahedra is

$$\frac{N_3}{N} = \frac{n_{3,o}}{4} = \frac{24}{35}\pi^2 \tag{27}$$

and

$$n_{1,o} = 2 + \frac{48}{35}\pi^2 = 15.53546... \qquad \frac{N_1}{N} = \left(1 + \frac{24}{35}\pi^2\right) \tag{28}$$

$$n_{2,o} = \frac{144}{35}\pi^2 = 40.6064... \qquad \frac{N_2}{N} = \frac{48}{35}\pi^2 \tag{29}$$

The average volume of a tetrahedron in natural units is the reciprocal of N_3/N

$$\ell_3 = \frac{35}{24\pi^2} = 0.1478... \tag{30}$$

From (19) one also computes the average area of a 2-simplex

$$\ell_2 = \left(\frac{3}{4\pi}\right)^{2/3} \frac{875}{243}\frac{1}{\pi}\Gamma\left(\frac{2}{3}\right) = 0.5973... \tag{31}$$

and the mean length of a one simplex, i.e. the average distance between neighbors

$$\ell_1 = \left(\frac{3}{4\pi}\right)^{1/3} \frac{1,715}{2,304}\Gamma\left(\frac{1}{3}\right) = 1.2371 \tag{32}$$

with the limitations explained above.

For the dual lattice we have

$$\sigma_3 = 1 \tag{33}$$

The average area is obtained by following the same steps as in two dimensions, with the result

$$\text{average cell area} = \left(\frac{4\pi}{3}\right)^{1/3} \frac{8}{3}\Gamma\left(\frac{2}{3}\right) = 5.821 \tag{34}$$

The average area of a face σ_2 is obtained by dividing by the connectivity $n_{1,o}$

$$\sigma_2 = \frac{\left(\frac{4\pi}{3}\right)^{1/3}\frac{4}{3}\Gamma(2/3)}{1 + \frac{24}{35}\pi^2} = 0.3747... \tag{35}$$

Similar calculations in four dimensions[1] yield Table IV

The average coordination grows very fast being al-

Simplex	direct lattice Number	Number incident on a point	Average length/area/volume	Dual lattice Average volume/area/length
point	$N_o = N$	$n_{o.o} = 1$		$\sigma_3 = 1$
segment	$N_1 = \left(1 + \frac{24}{35}\pi^2\right) N$	$n_{1.o} = 2 + \frac{48}{35}\pi^2$	$\ell_1 = \left(\frac{3}{4\pi}\right)^{1/3} \frac{1715}{2304}\Gamma(1/3)$	$\sigma_2 = \frac{\left(\frac{4\pi}{3}\right)^{1/3} \frac{4}{3}\Gamma\left(\frac{2}{3}\right)}{1 + \frac{24}{35}\pi^2}$
triangle	$N_2 = \frac{48}{35}\pi^2 N$	$n_{2.o} = \frac{144}{35}\pi^2$	$\ell_2 = \left(\frac{3}{4\pi}\right)^{1/3} \frac{875}{243} \frac{1}{\pi}\Gamma\left(\frac{2}{3}\right)$	
tetrahedron	$N_3 = \frac{24}{35}\pi^2 N$	$n_{3.o} = \frac{96}{35}\pi^2$	$\ell_3 = \frac{35}{24\pi^2}$	

Table III - Cell data in 3-dimensions

	direct lattice			dual lattice
dimension of simplex	Number	Number incident on a point	Average d-volume	Average d-volume
0	$N_0=N$	$n_{0,0}=1$		$\sigma_4=1$
1	$N_1=\frac{170}{9}N$	$n_{1,0}=\frac{340}{9}$	$\ell_1=\left(\frac{2}{\pi^2}\right)^{1/4}\frac{3^3}{7\times11}\,\frac{16!!}{15!!}\,\frac{1}{\pi}\Gamma\left(\frac{1}{4}\right)$	$\sigma_3=\frac{9}{225}(2\pi^2)^{1/4}\,\frac{1}{\pi}\Gamma\left(\frac{3}{4}\right)$
2	$N_2=\frac{590}{9}N$	$n_{2,0}=\frac{590}{9}$	$\ell_2=2^{1/2}\,\frac{2^7}{11^2\times13}\,\frac{17!!}{16!!}\Gamma\left(\frac{1}{2}\right)$	
3	$N_3=\frac{715}{9}N$	$n_{3,0}=\frac{2860}{9}$	$\ell_3=\left(\frac{2}{\pi^2}\right)^{3/4}\frac{2^7 7^2}{3\times13^2\times57}\,\frac{18!!}{17!!}$	
4	$N_4=\frac{286}{9}N$	$n_{4,0}=\frac{1430}{9}$	$\ell_4=\frac{9}{284}$	

Table IV - Cell data in four dimensions

ready $\frac{340}{9} = 37.778$ in dimension four as opposed to 8 on a regular hypercubic lattice. The average volume of the boundary of a cell is according to Meijering

$$\text{volume of cell boundary} = (2\pi^2)^{1/4} \frac{4}{5} \frac{1}{\pi} \Gamma\left(\frac{3}{4}\right) \tag{36}$$

so that the average volume of an hyperface is

$$\sigma_3 = \frac{9}{425} (2\pi^2)^{1/4} \frac{1}{\pi} \Gamma\left(\frac{3}{4}\right) \tag{37}$$

C- Some Results in Arbitrary Dimension

Since each d-1 simplex belongs to 2 d-simplices, and each d simplex has d+1 d-1-simplices

$$2N_{d-1} = (d+1)N_d \qquad n_{d-1,o} = \frac{d}{2} n_{d,o} \tag{38}$$

Euler's relation applied to the surface of a Voronoï cell homeomorphic to a sphere gives

$$n_{1,o} - n_{2,o} + \ldots -(-1)^{d-1} n_{d,o} = +1-(-1)^{d} \tag{39}$$

More generally one has

$$\sum_{k=m+1}^{d} (-1)^{k-m-1} n_{k,m} = 1-(-1)^{d-m} \tag{40}$$

Equivalently, since

$$n_{k,m} = \frac{N_k}{N_m} C_{k+1}^{m+1} \qquad d \geq k \geq m \tag{41}$$

$$\sum_{k=m}^{d} C_{k+1}^{m+1} (-1)^{d-k} N_k = N_m \tag{42}$$

The case m=d is trivial, m=d-1 and m=d-2 give equivalent information, etc... Hence (40) or (42) leave $\frac{d}{2}-1$ (d even) or $\frac{d-1}{2}$ (d-odd) unknown. This means no unknown in two dimensions, 1 unknown in three and four dimensions, etc..., in agreement with the previous sections.

We turn to the computation of the average number of d-simplices incident on a point $n_{d,o}$, given by

$$n_{d,o} = \frac{1}{d!} \int d^d x_1 \ldots d^d x_d \, e^{-v} \tag{43}$$

v = (hyper) volume of circumscribed sphere to simplex $(M_o, M_1, \ldots, M_d)$ with coordinates $(\vec{x}_o = 0, \vec{x}_1, \ldots, \vec{x}_d)$, of radius R, i.e. $(R^d/d) S_d$, where S_d is the "area" of the unit sphere in d-dimensional space

$$S_d = \frac{2\pi^{d/2}}{\Gamma(d/2)} \tag{44}$$

Taking the origin at the center of the sphere with $\hat{n}_o, \ldots, \hat{n}_d n+1$ unit vectors along $OM_o, OM_1, \ldots, OM_d$

$$\begin{aligned} n_{d,o} &= \int_o^\infty R^{d^2-1} \, dR \, e^{-\frac{R^d}{d} S_d} \, S_d^{d+1} \, \langle w \rangle \\ &= S_d \, d^{-2} \, d! \, \langle w \rangle \end{aligned} \tag{45}$$

We have shown in [17] that

$$\langle w \rangle = \frac{d+1}{d} \, \frac{(S_{d-1})^{d+1}}{((d-1)S_d)^d} \, \frac{\Gamma\left(\frac{1}{2}\right) \Gamma\left(\frac{d^2-1}{2}\right)}{\Gamma\left(\frac{d^2}{2}\right)} \tag{46}$$

Hence

$$n_{d,o} = \frac{2}{d} \, \frac{(d-1)!}{\Gamma\left(\frac{d+1}{2}\right)^2} \left[\frac{\Gamma\left(\frac{1}{2}\right)\Gamma\left(\frac{d}{2}+1\right)}{\Gamma\left(\frac{d+1}{2}\right)} \right]^{d-1} \frac{\Gamma\left(\frac{1}{2}\right)\Gamma\left(\frac{d^2+1}{2}\right)}{\Gamma\left(\frac{d^2}{2}\right)} \tag{47}$$

$$\underset{d\to\infty}{\sim} \frac{2}{d^{1/2}}\, e^{1/4}\, (2\pi d)^{\frac{d-1}{2}}$$

which can be compared with the results quoted for d=2, 3, 4. The average (volume) of the d-simplex is

$$\ell_d = \frac{d+1}{n_{d,0}} \tag{48}$$

Similarly we obtain the average total "area" of a Voronoï cell, call it P,

$$P = 2^{d+1}\,\frac{1}{d}\,\Gamma\left(2 - \frac{1}{d}\right)\frac{1}{\Gamma\left(d - \frac{1}{2}\right)}\left(\Gamma\left(\frac{d}{2}+1\right)\right)^{2-\frac{1}{d}} \tag{49}$$

and the average "area" of a d-1 simplex

$$\ell_{(d-1)} = \frac{d^2}{\pi^{1/2}}\,\frac{\Gamma\left(d+\frac{d-1}{d}\right)}{(\Gamma(d))^2}\left(\frac{\Gamma\left(\frac{d}{2}+1\right)}{\pi^{d/2}}\right)^{\frac{d-1}{d}}\frac{\Gamma\left(\frac{d+1}{2}\right)^d}{\Gamma\left(\frac{d}{2}+1\right)^{d-1}}$$
$$\left(\frac{\Gamma\left(\frac{d^2}{2}\right)}{\Gamma\left(\frac{d^2+1}{2}\right)}\right)^2\frac{\Gamma\left(\frac{d^2+d}{2}\right)}{\Gamma\left(\frac{d^2+d-1}{2}\right)} \tag{50}$$

Much remains to be done, not only in general dimension, but also with respect to correlations[22].

It would also be worthwile to investigate Poissonian lattices on other manifolds but Euclidean space. Little seems to be known in this direction.

II. ACTIONS AND FIELD EQUATIONS

The next step is to formulate classical and quantum field theory on these random lattices, and to analyse functions defined on various geometric objects. Two lattices have in fact been defined. In the direct lattice L we have 0, 1, 2,... simplices. The zero simplices are points i, we set $\ell_i \equiv 1$. The one-simplices are segments joining two "neighboring" points (ij) i.e. segments that cross a face of a cell, and $\ell_{ij}=\ell_{ji}$ is their length and so on... The dual lattice $\tilde{L}$ is made of d, d-1,... cells. We call i the d-cell to which the point i belongs and σ_j its volume, (ij) is a d-1 cell common to two neighboring cells i and j and $\sigma_{ij}=\sigma_{ji}$ its "area" etc...

Define L_0 as the set of σ-forms, i.e. functions defined on sites $i\mapsto\varphi_i$. Similarly L_1 is the set of antisymmetric 1 forms $\varphi_{ij}\equiv-\varphi_{ji}$ defined on 1-simplices; L_2 antisymmetric 2 forms $\varphi_{ijk}=-\varphi_{ijk}=\ldots$ defined on 2 simplices and so on. Furthermore let us call d-densities the function ψ_i defined on the (positively) oriented d-cells and the set of such densities $\tilde{L}_d$. The d-1 densities ψ_{ij} defined on oriented d-1 cells build $\tilde{L}_{d-1}$ and so on. The orientation on the cell (ij...) and on the simplex (ij...) are compatible, i.e. their product is unity.

The forms $\varphi_i, \varphi_{ij}, \varphi_{ijk}\ldots$ may be thought of as restrictions on the lattice of scalar, vector, antisymmetric tensor... fields defined in the continuum, the correspondence being

$$\varphi_i \leftrightarrow \varphi(x_i)$$

$$\varphi_{ij} \leftrightarrow \frac{1}{\ell_{ij}} \int_i^j dx^\mu \; \varphi_\mu(x) \tag{1}$$

$$\varphi_{ijk} \leftrightarrow \frac{1}{ijk} \iint_{\substack{\text{oriented}\\ \text{triangle } ijk}} dx^\mu \wedge dx^\nu \varphi_{\mu\nu}(x)$$

Between L_p and $\tilde{L}_{d-p}$ there exists a natural scalar product, and $\tilde{L}_{d-p}$ is the dual of L_p. We call this scalar product $\langle\varphi|\psi\rangle$ with

$$p=0 \qquad \langle\varphi|\psi\rangle = \frac{1}{1!} \sum_i \varphi_i \psi_i$$

$$p=1 \quad \langle\varphi|\psi\rangle = \frac{1}{2!} \sum_{\substack{ij \\ \{ij\}\ \text{neighbors}}} \varphi_{ij}\psi_{ij} \tag{2}$$

$$p=2 \quad \langle\varphi|\psi\rangle = \frac{1}{3!} \sum_{\substack{ij \\ \{ijk\}\ \text{neighbors}}} \varphi_{ijk}\psi_{ijk}$$

Moreover we can define a one to one correspondence betwen L_p and $\tilde{L}_{d-p}$ which we call duality, in such a way that each one of them becomes a Hilbert space as follows

$$\varphi \mapsto \psi$$

$$L_p \mapsto \tilde{L}_{d-p} \tag{3}$$

$$\ell_{ijk\ldots}\,\varphi_{ijk\ldots} = \frac{\psi_{ijk\ldots}}{\sigma_{ijk\ldots}} \ \text{(no summation)}$$

For any p this entails in terms of dimensions

$$[\psi] = [\varphi]\,[\text{length}]^d$$

The dimensional factor is the volume of a cell for p=0. For p=1 it is d times the sum of the volumes of two pyramides of apex i and j with base σ_{ij} and so on. We denote the correspondence (3) $\psi=\tilde{\varphi}$ or $\varphi=\tilde{\psi}$.

We define next on L_p a potential (or mass) term as

$$V(\varphi) = \frac{1}{2}\langle\varphi|\tilde{\varphi}\rangle \tag{4}$$

i.e.

$$p=0 \qquad V(\varphi) = \frac{1}{2}\sum_i \sigma_i \varphi_i^2$$

$$p=1 \qquad V(\varphi) = \frac{1}{2} \sum_{(ij)} \ell_{ij}\sigma_{ij}\varphi_{ij}^2 \tag{5}$$

$$p=2 \qquad V(\varphi) = \frac{1}{2} \sum_{(ijk)} \ell_{ijk}\sigma_{ijk}\varphi_{ijk}^2$$

To construct kinetic terms we need the analogs of gradient and divergence. The gradient (or exterior derivative) is an operator d from L_p to L_{p+1} (giving o on L_d) such that

$$L_p \overset{d}{\mapsto} L_{p+1}$$

$$(d\varphi)_{ij} = \frac{\varphi_i - \varphi_j}{\ell_{ij}}$$

$$(d\varphi)_{ijk} = \frac{\ell_{ij}\varphi_{ij} + \ell_{jk}\varphi_{jk} + \ell_{ki}\varphi_{ki}}{\ell_{ijk}} \tag{6}$$

$$(d\varphi)_{i_o \ldots i_q} = \sum_{s=0}^{q} \frac{(-1)^s \ell_{i_o \ldots \hat{i}_s \ldots i_q} \varphi_{i_o \ldots \hat{i}_s \ldots i_q}}{\ell_{i_o \ldots i_q}}$$

The operator d satisfies

$$d^2 = 0 \tag{7}$$

Using (2) one can transpose d into d^T, which operates as

$$\tilde{L}_p \overset{d^T}{\mapsto} \tilde{L}_{p+1} \tag{8}$$

and satisfies

$$\langle \varphi | d^T \psi \rangle = \langle d\varphi | \psi \rangle \tag{9}$$

Explicitly

$$\tilde{L}_{d-1} \mapsto \tilde{L}_d \qquad (d^T\psi)_i = \sum_{j(i)} \frac{\psi_{ij}}{\ell_{ij}} \tag{10}$$

$$\tilde{L}_{d-2} \mapsto \tilde{L}_{d-1} \qquad (d^T\psi)_{ij} = \sum_{k(ij)} \frac{\ell_{ij}\psi_{ijk}}{\ell_{ijk}}$$

and

$$d^{T^2} = 0 \tag{11}$$

The divergence operator d^* is obtained by shifting back d^T using the duality map (3) according to the diagram

$$\begin{array}{ccc} L_p & \xrightarrow{\text{duality}} & \tilde{L}_{d-p} \\ d^* \Big\Downarrow & & \Big\downarrow d^T \\ L_{p-1} & \xrightarrow[\text{duality}]{} & \tilde{L}_{d-p+1} \end{array} \tag{12}$$

For instance d^* maps L_1 on L_o. We start with φ_{ij}, get from duality $\tilde{\varphi}_{ij} = \ell_{ij}\sigma_{ij}\varphi_{ij}$, apply d^T, to obtain

$$(d^T\psi)_i = \sum_{j(i)} \frac{\varphi_{ij}}{\ell_{ij}} = \sum_{j(i)} \sigma_{ij}\varphi_{ij} .$$

Returning by duality to L_o yields

$$(d^*\varphi)_i = \frac{1}{\sigma_i} \sum_{j(i)} \sigma_{ij}\varphi_{ij} \tag{13}$$

which is seen to be the discrete analog of the divergence. Again

$$d^{*^2} = 0 \tag{14}$$

and d^* annihilates L_o.

One we can also compute d^{T^*} with square equal to zero. A kinetic term is defined as

$$K(\varphi) = V(d\varphi) = \frac{1}{2} \langle d\varphi | \widetilde{d\varphi} \rangle$$

$$= \frac{1}{2} \sum_{i_o,\ldots,i_p} \frac{\sigma_{i_o \ldots i_p}}{\ell_{i_o \ldots i_p}} \left(\sum_{s=0}^{p} (-1)^s \ell_{i_o \ldots \hat{i}_s \ldots i_p} \varphi_{i_o \ldots \hat{i}_s \ldots i_p} \right)^2 \tag{15}$$

For scalar and vectors

$$\varphi \in L_o \quad K(\varphi) = \frac{1}{2} \sum_{(ij)} \frac{\sigma_{ij}}{\ell_{ij}} (\varphi_i - \varphi_j)^2$$

$$\varphi \in L_1 \quad K(\varphi) = \frac{1}{2} \sum_{(ijk)} \frac{\sigma_{ijk}}{\ell_{ijk}} (\ell_{ij}\varphi_{ij} + \ell_{jk}\varphi_{jk} + \ell_{ki}\varphi_{ki})^2 \tag{16}$$

The dimension of K is

$$[K] = [\varphi]^2 \, [\text{length}]^{d-2} \tag{17}$$

as in the continuum theory.

To obtain (free massless) field equations we minimize the kinetic term K considered as an action

$$\delta K = \langle . | \widetilde{\delta\varphi} \rangle = 0$$

The field equations are

$$d^* d\varphi = 0 \tag{18}$$

The dimension of the operator d^*d is $[\text{length}]^{-2}$ and it transforms L_p into L_p.

A- Scalar Fields

To see the content of (18) consider the case of scalars. The operator d^*d can then be identified with the Laplacian Δ, up to a sign, $(d^*d + dd^*) = (d+d^*)^2 = -\Delta$, and

$$(-\Delta\varphi) = (d^{*}d\varphi)_i = \frac{1}{\sigma_i}\sum_{j(i)} \sigma_{ij} \,.\, \frac{\varphi_i - \varphi_j}{\ell_{ij}} \qquad (19)$$

The right hand side can be rewritten

$$(2d) \sum_{j(i)} \frac{\sigma_{ij}\ell_{ij}}{2d\sigma_i} \,.\, \frac{(\varphi_i - \varphi_j)}{\ell_{ij}{}^2}.$$

For fixed i, $p_{j,i} = \dfrac{\sigma_{ij}\ell_{ij}}{2d\sigma_i}$ when summed over j adds up to one and can be identified with hopping probabilities $p_{j,i}$ from site i to site j

$$\left[-\frac{1}{2d}\Delta\varphi\right]_i = \sum_{j(i)} p_{j,i} \frac{(\varphi_i - \varphi_j)}{\ell_{ij}{}^2} \qquad (20)$$

Because ℓ_{ij} is not bounded from below, as on a regular lattice, we may expect that the spectrum of $-\Delta$ is not bounded. As emphasized by the authors of Ref. [1] harmonic functions i.e. solutions of $\Delta\varphi=0$ are of the form

$$\varphi_i = a + \vec{k}.\vec{x}_i \qquad (21)$$

where $\vec{k}$ is a constant vector as in the continuum. Indeed since $\ell_{ij} = |\vec{x}_i - \vec{x}_j|$

$$\sum_{j(i)} \sigma_{ij} \frac{\vec{k}.(\vec{x}_i - \vec{x}_j)}{\ell_{ij}} = \vec{k}. \sum_{j(i)} \sigma_{ij} \frac{x_i - x_j}{|x_i - x_j|} =$$

$$- \int_{\substack{\text{surface}\\ \text{of cell}}} d\sigma\, \vec{k}.\vec{n} = 0 \qquad (22)$$

from Gauss's theorem. If one studies a diffusion equation of the form (D is a diffusion constant)

$$\left(\frac{1}{D}\frac{\partial}{\partial t}\varphi + d^{*}d\varphi\right) = 0 \qquad (23)$$

The integral over φ i.e. $Q(\varphi) = \langle\varphi|1\rangle = \sum_i \sigma_i \varphi_i$, is a conserved quantity, since

$$\frac{1}{D}\dot{Q} = -\sum_i \sum_{j(i)} \frac{\sigma_{ij}}{\ell_{ij}} (\varphi_i - \varphi_j) = 0$$

The solution of equation (23) can be obtained as a limit of a sum over paths

$$\varphi_j(t) = \sum_i (j|e^{-Dtd^*d}|i)$$

$$= \lim_{n\to\infty} \left(j\left|\left(1 - \frac{Dt}{n} d^*d\right)^n\right| i\right) \qquad (24)$$

$$= \lim_{n\to\infty} \sum_{\substack{\text{path of} \\ n \text{ steps}}} C_{path}$$

A path of n steps is a set $\omega_{ij} = \{i\, k_1 k_2 \ldots k_{n-1}\}$ with the constraint that $k_{a+1} \equiv k_a$ or k_{a+1} is a neighbor of k_a. Correspondingly the contribution C_{path} is a product of factors, one for each step, such that

$$\text{if } k_{a+1} = k_a \quad \text{factor} \left(1 - \frac{Dt}{n} \sum_{q(k_a)} \frac{\sigma_{qk_a}}{\sigma_{k_a} \ell_{qk_a}}\right)$$

$$\text{if } k_{a+1} \neq k_a \quad \text{factor} \ \frac{Dt}{n} \frac{\sigma_{k_{a+1},k_a}}{\sigma_{k_a} \ell_{k_{a+1},k_a}} \qquad (25)$$

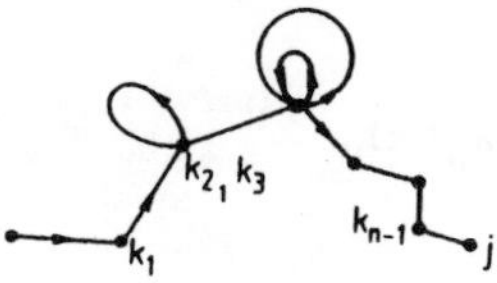

Fig. 4 : A path from i to j.

For n going to infinity, since all the factors are positive

barring a very improbable situation where some $\sum_{j(i)} \frac{\sigma_{ji}}{\sigma_i \ell_{ij}}$ would be arbitrarily large, the kernel $(j|e^{-Dtd^*d}|i)$ is a positive number as in the continuum. If φ is initially positive (or zero) it remains so and if its total "charge" $Q(\varphi)$ is normalized at t=0 it will remain so at any time. It is possible to interpret it as a probability density in the cell i with $\sigma_i\varphi_i$ the total probability in the cell i.

The links (ab) where $\frac{\sigma_{ab}}{\sigma_a \ell_{ab}}$ gets abnormally large as compared to its mean value, play a particular role in the sum over paths. They will increase the tendancy for the paths to leave a and to visit b. We will return to these considerations when we study the average spectrum of the Laplacian.

B- Vector Fields

Returning to equations (18) we observe that for 1,2,...-forms d^*d is not the Laplacian anymore, very much as Maxwell's equations (the case p=1) do not entail without a special choice of gauge (the Lorentz gauge) that the vector potential is harmonic. When p=1 we can write (18) as

$$(d^*d\varphi)_{ij} = \frac{1}{\sigma_{ij}} \sum_{k(ij)} \sigma_{ijk} \frac{\ell_{ij}\varphi_{ij}+\ell_{jk}\varphi_{jk}+\ell_{ki}\varphi_{ki}}{\ell_{ijk}} \tag{26}$$

To get the meaning of (26) assume for instance that the dimension is three and that φ_{ij} results from an underlying vector field $A_\mu(x)$ defined in the continuum, then

$$\varphi_{ij} \leftrightarrow \frac{1}{\ell_{ij}} \int_i^j dx^\mu \, A_\mu(x)$$

$$\frac{\ell_{ij}\varphi_{ij}+\ell_{jk}\varphi_{jk}+\ell_{ki}\varphi_{ki}}{ijk} \leftrightarrow \frac{1}{\ell_{ijk}} \oint_{\substack{\text{boundary}\\ \text{of } ijk}} dx^\mu A_\mu(x)$$

$$= \frac{1}{\ell_{ijk}} \iint_{(ijk)} dx^\mu \wedge dx^\nu (\partial_\mu A_\nu - \partial_\nu A_\mu)$$

On the r.h.s. one has the flux of the corresponding

magnetic field B through the (oriented) triangle ijk i.e. $\vec{B}.\vec{n}$ where $\vec{n}$ is the unit normal. The dual of (ijk) is a cell edge with length σ_{ijk}. The oriented edge is represented by the vector $\sigma_{ijk}\vec{n} = \vec{\sigma}_{ijk}$. Therefore

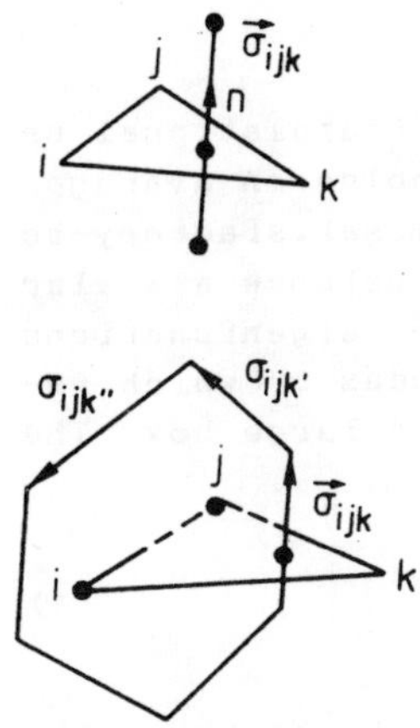

$$(d^{*}d\varphi)_{ij} \leftrightarrow \frac{1}{\sigma_{ij}} \sum_{k(ij)} \vec{B}.\vec{\sigma}_{ijk}$$

Hence we find the circulation of B around the face (ij) with area σ_{ij}

$$\frac{1}{\sigma_{ij}} \sum_{k(ij)} \vec{B}.\vec{\sigma}_{ijk} = \frac{1}{\sigma_{ij}} \oint_{(ij)} \vec{B}.\vec{d\ell}$$

$$= \frac{\iint \vec{\nabla} x \vec{B}.\vec{m} d^2\sigma}{\iint d^2\sigma} \sim \vec{\nabla} x \vec{B}.\vec{m}_{ij}$$

with $\vec{m}_{ij}$ the unit normal to the face (ij).

Fig. 5 : Geometry for Maxwell's equation.

In general dimension one finds similarly a discrete version of Maxwell's equations $\partial^{\mu}F_{\mu\nu} = \Delta A_{\nu} - \partial_{\nu}(\partial.A)$.

As emphasized before this is not the Laplacian on vector fields. Indeed from the expression for $\partial^{\mu}F_{\mu\nu}$ we get $\Delta A_{\nu} = \partial^{\mu}(\partial_{\mu}A_{\nu} - \partial_{\nu}A_{\mu}) + \partial_{\nu}(\partial.A)$. In general terms this reads div x grad + grad x div. More abstractly

$$-\Delta = d^{*}d + dd^{*} = (d+d^{*})^{2} \tag{27}$$

As a result one can formulate a Dirac-Kahler[23] equation on a collection $\phi \equiv \{\varphi_i, \varphi_{ij}, \varphi_{ijk}, \ldots\}$ of 0,1,...d forms of the type

$$(d+d^{*})\Phi = 0 \qquad (28)$$

Further generalization to include interactions or gauge fields[1] can similarly be handled.

III. SPECTRUM OF THE LAPLACIAN

On a random lattice we do not have translational or rotational invariance. The latter only holds on average. There are various approaches, none of them satisfactory to replace Fourier analysis. If we study for instance a scalar function φ_i we might expand it in terms of eigenfunctions of the Laplace operator labelled by an index α (which replaces momentum). Assume this is done in a large box. The eigenfunctions are normalized according to

$$\langle\varphi^{(\alpha)}|\varphi^{(\beta)}\rangle \equiv \sum_i \sigma_i \varphi_i^{(\alpha)}\varphi_i^{(\beta)} = \delta^{\alpha\beta} \qquad (1)$$

this being the scalar product with respect to which the Laplacian is symmetric (up to boundary terms). An arbitrary φ reads

$$\varphi_i = \sum_\alpha \gamma_\alpha \varphi_i^{(\alpha)}$$

$$\gamma_\alpha = \langle\varphi|\tilde{\varphi}^{(\alpha)}\rangle \equiv \sum_i \sigma_i \varphi_i \varphi_i^{(\alpha)} \qquad (2)$$

The Laplacian is diagonal

$$(-\Delta\varphi)_i = \sum_\alpha \gamma_\alpha \omega_\alpha \varphi_i^{(\alpha)} \qquad (3)$$

and the $\varphi^{(\alpha)}$ replace the plane waves. In (3) ω_α is the eigenvalue corresponding to the mode $\varphi^{(\alpha)}$, both of which are still stochastic variables.

This leads to the hard problem of finding the most likely distribution of eigenvalues and corresponding structure of eigenfunctions.

On a large lattice of N sites, α runs over N values. As N goes to infinity denote by $N\,\rho(\omega)\,d\omega$ the number of eigenvalues between ω and $\omega+d\omega$

$$\int_0^\infty d\omega\ \rho(\omega) = 1 \tag{4}$$

We expect that the thermodynamic limit ensures that $\rho(\omega)$ is already the most likely distribution. As $\omega\to0$ assuming that plane waves are almost eigenstates, suggests that $d\omega\rho(\omega) \sim \frac{d^d k}{(2\pi)^d}$, $\omega=k^2$, i.e.

$$\rho(\omega) \underset{\omega\to0}{\mapsto} = \frac{\omega^{\frac{d}{2}-1}}{(4\pi)^{d/2}\Gamma(d/2)} \tag{5}$$

To lend credence to (5) consider

$$\varphi_i(\vec{k}) = e^{i\vec{k}.\vec{x}_i} \tag{6}$$

and the approximate "eigenvalue", still a function of the site i

$$\omega_i(\vec{k}) = \varphi_i(\vec{k})^*(-\Delta\varphi(\vec{k}))_i = \frac{1}{\sigma_i}\sum_{j(i)}\frac{\sigma_{ij}}{\ell_{ij}}\left[1-e^{i\vec{k}.(\vec{x}_j-\vec{x}_i)}\right] \tag{7}$$

From (II.21) $\omega_i(k)$ is of order k^2 and

$$\overline{\omega}(\vec{k}) = \frac{\sum_i\sigma_i\omega_i(\vec{k})}{\sum_i\sigma_i} \tag{8}$$

can be interpreted as the average of $(-\Delta)$ in the state $\varphi(\vec{k})$ i.e. the diagonal elements in the plane wave basis. For $N\mapsto\infty$ we have no preferred direction in the lattice so (8) can be averaged over the directions of $\vec{k}$

$$\bar{\omega}(k) = \sum_{p=1}^{\infty} (-1)^{p-1} \frac{k^{2p}}{2^{2p}} \frac{1}{p!} \frac{\Gamma\left(\frac{d}{2}\right)}{\Gamma\left(\frac{d}{2}+p\right)} \left\langle \int d\sigma\, \ell^{2p-1} \right\rangle \tag{9}$$

The meaning of the last bracket is

$$\left\langle \int d\sigma\, \ell^{2p-1} \right\rangle = \frac{1}{N} \sum_{i} \sum_{j(i)} \sigma_{ij} \ell_{ij}^{2p-1} \tag{10}$$

It can be computed following Meijering's argument to yield

$$\bar{\omega}(k) = \frac{2^d}{d\pi^{1/2}} \Gamma\left(1+\frac{d}{2}\right)^2 \sum_{p=1}^{\infty} \left[\frac{-\Gamma\left(1+\frac{d}{2}\right)^{2/d}}{\pi}\right]^{p-1} \frac{\Gamma\left(p+\frac{d-1}{2}\right)\Gamma\left(2+\frac{2(p-1)}{d}\right)}{\Gamma\left(p+\frac{d}{2}\right)\Gamma(p+d-1)} \frac{K^{2p}}{p!} \tag{11}$$

when d=1, 2 this reduces to

$$\begin{aligned} d=1 \qquad & \bar{\omega}(k) = \ell n(1+k^2) \\ d=2 \qquad & \bar{\omega}(k) = 2\pi \left[1 - e^{-\frac{k^2}{2\pi}} I_o\left(\frac{k^2}{2\pi}\right)\right] \end{aligned} \tag{12}$$

with $I_o(z)$ the modified Bessel function

$$I_o(z) = \sum_{n=0}^{\infty} \left(\frac{z^2}{4}\right)^n \frac{1}{(n!)^2} \underset{z\mapsto\infty}{\sim} \frac{e^z}{\sqrt{2\pi z}} \left(1 + \frac{1}{8z} + \ldots\right) \tag{13}$$

The general formula as well as the explicit forms (12) show that independently of d, as k goes to zero $\bar{\omega}(k) \mapsto k^2$.

The function $\bar{\omega}(k)$ being the average diagonal term of the Laplacian in the plane wave basis is appropriate for an estimate of the lower end of the spectrum. Its meaning in uncertain as k increases. The dimension d=1 is marginal. Up to d=1, $\bar{\omega}(k)$ increases without bound as $k\mapsto\infty$ while for d>1 it is bounded. For d=2, $\bar{\omega}(k) \mapsto 2\pi\left(1 - \frac{1}{k} + \ldots\right)$ as $k\mapsto\infty$. In general by dropping the oscillatory term in (7)

$$\lim_{k\mapsto\infty} \bar{\omega}(k) \equiv \omega_\infty = \pi d \frac{\Gamma\left[2\left(1 - \frac{1}{d}\right)\right]}{\left[\Gamma\left(1 + \frac{d}{2}\right)\right]^{2/d}} \quad d>1 \tag{14}$$

The spectrum of $(-\Delta)$ contains at least the interval $[0,\omega_\infty]$ so that in one dimension we have a proof that it covers all the positive axis.

We can estimate the large ω behavior of $\rho(\omega)$ on the basis of the path expansion for the diffusion equation (section III). What is suggested is that for short times i.e. ω large, we can think of ultralocalized wave functions corresponding to links where σ_{ij}/ℓ_{ij} was exceptionally large. For those links a bold approximation is that the wave function φ^α is essentially concentrated on the two neighboring points i and j. As a result

$$\begin{aligned} \omega\varphi_i &= \frac{1}{\sigma_i}\frac{\ell_{ij}}{\sigma_{ij}}(\varphi_i-\varphi_j) \\ \omega\varphi_j &= \frac{1}{\sigma_j}\frac{\ell_{ij}}{\sigma_{ij}}(\varphi_j-\varphi_i) \\ \omega &= \frac{\sigma_{ij}}{\ell_{ij}}\left(\frac{1}{\sigma_i} + \frac{1}{\sigma_j}\right) \end{aligned} \tag{15}$$

It is assumed of course that the right hand side is (with very small probability) very large. A natural assumption is that this results from the fact that i and j are exceptionnally close hence ℓ_{ij} very small. If this is the case we may think of the two neighboring cells as a unique cell, call it α cut by the (ij) bissecting plane up to corrections, i.e. a random plane through the center of the cell

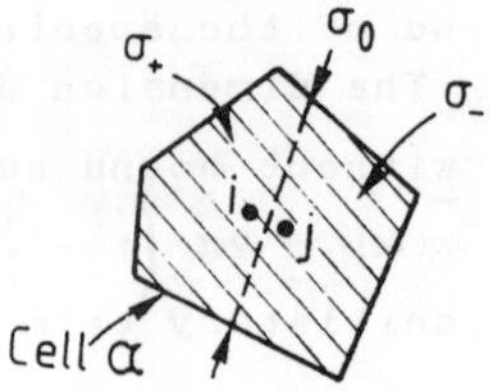

Fig. 6 : Bringing two points very close to each other.

α, cutting it in two parts of volumes σ_+ and σ_-, having an "area" σ_0 in the cell. Then, for $\omega\to\infty$, we expect

$$\rho(\omega) \underset{\omega\to\infty}{\sim} \left\langle \left[\sigma_0\left(\frac{1}{\sigma_+} + \frac{1}{\sigma_-}\right)\right]^d \right\rangle \frac{2\pi^{d/2}}{\Gamma\left(\frac{d}{2}\right)} \frac{1}{\omega^{d+1}} \tag{16}$$

The average is both on the direction of the plane and the various cells. This fast decrease with ω is in agreement with the fact that the integral of $\rho(\omega)$ is one. This is to be contrasted with the continuum case, where there are infinitely many modes per unit volume and therefore the integral of ρ is infinite.

A tail in the spectrum of the type (16) does not occur on a regular lattice. Our argument shows that it is related to localization effects. This leads into the much more delicate question of the structure of wave functions. At $\omega\to\infty$ wave functions are localized, while as $\omega\to 0$, this is not so. We shall show that in one dimension we have a finite localization length no matter how small ω is. Current wisdom would suggest that d=2 might be a lower critical dimension for this phenomenon (how the wave functions decrease at large distances is yet another question). For d>2 there might exist a range $[0,\omega_c]$ in which wave functions extend to infinity, ω_c being a localization threshold.

It is important to get some understanding of this point in order to proceed to interacting systems, where the interplay of two length scales of localization and of ordering might lead to complex results.

Let us add two remarks. From (14) one can derive a sum rule on the spectrum

$$\langle\omega\rangle = \int_0^\infty d\omega\ \omega\rho(\omega) = \omega_\infty = \pi d \frac{\Gamma\left(2-\frac{2}{d}\right)}{\Gamma\left(1+\frac{d}{2}\right)^{2/d}} \tag{17}$$

This is infinite for d=1, but eventually grows like d for large d due to the suppression of small and large frequencies.

One can also derive the following inequality

$$\frac{\langle\omega^2\rangle}{\langle\omega\rangle 2} - 1 \geq \frac{1}{q} \tag{18}$$

where q is the average coordination. This is an equality on any regular lattice. It could perhaps be that (18) becomes an equality as d tends to infinity.

With B. Derrida and E. Gardner we have studied[15] the case d=1 in some detail using a simplified version of the Laplacian, namely

$$\omega\varphi_n = -\left\{\frac{\varphi_{n+1}-\varphi_n}{\ell_n} - \frac{\varphi_n-\varphi_{n-1}}{\ell_{n-1}}\right\} \tag{19}$$

The random points have been ordered and ℓ_n denote the successive independent Poissonian distributed intervals (I-2). Our results were given in terms of small and large ω expansions, for ρ(ω), the density of eigenvalues and L(ω) the localization length. For small ω

$$\rho(\omega) = \frac{1}{2\pi\omega^{1/2}}\left(1 - \frac{1}{128}\omega + O(\omega^2)\right) \tag{20}$$

$$L(\omega) = \frac{8}{\omega} - \frac{1}{8} + O(\omega) \tag{21}$$

in agreement with the above discussion. No matter how small ω is, L(ω) is finite, but of course diverges as ω approaches zero. The leading correction to the behaviour of ρ(ω) is very small. In the large ω regime, in agreement with expectations, ρ(ω) behaves as ω^{-2} and L(ω) tends to vanish as (C is related to Euler's constant γ through $C = e^\gamma$)

$$\rho(\omega) = \frac{2}{\omega^2} + \frac{4}{\omega^3}\left(\ell n\frac{\omega}{2C} - \frac{3}{2}\right) + \ldots \tag{22}$$

$$\frac{1}{L(\omega)} = \ell n \frac{\omega}{C} - \frac{2}{\omega}\left(\ell n \frac{\omega}{2C}+1\right) - \frac{1}{\omega^2}\left[\left(\ell n \frac{\omega}{2C}-1\right)^2 + \eta - 4 - \pi^2\right] + \ldots$$

$$\eta = 0.13070\ldots \qquad (23)$$

The appearance of logarithms is characteristic of a Poissonian distribution for the ℓn with no cut off at small values.

These expressions are in good agreement with a numerical simulation as shown on figure 7.

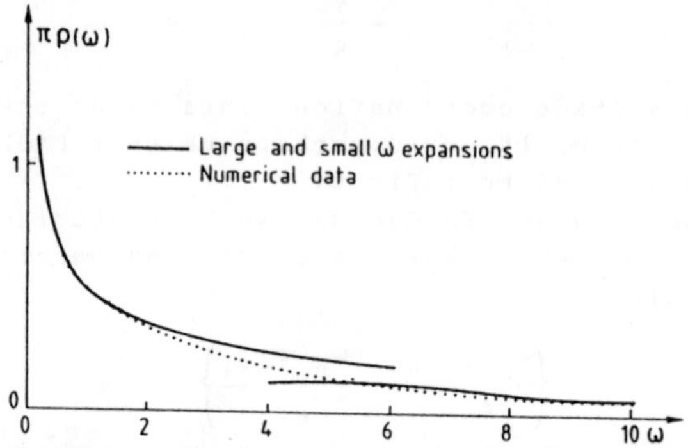

Fig. 7 : The density of states $\rho(\omega)$ for the random one dimensional Laplacian.

One can also study Green functions at arbitrary separations. Since averages of products are different from product of averages, the "free field" theory is in effect an interacting one. In particular one finds a short range attraction. In any case the one dimensional case illustrates the points made in the general discussion.

IV. RANDOM SURFACES

In this section I discuss work done with M. Bander[18,19] on random triangulated surfaces.

It has been advocated by a number of authors[7-13], that a theory of random surfaces might play a role in a diversity of situations, ranging from interfacial effects to string models and gauge theories.

One may wish to consider a surface as an abstract

object endowed with internal properties like connectedness and its topological generalization (Betti numbers), metric, curvature of various other bundle structure that it can support. Or one may think of its embeddings in larger manifolds, in particular Euclidean spaces, with new topological invariants attached to its complementary. An important circumstance is when one deals with an interface in $\mathbb{R}^3$ between distinct material phases.

An other new aspect is the non trivial structure of boundary problems. There is a clear distinction between curves bounded by structureless sets of points and surfaces bounded by curves. Typical of such questions is the Plateau problem of minimal surfaces, or the study of Wilson loops in gauge theories.

A third aspect has to do with parametrization and correlatively the dynamical generation of surfaces. A curve may be viewed as the evolution of a point, while a surface is generated by a string with infinitely many degrees of freedom. String theories are notoriously difficult and have given rise to a large body of literature[24]. On the other hand, one may want to think of the surface without prejudice about any coordinate choice, i.e. insist on reparametrization invariance, or general covariance of its physical properties.

The combination of all these difficulties has prevented up to now to develop a simple reference model with easily computable properties (at least some of them). A number of ingenious attempts have been made, but a clear picture of their inter-relationships is still missing. Recent developments in string theory lie outside the scope of this presentation.

A- Brownian Curves

Let us first recall a down to earth definition of Hausdorff (or scaling) dimension on the example of random curves.

An actual realization of a closed curve can be viewed as a map from the circle to Euclidean space $\mathbb{R}^d$. The parameter space, the circle, enables one to assign to each point of the curve a "time" or angle, and to distinguish among possible multiple points on the image. We may be interested in properties specific to the image and independent of any particular map, assumed at least to be continuous. A complete description requires the specification of the allowed maps. A physicist's point of view might be more concrete, with the curve being thought as an idealization for a material object, be it for instance a (closed) long molecule, a defect line in an otherwise ordered medium, a particle trajectory in a dense material... In such instances, the infinitesimal (or short distance) structure is of little relevance to the large scale properties. One should then be allowed, in certain circumstances, to replace the continuous aspect by a discrete one (a very fine mesh) without altering in a noticeable way the overall picture. This we do, for instance, by replacing the continuous parameter space (the circle) by a sequence of densely packed point labelled from 0 to N-1 (with N identified with 0), N very large, to which we assign N points in $\mathbb{R}^d$: $\vec{x}_0, \vec{x}_1, \ldots, \vec{x}_{N-1}$ and we think of the curve as a linear interpolation between $\vec{x}_i$ and $\vec{x}_{i+1}$.

A statistical model on these "curves" requires two ingredients which are really not independent. The first is a (relative) statistical weight for each curve e^{-S}, where S is a dimensionless action, or an energy divided by kT. The second, is to find a mean to distinguish and count the curves (this is the entropy). As the curves are embedded in $\mathbb{R}^d$ which has a metric structure, these prescriptions are required to respect Euclidean invariance. The arbitrariness is further reduced by a locality requirement (short range interactions). The most stringent form assumes S to be a sum of contributions from a successive pair of neighboring points

$$S = \sum_{i=0}^{N-1} L(\ell_{i,i+1}) \qquad \ell_{i,i+1} = |\vec{x}_i - \vec{x}_{i+1}|$$

The measure can be taken as the product measure over all but one $\vec{x}_i$. We indicate this by π'

$$d\mu = \pi' \, d^d x_i e^S$$

With $\vec{y}_1 = \vec{x}_1 - \vec{x}_0, \ldots, \vec{y}_{N-1} = \vec{x}_{N-1} - \vec{x}_{N-2}$, the measure factorises into a product, except for the term $L(\ell_{N-1,0})$. We add $\vec{y}_N$ in such a way that $\sum_{i=1}^{N} \vec{y}_p = 0$. Standard choices for $L(\ell)$ are $\propto \frac{\ell}{a}$ or $\propto \frac{\ell^2}{a^2}$ where a is an arbitrary unit of length. The first choice is attractive, since S is proportional to the total length of the curve. We assume all moments of the measure $d^d y\, e^{-L(|\vec{y}|)}$ to be finite. Random curves such that this would not be fulfilled would fall in a very different class.

A universal property of such models is the following. Choose the arbitrary position of the center of mass to be the origin

$$\frac{1}{N} \sum_{0}^{N-1} \vec{x}_i = 0$$

Relative to this origin, the mean square radius R_N^2 is

$$R_N^2 = \frac{a_N^2}{12} (N+1), \qquad \langle \vec{y}_i^{\,2} \rangle = a_N^2$$

To keep the average extent finite in the limit $N \mapsto \infty$, requires to adjust the bare parameters (in the Lagrangian) in such a way that $a_N \sim \frac{1}{\sqrt{N}}$. This is typical of a Brownian curve, and make it look more like a two dimensional manifold than an ordinary regular curve. Indeed if one approximates a manifold of dimension δ by a set of N points in some regular fashion the distances between neighbors will scale as $a_N \sim \frac{1}{N^{1/\delta}}$. This motivates a more rigorous definition of the Hausdorff dimension which leads to $\delta_H = 2$ for a Brownian curve. What lies at the heart of the matter here, is that, apart from the overall constraint of being closed, the curve was constructed from independent increments. Their mean square add, leading to the above conclusion. Details of the short range structure are immaterial, as is the dimension of the imbedding space. All these are lumped together in a_N^2.

We cannot expect such simple properties for higher dimensional manifolds. For surfaces, extreme suggestions have been made with the Hausdorff dimension being 4 or infinity, pointing to the fact that either one were not using the same concept, or one were discussing utterly different models.

B- Piecewise Linear Triangulated Surfaces

We discuss here two dimensional (compact, orientable) manifolds of fixed topology. First one picks a compact model (or parameter space) which fixes the topology. It is characterized by its genus, or number of handles, g, related to the Euler characteristics χ through $\chi=2-2g$. Then we discretize and triangulate it, for instance introducing a metric and using a Voronoi construction. This triangulation introduces the notion of nearest neighbor pairs (or links) and elementary 2-simplices or triangles. If $N_0 \equiv N$ is the number of sites, N_1 of links, N_2 of simplices, then $N_0-N_1+N_2=\chi$. Specific to a triangulation one has $2N_1=3N_2$, since each link belongs to two triangles and each triangle has three sides, thus

$$\begin{aligned} N_0 &= N \\ N_1 &= 3(N-\chi) \\ N_2 &= 2(N-\chi) \end{aligned} \tag{1}$$

A troublesome question is the one of inequivalent triangulations, even for given $N_0=N$, that is as abstract (unlabelled) graphs. We shall assume that one is selected for a growing sequence of N's. There is also the possibility to sum for each N over all possible inequivalent choices adding further entropy to the system[13].

We then map the N points in $\mathbb{R}^d$ and interpolate linearly between the images of neighbors. This produces in $\mathbb{R}^d$ a set of points, links (linear segments), planar faces (triangles), which may of course have numerous self intersections, which we do not take into account. For these sets we want to introduce a measure with the same requirements as previously.

The metric on $\mathbb{R}^d$ induces on the image surface, lengths and areas for the links and triangles as well as angles. The triangles are Euclidean, so that their internal angles add to π. For each triangle we can therefore split unity into $1 = \frac{1}{\pi}(\theta_1+\theta_2+\theta_3)$ with each θ between 0 and π. Summing over triangles gives N_2. But we can rearrange the sum by collecting all θ's pertaining to a vertex, then summing over vertices. Call θ_i the sum at each vertex. Then $N_2 = \sum_i \theta_i/\pi$. From (1) this is $2N-2\chi=-2\chi+\sum_i 1$. Identification leads to

$$\chi = \sum_i \left(1 - \frac{\theta_i}{2\pi}\right) \tag{2}$$

the discrete form of the Gauss-Bonnet formula in terms of deficit angles. When the deficit angle vanishes at a vertex the corresponding triangles fit in flat space. We have the identification: deficit angle $\Leftrightarrow$ curvature $\Leftrightarrow$ frustration from planar situation. This can be understood on the example of a sphere, of radius be r, with curvature $1/r^2 \equiv R$. For a spherical triangle with inner angles $\alpha_1, \alpha_2, \alpha_3$ the area A is given by $RA=(\alpha_1+\alpha_2+\alpha_3-\pi)$. The total amount that a tangent vector has rotated in one circumnavigation (rounding vertices) is $\theta = \sum_1^3(\pi-\alpha_i) = 2\pi-RA$, the total angular deficit. The limiting value as the spherical area shrinks to zero and R scales as 1/A, is the angular defect. On the triangulated surface, curvature is concentrated at the vertices. Incidentally with our normalization the continuum version of the Gauss Bonnet formula reads $\chi = \frac{1}{2\pi}\int dA\, R$, i.e. 2 for a sphere. For surfaces, curvature is a scalar concept and (2) relates geometry and topology. One observes a disymmetry between positive and negative deficits: θ_i is positive, so $1 - \frac{\theta_i}{2\pi}$ runs from 1 to $-\infty$ (more precisely between 1 and $-(q_i/2-1)$ if q_i triangles meet at the point i). Angular-wise we cannot have more than a 2π deficit, but we can have as much as we want of extra matter.

In higher dimensions, for piece-wise linear triangulated compact manifolds (dimension D) expression (2) generalizes in two ways. The notion of deficit angle extends easily and represents the frustration in being able to paste together in $\mathbb{R}^d$ all D-simplices incident on a D-2 simplex of "volume" a. But now the "direction" of the (D-2) simplex matters. Curvature is not more a scalar but a tensor. Regge[6] has shown that if we label $\{\alpha\}$ the D-2 simplices, the Euclidean action for Einstein's gravity in discretized form is proportional to

$$S_E = \sum_{\alpha} a_{\alpha}\left(1-\frac{\theta_{\alpha}}{2\pi}\right) \tag{3}$$

a rather remarkable formula.

An other generalization is to ask for an expression of the Euler characteristics χ in terms of curvature, when D is larger than 2 (and even, otherwise it vanishes) which would generalize (2). To get as simple and explicit a formula in terms of deficit angles is not easy, as discussed by Cheeger, Müller and Schrader[25]. A partial unsatisfactory answer is as follows. We want to dissect the number of p dimensional simplices occuring in $\chi = \sum_{p=0}^{D} (-1)^p N_p$ in contributions from its vertices. This is obtained by noticing that the exterior normal of a p simplex sweeps the S_{p-1} sphere, if we round off corners, in such a way that we can assign to each vertex i the corresponding fraction $\varphi_i^{(o)}$ (normalized angle) of S_{p-1}. The sphere S_0 has two points, so for a link we assign $\frac{1}{2}$ to each end point, and for a point itself $\varphi=1$. In the sum for χ we collect for each vertex i the contributions from the different p simplices incident on i and call it again $\varphi_i^{(p)}$, thus getting

$$\chi = \sum_i \left(\sum_{p=0}^{D} (-1)^p \varphi_i^{(p)}\right) \tag{4}$$

While (4) reproduces (2) in the case D=2, it has serious drawbacks otherwise. First one fails to see that χ vanishes if D is odd. Also, if regions of the manifold are flat, their contribution is not seen to vanish, as one could expect. Some improvements can however be made which don't seem to have the elegance of the corresponding continuum formula, nor of (2). Of course the Euler characteristics is not the only topological invariant of higher dimensional manifolds.

C- Free Field

Given a triangulated surface embedded in flat space $\mathbb{R}^d$, lengths, areas and angles being defined, we have a metric on this structure. We want to write the action for free fields, and the corresponding classical equations and quantum mechanical path integrals, generalizing the expresions given in II. We start with a massless scalar field defined on vertices and look for a quadratic form which approximates the kinetic term on a continuous surface, i.e.

$$S_{cont.} = \frac{1}{2} \int d^2\alpha \sqrt{g}\, g^{ab} \partial_a \varphi\, \partial_b \varphi \qquad (5)$$

Here α stands for the parameters, g^{ab} is the inverse of the metric giving the length square $ds^2 = g_{ab}\, d\alpha^a d\alpha^b$, $g = \det g_{ab} > 0$. Take on each (flat) triangle, with vertices $\vec{x}_1$, $\vec{x}_2$, $\vec{x}_3$ in $\mathbb{R}^d$, barycentric coordinates

$$\vec{x}(\alpha) = \alpha^1 \vec{x}_1 + \alpha^2 \vec{x}_2 + \alpha^3 \vec{x}_3 \qquad \alpha_i \geq 0 \quad \Sigma\alpha_i = 1 \qquad (6)$$

and extend the field linearly through

$$\varphi(\vec{x}(\alpha)) = \alpha^1 \varphi_1 + \alpha^2 \varphi_2 + \alpha^3 \varphi_3 \qquad (7)$$

given its values at the three vertices. This is the natural harmonic extension inside the triangle. We then apply (1), with the metric inherited on each triangle. It follows that

$$S_{discrete} = \sum_{(ijk)} \frac{1}{8} \frac{[\varphi_i(\vec{x}_j - \vec{x}_k) + \varphi_j(\vec{x}_k - \vec{x}_i) + \varphi_k(\vec{x}_i - \vec{x}_j)]^2}{\ell_{ijk}} \qquad (8)$$

The sum is over the triangles, with area ℓ_{ijk}

$$\ell^2_{ijk} = \frac{1}{16}(\ell_{ij}+\ell_{jk}+\ell_{ki})$$

$$(\ell_{ij}+\ell_{jk}-\ell_{ki})(\ell_{ij}-\ell_{jk}+\ell_{ki})(-\ell_{ij}+\ell_{jk}+\ell_{ki}) \quad (5)$$

Translation or rotation of the surface, or shift of φ_i by a constant, does not affect (8). It is possible to give to (8) a form similar to the one for a flat random lattice, in spite of the fact that a dual lattice is missing. Very much as one can define $\ell_i \equiv 1$, ℓ_{ij}, ℓ_{ijk} for the direct lattice, we introduce corresponding quantities $\sigma_{ijk}=1$, σ_{ij}, σ_i related to a virtual dual lattice as follows. Pick one of the triangles, call it (123) with interior angles θ_1, θ_2, θ_3 between 0 and π. The corresponding contribution to (8) can be rewritten

$$S_{123} = \frac{1}{4}[\cot g\theta_1 \, (\varphi_2-\varphi_3)^2+\cot g\theta_2 \, (\varphi_3-\varphi_1)^2+\cot g\theta_3 \, (\varphi_1-\varphi_2)^2]$$

as a sum of squares with algebraic dimensionless coefficients. The whole expression is of course positive. If R denotes the radius of the circumscribed circle to the triangle

$$\frac{1}{4} \cot g \, \theta_i = \frac{R \cos \theta_i}{2\ell_{23}}$$

The quantity $R \cos \theta_i$ is the (algebraic) distance from the center 0 of the circumscribed triangle to the edge (jk) and it is of the sign of $\frac{\pi}{2} - \theta_i$, i.e. positive or negative according to wether i and 0 are on the same side or not of the chord jk. Two and only two triangles share a given link (23). Define σ_{23} as the algebraic length of the virtual link dual to (23) as

$$\sigma_{23} = \sigma^+_{23} + \sigma^-_{23} = R \cos \theta_i + R' \cos \theta'_1 \quad (10)$$

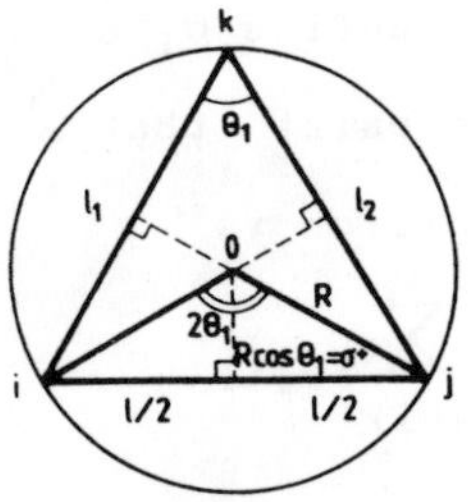

Figure 8

From the lengths ℓ_{ij} alone one can construct the coefficients $\frac{\sigma_{ij}}{\ell_{ij}}$. Each one is a sum of two terms pertaining to adjacent triangles. One of them is given by

$$a = \frac{\sigma^{+}}{\ell} = \frac{\cos\theta}{2\sin\theta} = \frac{1}{2}\frac{\ell_1^2+\ell_2^2+\ell^2}{\sqrt{[(\ell_1+\ell_2)^2-\ell^2][\ell^2-(\ell_1-\ell_2)^2]}}$$

provided of course that triangular inequalities are satisfied. To get $\frac{\sigma_{ij}}{\ell_{ij}}$ just add two such values for adjacent triangles.

The definition (10) agrees with the one given in the random flat case, where σ_{23} was positive, and equal to the length of the edge common to two adjacent Voronoi cells. Positivity arose from the circumstance that a circle circumscribed to a triangle contained no other point of the triangulation. In the general case, the quantities $\sigma_{ij}=\sigma_{ij}$ can unfortunately be negative. Collecting the contribution of edges

$$S_{discrete} = \frac{1}{2}\sum_{(ij)}\frac{\sigma_{ij}}{\ell_{ij}}(\varphi_i-\varphi_j)^2 \tag{11}$$

The positivity of S (non negativity) entails certain inequalities, like

$$\sum_{j(i)}\frac{\sigma_{ij}}{\sigma_{ij}} \geq 0 \tag{12}$$

We have now σ_{ij} defined for each virtual dual link and we set $\sigma_{ijk}=1$. It remains to define σ_i's as areas of virtual dual cells, with the constraint that $\sum_i \sigma_i = \sum_{(ijk)} \ell_{ijk} =$ total area. A natural definition is

$$\sigma_i = \frac{1}{4} \sum_{j(i)} \sigma_{ij}\ell_{ij} \tag{13}$$

One verifies that the abovve requirement is satisfied and (13) reduces in flat space to the area of a Voronoi cell. Here σ_i is in general algebraic. Inequality (12) goes in the right direction to show that if the ℓ_{ij}'s are not too different, σ_i is likely to be positive.

For the time being we shall assume that no σ is zero. A small deformation of the surface would restore this condition.

We can repeat the construction of the operators d and d^* defined in II and analogous to gradient (and curl) and divergence. Apart from scalar fields φ_i, we introduce vector fields $\varphi_{ij}=-\varphi_{ij}$ associated to links, and antisymmetric tensor fields (pseudo scalars) $\varphi_{ijk}=(-1)^P\varphi_{PiPjPk}$ associated to triangles. Then we set

$$\varphi_i \mapsto (d\varphi)_{ij} = \frac{\varphi_i-\varphi_j}{\ell_{ij}} \qquad (\ell_i \equiv 1)$$

$$\varphi_{ij} \mapsto (d\varphi)_{ijk} = \frac{\ell_{ij}\varphi_{ij}+\ell_{jk}\varphi_{jk}+\ell_{ki}\varphi_{ki}}{\ell_{ijk}}$$

$$\varphi_{ijk} \mapsto 0$$

$$\varphi_i \mapsto 0 \tag{14}$$

$$\varphi_{ij} \mapsto (d^*\varphi)_i = \frac{1}{\sigma_i} \sum_{j(i)} \sigma_{ij}\varphi_{ij}$$

$$\varphi_{ijk} \mapsto (d^*\varphi)_{ij} = \frac{1}{\sigma_{ij}} \sum_{k(ij)} \varphi_{ijk} \qquad (\sigma_{ijk} \equiv 1)$$

These operations lead to a definition of Laplacians. For scalars

$$(-\Delta\varphi)_i = (d^*d\varphi)_i = \frac{1}{\sigma_i}\sum_{j(i)}\frac{\sigma_{ij}}{\ell_{ij}}(\varphi_i-\varphi_j) \qquad (15)$$

which enables one to "integrate by parts" in (11) with the result

$$S_{discrete} = \frac{1}{2}\sum_i \sigma_i\varphi_i\ (-\Delta\varphi)_i \qquad (16)$$

The Laplacian Δ has dimension $(\text{length})^{-2}$ while S is dimensionless, as is the field φ. On a compact surface $S_{disc.}>0$ as soon as two adjacent φ_i's are unequal; the only harmonic functions are constants, and $-\Delta$ is a non negative operator with respect to the (possible indefinite) square norm $\sum\sigma_i\varphi_i^2$.

For pseudoscalar fields one can write a formula analogous to (11) namely

$$\frac{1}{2}\sum_{(ijk)}\ell_{ijk}\varphi_{ijk}(dd^*\varphi)_{ijk} = \frac{1}{2}\sum_{(ij)}\frac{\ell_{ij}}{\sigma_{ij}}(\varphi_{ijk}-\varphi_{jik'})^2 \qquad (17)$$

where (ijk) and (jik') are the two triangles adjacent to the link ij with compatible orientations. Typically ℓ's and σ's appear interchanged as compared to (11). Assuming all σ's positive, then again an harmonic pseudo scalar is a constant multiple of the orientation $\eta_{ijk}(=\pm1)$. For vector fields φ_{ij}, harmonicity is equivalent to being both divergenceless and curl-less

$$[(dd^*+d^*d\varphi]_{ij} = 0 \leftrightarrow \left\{(d\varphi)_{ijk}=0,\ (d^*\varphi)_i=0\right\} \qquad (18)$$

Indeed if φ_{ij} is harmonic, $\psi_i=(d^*\varphi)_i$ satisfies $d^*d\psi=d^*dd^*\varphi=-d^{*2}d\varphi=0$, so ψ is constant and

$$\sum\sigma_i\psi_i=\psi\text{xArea}=\sum_i\sigma_i(d^*\varphi)_i=\sum_i\sum_{j(i)}\sigma_{ij}\varphi_{ij}=0,$$

hence $\psi=0$. So $d^*d\varphi=0$ and $dd^*(d\varphi)=0$, hence $d\varphi$ is harmonic, therefore a constant multiple $\tilde{\psi}$ of the orientation

$$\tilde{\psi}\times\text{Area}=\sum_{(ijk)} \eta_{ijk}\ell_{ijk}(d\varphi)_{ijk} =$$

$$\sum_{(ijk)} \eta_{ijk}(\ell_{ij}\varphi_{ij}+\ell_{jk}+\varphi_{jk}+\ell_{ki}+\varphi_{ki})=0.$$

So (18) is fully justified. Incidentally (14) also shows that $d^*\psi=0$. The usual counting of harmonic vector fields still holds in this discrete context. There are N_1 linearly independent φ_{ij}'s. From (18) the number of conditions is $(N_0-1)+(N_2-1)$ (since $\sum_{ijk} \ell_{ijk}\psi_{ijk}(d\varphi)_{ijk}=0$ and $\sum_i \sigma_i(d^*\varphi)_i=0$). Therefore there are $N_1-(N_0+N_2-2)=2-\chi=2g$ linearly independent solutions to (18).

We continue to assume the σ's to be positive and promote free field theory from classical to quantum mechanical by constructing the corresponding path integral. This requires defining the a priori measure on fields.

The a priori measure is not the product of Lebesgue measures of the φ_i's, or else we would miss the two dimensional "dilatation anomaly" of the continuous case. Since we are interested in comparing results for different surfaces, this difference matters. The point is that we require the free field path integral to be related to the determinant of $-\Delta$ in the subspace orthogonal to its zero eigenvalue mode, call it $\det'(-\Delta)$ or equivalently to the product of all its positive eigenvalues E_n, $1\leq n\leq N-1$, $E_0=0$. Let $\psi^{(n)}$ denote the corresponding eigenfunctions of $-\Delta$, and expand the field φ in eigenmodes

$$\varphi_i = \sum_n c_n \psi_i^{(n)} \tag{19}$$

Then

$$S = \frac{1}{2}\sum_n E_n \varphi_n^2 \tag{20}$$

$$Z = \frac{1}{\left(\prod_{1}^{n-1} E_n\right)^{1/2}} = \frac{1}{(\det' -\Delta)^{1/2}}$$

$$= \int \prod_{0}^{N-1} \frac{dc_n}{\sqrt{2\pi}} \, \delta(c_o) \, e^{-S_{disc}} \qquad (21)$$

The ψ's are normalized through $\delta^{n_1 n_2} = \sum_i \sigma_i \psi_i^{(n_1)} \psi_i^{(n_2)}$.

Hence $\prod_{0}^{N-1} dc_n = (\prod_{0}^{N-1} \sigma_i)^{1/2} \prod_i d\varphi_i$. If A is the total area, then $\psi_i^{(o)} = \frac{1}{\sqrt{A}}$ and $c_o = \frac{1}{\sqrt{A}} \sum_i \sigma_i \varphi_i$. Thus

$$Z = \sqrt{2\pi} \int \prod \frac{d\varphi_i}{\sqrt{2\pi}} \, (\prod \sigma_i)^{1/2} \; e^{-S(\varphi)} \delta \left(\frac{1}{\sqrt{A}} \sum \sigma_i \varphi_i \right) \qquad (22)$$

and this can readily be extended to include an external source. In flat space the extra factor plays of course no role, but it is seen to restore sensitivity to scale transformations, if only in the crudest sense, when we dilate all lengths by a constant factor. An estimate of this finite integral can only be done numerically.

D- Conformal Anomaly

In a continuous theory, the action is conformally invariant. Locally any two dimensional metric can be written (isothermal coordinates)

$$g_{ab} = \rho^2 \delta_{ab} \qquad \sqrt{g} = \rho^2 \qquad (23)$$

As an example on a sphere of radius r by projection on a plane

$$ds^2 = \rho^2 \, d\vec{\alpha}^2 \qquad \rho = \frac{1}{1+\frac{\vec{\alpha}^2}{4r^2}} \tag{24}$$

valid on all sphere except one point. From (23)

$$S_{cont} = \frac{1}{2}\int d^2\alpha \sum_{a=1,2} \partial_a\varphi\partial_a\varphi \tag{25}$$

which makes obvious the independence over the local scale parameter ρ. On the other hand the Laplacian is

$$\Delta\varphi = \frac{1}{\rho^2}\sum_{a=1}^{2} \partial_a^2\varphi \tag{26}$$

and does depend on ρ. The free field path integral is required to be renormalized version of the determinant of $-\Delta$ to the power $-1/2$ with the zero eigenvalue omitted. The ultraviolet difficulty is related to the continuum infinite number of modes. Following[26], consider the variation of Z under a local variation of scale

$$\delta\, \ell n\, Z = \delta\, \ell n \int \pi' \frac{d\varphi(m)}{\sqrt{2\pi}}\, e^{-\frac{1}{2}\Sigma E_n\varphi_{(n)}^2} \tag{27}$$

The field has been expanded in eigenmodes of the Laplacian with amplitudes $\varphi_{(n)}$ so that the action $\Sigma\, E_n\varphi_{(n)}^2$ is invariant under changes of ρ, and π' indicates that the zero mode is omitted. Call the corresponding amplitude $\varphi_{(o)}$ $(E_o=0)$. As ρ varies, for a fixed field φ, both the $\varphi_{(n)}$ and the E_n vary. Therefore

$$\delta\, \ell n\, Z = \left\langle \left(\sum_1^{\infty} \frac{\partial\delta\varphi_{(n)}}{\partial\varphi_{(n)}} - \frac{\partial\delta\varphi_{(o)}}{\partial\varphi_{(o)}} \right) \right\rangle \tag{28}$$

arising from the Jacobian of the transformation from $\varphi_{(n)}+\delta\varphi_{(n)}$ to $\varphi_{(n)}$. Since the field is invariant

$$0 = \delta\varphi = \sum_{0}^{\infty} \delta\varphi_{(n)}\psi^{(n)} + \varphi_{(n)}\delta\psi^{(n)}$$

and

$$\frac{\partial\delta\varphi_{(n)}}{\partial\varphi_{(n)}} = -\int d^2\alpha\ \rho^2\psi^{(n)}\delta\psi^{(n)} = \int d^2\alpha\ \rho\delta\rho\ \psi^{(n)2}$$

The quantity in (28) does not require to be computed in the mean, since it is $\varphi_{(n)}$ independent. Consequently

$$\delta\ \ell n\ Z = \int d^2\alpha\ \rho\delta\rho \sum_{0}^{\infty} \psi^{(n)2} - \frac{1}{2A}\ \delta A \tag{29}$$

The second term results from the zero mode subtraction. The first one is ultraviolet infinite, due to the infinity of modes, and requires a subtraction. It is regularized as

$$\sum_{0}^{\infty} \psi^{(n)2}(\alpha) \underset{s\mapsto 0}{\mapsto} \sum_{0}^{\infty} \psi^{(n)2}(\alpha)\ e^{-sE_n} = U(s;\alpha,\alpha) \tag{30}$$

where $U(s;\alpha,\beta)$ is the heat kernel

$$U(s;,\beta) = U(s;\beta,\alpha) = \sum_{0}^{\infty} \psi^{(n)}(\alpha)\psi^{(n)}(\beta)\ e^{-sE_n} \tag{31}$$

$$\left(\frac{\partial}{\partial s} - \Delta\right) U(s;\alpha,\beta) = 0 \qquad s > 0$$
$$\lim_{s\mapsto 0} U(s;\alpha,\beta) = \delta(\alpha,\beta) \tag{32}$$

The invariant Dirac distribution $\delta(\alpha,\beta)$ satisfies

$$\varphi(\alpha) = \int d^2\beta\ \rho^2(\beta)\ \delta(\alpha,\beta)\varphi(\alpha,\beta) \tag{33}$$

and is singular when $\beta\mapsto\alpha$.

We consider

$$\delta\, \ell n\, Z_{reg}(s) = \int d^2\alpha\, \rho^2 \frac{\delta\rho}{\rho} U(s;\alpha,\alpha) - \frac{1}{2A}\delta A \qquad (34)$$

and ask for the small s behaviour. We expect terms of order $\frac{1}{s}$ and finite ones as $s\mapsto 0$. In flat space

$$U_{flat}(s;\alpha,\beta) = \frac{e^{-\frac{1}{4s}(\alpha-\beta)^2}}{4\pi s} \qquad (35)$$

and $U_{flat}(s;\alpha,\alpha) = \frac{1}{4\pi s}$, independently of α. In curved space, and for dimensional reasons, $U_{curved}(s;\alpha,\alpha) = \frac{1}{4\pi s}(1+a\, R_\alpha s+\ldots)$ where R_α is the curvature at α. It is sufficient to do the calculation for a sphere with the curvature R equal to r^{-2}. To first order in R, using (2)

$$U_{sphere}(s;\alpha,0) = \frac{e^{-\frac{\alpha^2}{4s}}}{4\pi s}\left[1+\frac{Rs}{3}\left(1+\frac{1}{4}\frac{\alpha^2}{s}+\frac{1}{8}\frac{\alpha^4}{s^2}\right)+\ldots\right] \qquad (36)$$

and

$$\begin{aligned} U_{curved}(s;\alpha,\alpha) &= \frac{1}{4\pi s}\left[1+\frac{R_\alpha s}{3}+\ldots\right] & \text{(a)} \\ &= \frac{1}{4\pi s} - \frac{1}{12\pi}\Delta_{curv.}\ell n\, \rho & \text{(b)} \end{aligned} \qquad (37)$$

The second expression is checked in the case of the sphere. Consequently

$$\begin{aligned} \delta\, \ell n\, Z_{reg} &= \left(\frac{1}{8\pi s}-\frac{1}{2A}\right)\delta A + \frac{1}{12\pi}\int d^2\alpha\, \rho^2(\delta\, \ln\rho)(-\Delta_{curv}\ell n\,\rho) \\ &= \delta\left[\frac{A}{8\pi s}-\frac{1}{2}\ln\frac{A}{A_0}+\frac{1}{24\pi}\int d^2\alpha\, \rho^2\, \ell n\,\rho(-\Delta_{curv})\ell n\,\rho\right] \end{aligned} \qquad (38)$$

Equation (37a) could be slightly improved to read for small s and in the vicinity of a point α

$$U_{curved}(s;\alpha,\beta) = \frac{e^{-d^2_{\alpha\beta}/4s}}{4\pi s}\left[1+R_\alpha\frac{s}{3}+R_\alpha\frac{d^2_{\alpha\beta}}{12}+\ldots\right]$$

ignoring derivatives of the curvature. Here $d_{\alpha\beta}$ is the geodesic distance between α and β. The factor $\frac{1}{4\pi s}$ in the flat case can be interpreted as the inverse area of a circle of radius $r_{eff}=2\sqrt{s}$ as is reasonable in a Brownian motion interpretation of U as a probability density. On a curved surface a circle of same radius (assumed to be small) has an area $2\pi R^{-1}(1-\cos\theta)\approx 2\pi R^{-1}\left(\frac{\theta^2}{2}-\frac{\theta^4}{24}\right)$ with $\theta=R^{+1/2}2\sqrt{s}$, i.e. $4\pi s\left(1-\frac{Rs}{3}\right)$. The inverse of this area expanded in powers of $Rs\ll 1$ yields (37a), a neat way of understanding this expression.

When discussing (38) we realize a shortcoming: ρ can only be defined in a coordinate patch, which means sensitivity to boundary conditions, if we want to interpret $\ell n\ Z_{reg}-A\left(\frac{1}{s}-\frac{1}{s_0}\right)$ as $\ell n\ Z_{renormalized}$. This is exemplified in the case of a sphere with ρ given by (24). While it is tempting to relate $\int d^2\alpha\ \rho^2\ell n\ \rho(-\Delta_{curved})\ell n\ \rho = \int d^2\alpha\ \ell n\ \rho(-\Delta_{flat})\ell n\ \rho$ to $\int d^2\alpha \sum_{a=1}^{2}(\partial_a \ell n\ \rho)^2$ the first integral is finite and the second is not. With our convention

$$-\Delta_{curved}\ell n\ \rho = R \tag{39}$$

showing why for a compact surface with non vanishing Euler characteristics

$$\chi = \frac{1}{2\pi}\int d^2\alpha\ \rho^2 R \tag{40}$$

$\ell n\ \rho$ cannot be defined everywhere as a non singular function. In (40) the interpretation is that $d^2\alpha\ \rho^2$ is an invariant element of area, and several coordinate patches may be needed to compute the integral. We do not have these difficulties, on a torus, $\chi=0$, and we therefore assume such a simplification from now on. These subtleties do not play any role in a variation of $\delta \ell n Z_{ren.}$ when $\delta\rho$ is contained in a coordinate patch. We recall that the non local term $\ell n \frac{A}{A_0}$ is due to the fact that we consider compact surfaces.

Define $\tilde{G}_{curved}(\alpha,\beta)$ as a subtracted propagator -again because of the zero mode problem- as

$$\tilde{G}(\alpha,\beta) = \sum_{1}^{\infty} \frac{1}{E_n} \psi^{(n)}(\alpha)\psi^{(n)}(\beta) \quad ; \quad -\Delta_{curv}\tilde{G}(\alpha,\beta) = \delta_{cur}(\alpha,\beta) - \frac{1}{A} \tag{41}$$

Then for $\chi=0$

$$\ell n \; \rho(\alpha) = \int d^2\beta \; \rho^2(\beta) \; \tilde{G}(\alpha,\beta) \; R(\beta) \tag{42}$$

and the r.h.s. is insensitive to the addition to $\tilde{G}(\alpha,\beta)$ of an arbitrary constant. We therefore define for $\chi=0$

$$\ell n \; Z_{ren} = \frac{A}{s_0} - \frac{1}{2} \; \ell n \frac{A}{A_0} + \frac{1}{24\pi} \iint d^2\alpha \; d^2\beta \; \rho^2(\alpha)\rho^2(\beta) R(\alpha) \tilde{G}(\alpha,\beta) R(\beta) \tag{43}$$

Each elementary piece of the surface is endowed with a curvature "charge"

$$dq(\alpha) = d^2\alpha \; \rho^2(\alpha) \; R(\alpha) \tag{44}$$

The total charge is zero

$$\int dq(\alpha) = 0 \tag{45}$$

and $\tilde{G}(\alpha,\beta)$ has a (classical) two-dimensional long range character of a Coulomb potential. For all its aesthetic appeal, equation (43) is not that great a simplification. Z was given in terms of the product of eigenvalues and it requires as much effort to compute $\tilde{G}$. On the other hand given a metric in the form (23), it is easy to find the variations of $\ell n\ Z_{ren}$. by simple integrations.

We have tried to find analogs of (38) and (43) starting directly from the discrete situation of the previous sections but failed to obtain "nice" expressions. This is perhaps not unrelated to localization problems on such a random lattice. However inspired by (43), the so called "Liouville action", we may obtain a natural discretization for it. First the analog of $\tilde{G}$ is the solution of the equation

$$\sum_{k(i)} \frac{\sigma_{ik}}{\ell_{ik}} \left(\tilde{G}_{(i,j)} - \tilde{G}_{(k,j)}\right) = \delta_{ij} - \sigma_i \tag{46}$$

To make things more symmetric, we add a small mass term to the Laplacian which shifts all eigenvalues by a constant positive amount. The lowest eigenmode with positive eigenvalue is orthogonal to the curvature for $\chi=0$ (call it the neutral case), hence the limit $m^2 \mapsto 0$ is well defined. We can replace $\tilde{G}$ by G such that

$$\sum_{k(i)} \frac{\sigma_{ik}}{\ell_{ik}} [G(i,j;m^2) - G(k,k;m^2] + m^2 \sigma_i G(i,j;m^2) = \delta_{ij} \tag{47}$$

and write in discretized form

$$S_{Liouville} = \lim_{m^2 \mapsto 0} \frac{1}{2} \sum_{i,j} q_i G(i,j;m^2)\ q_j$$
$$q_i = \left(1 - \frac{\theta_i}{2\pi}\right) \quad ; \quad \sum_i q_i = 0 \tag{48}$$

The limit $m^2 \mapsto 0$ has to be taken when the double sum is first computed. We can relax here the condition that the σ's be positive. If we define

$$\psi_i = \sum_{\substack{m^2 \mapsto 0 \\ j}} G(i,j;m^2)\ q_j \tag{49}$$

Then

$$S_{Liouville} = \frac{1}{2} \sum_i \sigma_i \psi_i (-\Delta\psi)_i \tag{50}$$

In the neutral case, (49) has a limit when $m^2 \mapsto 0$, ψ remains finite, hence $S_{Liouville}$ is positive.

E- Embedding in Large Dimension

There is not at the moment a consensus on good candidates for models of random surfaces (the most random ones !). Here we limit ourselves to repeat the analysis of the simplest one[11,12], and suggest some natural modification.

Take a torus ($\chi=0$). Choose a triangulation, and map it in R^d as explained above. The simplest statistical weight is, with a an arbitrary length scale,

$$d\mu = \pi' \; d^d x_i \; e^{-\beta_o S_o} \tag{51}$$

$$S_o = \frac{1}{a^2} \; A = \frac{1}{a^2} \sum_{(ijk)} \ell_{ijk} \tag{52}$$

This is the direct analog of the Brownian case, and we wish to estimate what happens when the number points gets large and the points dense on the torus. This is was has been done numerically and analytically (for large d) by Billoire Gross and Marinari[12]. Let us briefly recapitulate what happens. Use for instance the triangulation shown on Figure 9, with $n=n_1=n_2$, $N_o=n^2$, $N_1=3n^2$, $N_2=2n^2$. The base space has a discrete translational invariance, assumed unbroken. Set for simplicity $f_{ijk} = \beta_o \dfrac{\ell_{ijk}}{\alpha^2}$. Then from dilatation covariance we have

$$Z = \int \pi' \; d^d x_i \; e^{-\beta_o S_o} = \lambda^{d(n^2-1)} \int \pi' \; d^d x_i \; e^{-\lambda^2 \beta_o S_o} \tag{53}$$

$$\langle \beta_o S_o \rangle = (n^2-1) \frac{d}{2} \approx N \frac{d}{2} \qquad\qquad \langle f \rangle = \frac{d}{4}$$

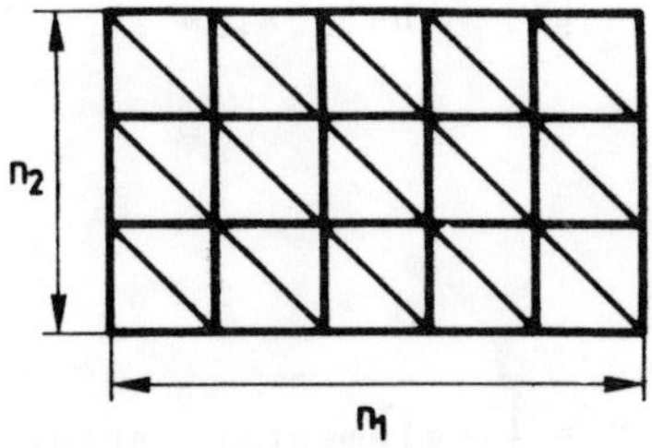

Figure 9

The limit $d\mapsto\infty$ is a good place to look for a valid saddle point approximation, to which one then proceeds. One rewrites

$$Z = \int \pi' \, d^d x_i \int \pi \frac{d\lambda_{ij}}{2\pi i} \, d\ell_{ij} \, e^{\sum_{(i)} \lambda_{ij}[(x_i - x_j)^2 - \ell_{ij}^2] - \beta_o S_o\{\ell_{ij}\}} \tag{54}$$

with λ_{ij} integrated over the imaginary axis (anticipating that the saddle point value will then be real) and assumes a translational invariant saddle point with $\lambda_{ij}^o \equiv \lambda$, $\ell_{ij}^o = \ell$. This leads to

$$Z \mapsto Z_o = \text{cst } e^{-S_{eff}(\lambda,\ell)}$$

$$S_{eff} = N\left[\frac{d}{2}\ \ell n\ \lambda - 3\lambda\ \ell^2 + 2f(\ell,\ell,\ell)\right] \tag{55}$$

where $f(\ell,\ell,\ell) = \beta_o \frac{\sqrt{3}}{4} \frac{\ell^2}{a^2}$ and the elementary area is effectively the one for an equilateral triangle. The variational equations to be satisfied by λ and ℓ reduce to

$$\lambda = \frac{d}{6\ell^2} \qquad f = \beta_o \frac{\sqrt{3}}{4} \frac{\ell^2}{a^2} = \frac{d}{4} \tag{56}$$

The second equation has of course to agree with (53). Therefore S_{eff}=cst + $\frac{Nd}{2}$ $\ell n\ \beta_o$ and $\beta_o \frac{dS_{eff}}{d} = \frac{Nd}{2} = \langle S_o \rangle$ as it should. With these values for λ and ℓ the model reduces to d-uncoupled gaussian free field in flat space for any question pertaining to the x_i's . In particular $\langle (x_i - x_j)^2 \rangle = \ell^2$ and

$$R_N^2 = \frac{1}{N} \langle \sum x_i^2 \rangle \tag{57}$$

$$= \frac{1}{\pi} \frac{\sqrt{3}}{4} \ell^2 \ln N = \frac{1}{\pi} \langle \text{elementary area} \rangle \ln N$$

The $\ln$ term results from the infrared behavior of the 2d massless propagator as the linear behavior in N for random curves arose from a similar infrared singularity of the one-dimensional propagator. The numerical factor on the r.h.s. of (56) is specific to triangulations. It behaves like d/4 for large d and is corrected by a factor (1+2/d+...) to first order in 1/d. This is in fair agreement with the observations made in [11] as shown in the following table giving $R_N^2 \dfrac{4\pi}{\sqrt{3}\,\ell^2} \cdot \dfrac{1}{\ln N}$

d	First order correction $1+\frac{2}{d}$	Numerical data [11]
3	1.667	1.73±0.04
4	1.500	1.43±0.03
6	1.333	1.27±0.01
12	1.167	1.16±0.03

It in unlikely that inequivalent triangulations of the torus would lead to a qualitatively different result, or that going to a surface with a different topology would change the $\ln N$ behavior also observed in numerical computation. This means that the image surface is strongly *collapsed* object in the mean.

Our previous discussion suggests to modify equations (51), (52) by taking as probability weight (forgetting about the $\ln A/A_0$ term, presumably irrelevant as compared to S_0)

$$d\mu = \prod_i d^d x_i \; e^{-\beta_0 S_0 - \beta_1 S_{Liouville}} \tag{58}$$

as a slight generalization of the above model.

A rough way to estimate the size of the Liouville term would be, in the same type of simulations that have been performed before, to compute $\langle q_i^2 \rangle$. Or else one could try to estimate it from the Gaussian model emerging from (54). We believe that it is order 1. Hence the Liouville term could become relevant even when $d \mapsto \infty$ provided we take β_1 of order d. This would then require to find different saddle point.

To see what is involved let us look at the Gaussian model implied by (54). Since the sum of angles of a given triangle is π and each of them has equal mean value it is indeed very reasonable that their average value is $\frac{\pi}{3}$. Then look at figure 10 which involves the points A, B, with A second neighbor to B, and compute ($N \mapsto \infty$)

$$\frac{\ell'^2}{\ell^2} = \frac{1}{(2\pi)^2} \int_{-\pi}^{+\pi}\int_{-\pi}^{+\pi}\int_{-\pi}^{+\pi} dk_1 dk_2 dk_3 \; \frac{3-\cos(k_1-k_2)-\cos(k_2-k_3)-\cos(k_3-k_1)}{3-\cos k_1-\cos k_2-\cos k_3} \delta(k_1+k_2+k_3) \tag{59}$$

$$= 6\sqrt{3}/\pi - 2 = 1.308$$

The points ABC fit in a plane and either O is in this plane or else OABC fits in a 3-dimensional linear subspace. If the picture were flat and regular (as on the regular triangulated plane) then $\frac{\ell'^2}{\ell^2} = 3$. Apparently this is not the case showing the importance of fluctuations.

We can also rewrite (58) using a Lagrange multiplier

$$d\mu = \frac{\pi' d^d x_i \pi_i' d\phi_i e^{-\beta_0 S_0 - \beta_1 S_{free\ field}(\phi) + i\sum_i \phi_i q_i}}{\int \pi_i' d\phi_i e^{-\beta_1 S_{free\ field}(\phi)}} \tag{60}$$

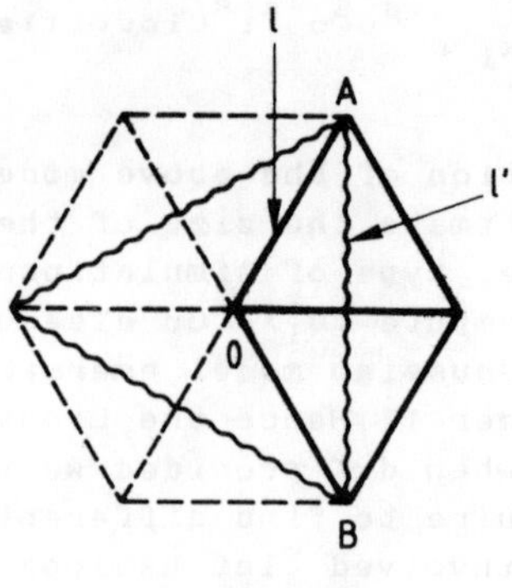

Figure 10

It is perhaps easier numerically to deal with (60) than with (58). At any rate they provide a non trivial discretization of the Liouville theory and it is perhaps worthwhile to investigate their relevance to a model of random surfaces. Surprisingly to leading order we have shown in [19] that the Liouville term does not modify any of the abovve conclusions. Of course this is only a very first step towards proposing a consistent scheme to study quantized gravity in a discretized version, which is one of the motivations to get a better understanding of random geometry.

REFERENCES

[1] Christ, N.H., Friedberg, R., Lee, T.D., Nucl. Phys. B202, 89 (1982) B210 (FS6) 310, 337 (1982).

[2] Collins, R., in "Phase Transitions and Critical Phenomena", Vol. 2, Domb, S. and Green M.S. eds, Academic Press, London (1972).

[3] Zallen, R., in "Fluctuation Phenomena", Montroll, E.W., Lebowitz, J.L., eds, North Holland, Amsterdam (1979).

[4] Ziman, J.M., "Models of Disorder", Cambridge University Press, (1979).

[5] Santalo, L.A., "Integral Geometry and Geometrical Probability" Addison-Wesley, Reading (1979).

[6] Regge, T., Nuovo Cimento, 19, 551 (1961).

[7] Parisi, G., Phys. Lett. B81, 957 (1979).

[8] Wallace, D., in "Recent Advances in Field Theory and Statistical Mechanics", Zuber, J.B. and Stora, R, eds, Les Houches 1983, North Holland, Amsterdam (1984).

[9] Polyakov, A.M., Phys. Lett. B103, 1207 (1981).

[10] Fröhlich, Proc. Sitges Conf. on Statistical Mechanics, Sitges, June 1984, Springer Berlin (1985).

[11] Billoire, A., Gross, D.J., Marinari, E., Phys. Lett. 139B, 239 (1984), Gross, D.J., Phys. Lett. 138B, 185 (1984).

[12] Duplantier, B., Phys. Lett. 141B, 239 (1984).

[13] David, F., Nucl. Phys. B257 (FS14) 45, 543 (1985).

[14] Itzykson, C., in "Non-Perturbative Field Theory and QCD", Trieste Worshop, Lengo, R., et al eds, World Scientific, Singapore (1983).

[15] Gardner, E., Itzykson, C., Derrida, B., J. Phys. A17, 1093 (1984).

[16] Drouffe, J.M., Itzykson, C., Nucl. Phys. B235 (FS11) 45 (1984).

[17] Itzykson, C., in "Progress in Gauge Field Theory, G't Hooft et al. eds, Cargese 1983, Plenum Press, New York (1984).

[18] Bander, M., Itzykson, C., in "Non Linear Equations in Classical and Quantum Field Theory", Proceedings Meudon-Paris 1983/1984. Sanchez, N. ed. Springer

Lecture Notes in Physics 226, Berlin (1985).
[19] Bander, M., Itzykson, C., Nucl. Phys. B257 (FS14) 531 (1985).
[20] Meijering, J.L., Philips Research report 8, 270 (1953).
[21] Rahman, A., J. Chem. Phys. 45, 2585 (1966).
[22] Hanson, H.G., J. Stat. Phys. 30, 591 (1983).
[23] Becher, P., Joos, H., Z. Phys. C15, 343 (1982).
[24] "String Theories", Jacob, M. ed. North Holland, Amsterdam (1974).
[25] Cheeger, J., Müller, W., Schrader, R., Comm. Math. Phys. 92, 405 (1982).
[26] Fukikawa, K., Phys. Rev. D21, 2848 (1980) D23, 2262 (1981).

STOCHASTIC APPROACH TO EUCLIDEAN QUANTUM FIELD THEORY *

(Stochastic Quantization)

P.K. Mitter

LPTHE**, Université Pierre et Marie Curie , Paris***

and

School of Mathematics****
Tata Institute of Fundamental Research
Homi Bhabha Road, Bombay 400 005, India

* Extended version of Lectures at the XVI GIFT International seminar: "New Perspections in Quantum Field theory", Jaca (Spain) June 3-8, 1985. To be published in the proceedings.

** Laboratoire Associé 280 au CNRS.

*** Permanent address: LPTHE, Université Pierre et Marie Curie
Tour 16, 4 Place Jussieu
75230 Paris Cedex 05, France.

**** Temporary address: 23.9.1985 - 30.8.1986

INTRODUCTION

Our aim in these lectures is to introduce the reader to a mathematically rigorous (and non-perturbative) version of stochastic quantization [11]. Let us briefly recall the essential idea behind this. As is by now well known [7,8,9] Euclidean (quantum) field theory with the Osterwalder-Schrader axioms engenders relativistic quantum field theory satisfying the Wightman axioms. In a dynamical approach the construction of Euclidean (quantum) field theories boils down to the construction of probability measures on some space of distributions. These measures are formally Gibbsian, and satisfy additional properties (reflection positivity,..., [7-9]). Stochastic quantization takes a particular dynamical point of view towards obtaining such measures. The idea is to construct Markov processes, in particular diffusion processes with values in a space of distributions, such that Euclidean field theory measures are realized as limiting distributions (invariant measures). In case of ergodicity we have a unique invariant measure. The lack of ergodicity is related to the problem of phase transitions.

A stochastic field theory model is specified by requiring that the diffusion process in question is a solution (in a sense to be specified) of a non-linear Langevin equation with drift formally proportional to the gradient of the Euclidean action, with the Wiener process as a driving process. The problem is complicated by the fact that (without cutoffs) we are in an infinite dimensional state space (that of distributions), and ultraviolet divergences of field theory require renormalization. To obtain renormalized measures we must work with renormalized Langevin equations and renormalized process. This programme has been successfully controlled by G. Jona-Lasinio and the author in [10] for the stochastic $P(\varphi)_2$ model in finite space-time volume Λ. In [10] we set up renormalized (generalized) Langevin equations and show that there exist weak solutions which are diffusion processes in $\mathcal{D}'(\Lambda)$. These Markov processes are ergodic with the finite volume $P(\varphi)_2$ Euclidean measure [7-9] as their unique invariant measure. The considerations are non-perturbative. Just as the construction of the $P(\varphi)_2$ model [7-9] was of fundamental importance in the development of the now imposing edifice of constructive quantum field theory, so also the stochastic $P(\varphi)_2$ model should have a paradigmatic role in the development of the stochastic approach to Euclidean quantum field theory. The goal of these lectures is to explain the content of [10].

One of the more interesting areas of application should be non-abelian gauge field theory where the continuum limit (in non-trivial cases) is not properly understood from the non-perturbative point of view. The dynamical view point of stochastic quantization may shed further light on the construction of gauge field theory measures. Indeed work is in progress in numerical Langevin simulation by G. Parisi and collaborators & K.G. Wilson and collaborators*. Interesting insights have also been obtained on the theoretical side [14,15]. None of these topics are touched upon on these lectures.

---

* private communication from Gerhard Mack.

We should also mention that stochastic statistical mechanical models have a long history. A list of references may be found in [13]. Prior to [10], the simpler $P(\varphi)_1$ model (stochastic Euclidean quantum mechanics) was controlled in [12].

In this written version of the lectures we have attempted to be pedagogic, given the fact that the subject is relatively young. In chapter I and II we give for the sake of the uninitiated reader a self-contained account of Brownian stochastic calculus and stochastic differential equations in $\mathbb{R}^N$, where N is to be interpreted as an ultraviolet cutoff in the context of field theory. We emphasize the theory of weak solutions. In chapter III, which is preparatory to stochastic field quantization, we consider a generalized Ornstein-Uhlenbeck process and drift perturbations in $\mathbb{R}^N$. We impose very weak (measurability) conditions on the drift and give prototype estimates to obtain weak solutions. These methods are generalized in Chapter IV, V and VI to the case of the stochastic $P(\varphi)_2$ model of [10]. The techniques are inspired by those of constructive field theory.

ACKNOWLEDGEMENTS

I wish to thank the organizers of the XVI International GIFT seminar for their truly generous hospitality and the opportunity to give these lectures. These notes were written up at the School of Mathematics, Tata Institute of Fundamental Research. I thank the National Board of Higher Mathematics (Government of India) for making my visit to India possible and TIFR for its hospitality.

I thank the CNRS (France) for according me leave of absence from Paris. Above all I thank M. Asorey, G. Jona-Lasinio and E. Seiler for many stimulating discussions. To G. Jona-Lasinio goes my special thanks for a fruitful collaboration leading to this work.

CHAPTER I

STOCHASTIC CALCULUS IN $\mathbb{R}^N$ AND APPLICATIONS

In this introductory chapter we summarize some standard properties of Brownian motion in $\mathbb{R}^N$ and the related I to stochastic calculus. In chapter II, III we give a brief introduction to a class of stochastic differential equations. This will serve as a background to the rigorous study of stochastic quantization of a model field theory taken up in chapters IV, V, VI and which necessarily takes place in infinite dimensions. Finite dimensional approximations (equivalently "ultraviolet cutoffs") nevertheless play a crucial role in intermediate steps.

§ 1.1 Brownian motion As is wellknown standard Brownian motion $\beta_t, 0 \leq t < \infty$, with values in $\mathbb{R}^N$ (the "state space") is a particularly simple Gaussian Markov process. Its transition probability i.e. the probability of the event that a standard Brownian particle starting at x passes through a (Borel) set $A \subset \mathbb{R}^N$ after time t, is given by:

$$P_x^{(W)}(\beta_t \in A) \equiv P_t^{(W)}(x, A) = \int_A p_t^{(W)}(x, dy) \qquad (1.1)$$

$$p_t^{(W)}(x, dy) = \frac{1}{(2\pi t)^{\frac{N}{2}}} e^{-\frac{\|x-y\|^2}{2t}} d^N y \qquad (1.2)$$

The super script 'W' stands for 'Wiener'. $\| \cdot \|$ is the standard Euclidean norm in $\mathbb{R}^N$.

The transition probabilities $p_t(x, dy)$ satisfy the Chapman-Kolmogoroff equation:

$$p_{t+s}(x,A) = \int_{\mathbb{R}^N} p_t^{(W)}(x,dy)\, p_s^{(W)}(y,A) \qquad (1.3)$$

which is a statement of the "Markov property". (1.3) is a well-known property of the 'heat kernel' appearing in (1.2). Related to Brownian motion we have the <u>heat semigroup</u>:

$$(T_t^{(W)} f)(x) = (e^{t(\Delta/2)} f.)(x) = \int_{\mathbb{R}^N} p_t^{(W)}(x,dy)\, f(y) \qquad (1.4)$$

$f \in L^2(\mathbb{R}^N)$, and the semigroup property follows from (1.3). Δ is the Euclidean Laplacian:

$$\Delta = \sum_{i=1}^{N} \frac{\partial^2}{\partial x_i^2}$$

The physical interpretation of $P_x^{(W)}(\beta_t \in A)$ was given earlier. To go further, let $0 \le t_1 < t_2 < \cdots < t_k < \infty$. We define 'joint probability distributions':

$$P_x^{(W)}(\beta_{t_1} \in A_1, \beta_{t_2} \in A_2, \ldots, \beta_{t_k} \in A_k)$$

$$= \int \cdots\cdots \int_{A_1 \times A_2 \times \cdots \times A_k} p_{t_1}^{(W)}(x, dy_1)\, p_{t_2 - t_1}^{(W)}(y_1, dy_2) \ldots p_{t_k - t_{k-1}}^{(W)}(y_{k-1}, dy_k) \qquad (1.5)$$

It is simply the probability of the event that a Brownian particle β_t starting at x, passes at subsequent times $t_1, t_2, \ldots, t_k$ through (respectively) the (Borel) subsets $A_1, A_2, \ldots, A_k$ of $\mathbb{R}^N$.

From (1.1, 1.2, 1.5) it is easy to deduce two characteristic properties of Brownian motion, viz for t>s:

$$P_x^{(W)} \{ (\beta_t - \beta_s) \in A \} = p_{t-s}^{(W)} (0, A) \qquad (1.6)$$

$$P_x^{(W)} \{ \beta_s \in A, (\beta_t - \beta_s) \in B \} = p_s^{(W)} (x, A)\, p_{t-s}^{(W)} (0, B) \;. \qquad (1.7)$$

Indeed (1.1 - 1.2) together with (1.6 - 1.7) can be considered to be defining properties of Brownian motion.

The elementary events whose probabilities are given by (1.5) are subsets of the space $\hat{\Omega}$ of <u>all</u> paths in $\mathbb{R}^N$ (this space is too big). Consider the smallest σ-algebra $\mathcal{F}$ generated by taking a countable number of unions, intersections and compliments of these elementary events (cylinder sets). Because of (1.3), Kolmogoroff showed [1] that $P_x^{(W)}$ has a countably additive extension as a probability measure on $(\hat{\Omega}, \mathcal{F})$. Let $E_x^{(W)}$ stand for integration (expectation) with respect to this measure. A theorem* due to Kolmogoroff [1] says that if there exists $\alpha, \delta > 0$ such that

$$E_x^{(W)} (\| \beta_t - \beta_s \|^{\alpha}) \leq C \, |t-s|^{1+\delta} \qquad (1.8)$$

then $P_x^{(W)}$ is realized as a (countably additive) probability measure on the much smaller space Ω of <u>continuous paths in $\mathbb{R}^N$</u>, $\Omega \equiv C^o([0,\infty), \mathbb{R}^N)$.

---

* This theorem holds for <u>any</u> Markov process satisfying (1.1, 1.3, 1.5) (and not just the Wiener process), with $p_t(x, dy)$ countably additive probability measure.

In our case an elementary computation shows that (1.8) is satisfied with $\alpha=4$, $\delta=2$. $(P_x^{(W)}, \Omega)$ is known as Wiener space, and $P_x^{(W)}$ is known as Wiener measure [N. Wiener was the first to prove this]. Henceforth we work in Wiener space.

It is easy to check the following proporties of Wiener expectations. We write $\beta_t = (\beta_t^1, \ldots, \beta_t^n)$ as a vector in $\mathbb{R}^N$. Then:

$$E_o^{(W)}(\beta_t^i) = 0 \quad \text{(mean)}$$

$$E_o^{(W)}(\beta_t^i \beta_s^j) = \delta^{ij} \min(t,s) \qquad \text{(covariance)}$$

$$E_o^{(W)}((\beta_t^i - \beta_s^i)(\beta_t^j - \beta_s^j)) = \delta^{ij} |t-s| \tag{1.9}$$

$$E_o^{(W)}((\beta_t^i - \beta_s^i)\beta_s^j) = 0, \quad t > s.$$

The last equality states that the forward increment of Brownian motion is independent of past events.

We say that a path β_t in a (bounded) interval $[0,T]$ is of bounded variation, if for partitions $Q_n: 0=t_o<t_1<t_2<\cdots<t_n=T$ of $[0,T]$, $t_j - t_{j-1} = \frac{T}{n}$, the sum:

$$V_n(Q_n) = \sum_{i=1}^{n} \|\beta_{t_i} - \beta_{t_{i-1}}\| \xrightarrow[n\to\infty]{} C < \infty .$$

It is an important fact about Brownian motion that, $P_o^{(W)}$ a.e (almost everywhere), paths in Ω are of unbounded variation (and hence also non-differentiable).

Proof: It suffices to prove that there is a subsequence $\{n_k\}$ such that $V_{n_k} \longrightarrow \infty$, $P_o^{(W)}$ a.e as $\{ n_k \} \uparrow \infty$.

Define:

$$U_n \equiv \sum_{i=1}^{n} \| \beta_{t_i} - \beta_{t_{i-1}} \|^2 \quad \text{and} \quad A_n = \sup_{1 \le i \le n} \| \beta_{t_i} - \beta_{t_{i-1}} \| .$$

Then, since β_t is continuous P^W a.e, we have

$$\lim_{n \longrightarrow \infty} A_n = 0. \qquad (*)$$

Now we have the obvious inequality:

$$U_n \le A_n V_n \qquad (**)$$

Note that

$$E_o^{(W)} | U_n - U_{n-1} | = E_o^{(W)} (\| \beta_{t_n} - \beta_{t_{n-1}} \|^2) = t_n - t_{n-1}$$

$$= \frac{T}{n} \underset{n \to \infty}{\longrightarrow} 0 .$$

Hence the sequence $\{ U_n \}$ converges in $L^1(P_o^{(W)}, \Omega)$ and thus we can always find a subsequence $\{ U_{n_k} \}$ which converges $P_o^{(W)}$ a.e. Since,

$$U_{n_k} \le A_{n_k} V_{n_k}$$

and the left-hand side converges to a non-zero element, whereas $A_{n_k} \longrightarrow 0$ as $n_k \uparrow \infty$ $P_o^{(W)}$ a.e., it follows that $V_{n_k} \longrightarrow \infty$, $P_o^{(W)}$ a.e., and we are done.

There is a great deal of detailed information on Brownian paths available in the literature [1] but the above suffices for the subsequent developments.

§ 1.2 Stochastic integrals over Brownian paths [1,2]

We want to give meaning to the following 'integral' as a random variable :

$$I_t = \sum_{i=1}^{N} \int_o^t f_i(\beta_s)\, d\beta_s^i = \int_o^t (f(\beta_s), d\beta_s) \qquad (1.10)$$

where $(.,.)$ is the scalar product in $\mathbb{R}^N$, and to begin with,

$$f(x) = (f_1(x), \ldots, f_N(x))$$

is a bounded once differentiable vector valued function on $\mathbb{R}^N$ with bounded derivative (C_b'). Our task is complicated by the fact that, as we have seen, β_s, although continuous, is $P^{(W)}$ a.e. non-differentiable and of unbounded variation. Thus we cannot hope to define it as a Lebesque-Stieltjes integral. Instead we opt for Ito's definition [1,2], and the ensuing powerful stochastic calculus.

To this end we consider a finite partition $0=t_o < t_1 < \cdots\cdots < t_{(2^n-1)} < t_{2^n} = t$ of the bounded interval $[0,t]$, $t_j = \frac{jt}{2^n}$, and an approximant to (1.10):

$$I_t^{(n)} = \sum_{j=1}^{2^n} (f(\beta_{t_{j-1}}), (\beta_{t_j} - \beta_{t_{j-1}})). \qquad (1.11)$$

Note the introduction of the "forward" increment of Brownian motion in (1.10), characteristic of Ito integration theory.

Using

$$E^{(W)}((\beta_t^i - \beta_s^i)(\beta_t^j - \beta_s^j)) = |t-s|\, \delta^{ij}$$

and $E^{(W)}((\beta_t^i - \beta_s^i)(\beta_u^j - \beta_v^j)) = 0$ for $t > s \geq u \geq v$

we obtain by elementary computation (left as an exercise)

$$E^{(W)}(|I_t^{(n)}|^2) = \sum_{j=1}^{2^n} E^{(W)}(\|f(\beta_{\frac{(j-1)t}{2^n}})\|^2)\frac{1}{2^n} \qquad (*)$$

$$\leq \|f\|_\infty^2, \quad \|f\|_\infty = \sup_{x \in \mathbb{R}^N} \|f(x)\| < \infty$$

since f is a bounded continuous function.

Moreover, by a similar computation,

$$E^{(W)}(|I_t^{(n+1)} - I_t^{(n)}|^2) = \sum_{j=0}^{2^{n+1}-1} E^{(W)}(\|f(\beta_{\frac{(2j+1)t}{2^{n+1}}}) - f(\beta_{\frac{2jt}{2^{n+1}}})\|^2)\frac{1}{2^{n+1}}.$$

Since f is in C_b^1, by using the mean value theorem,

$$\|f(x) - f(y)\| \leq C\|x-y\|.$$

Hence

$$E^{(W)}(|I_t^{(n+1)} - I_t^{(n)}|^2) \leq \frac{C}{2^{n+1}} \sum_{j=0}^{2^{n+1}-1} E^{(W)}(\|\beta_{\frac{(2j+1)t}{2^{n+1}}} - \beta_{\frac{2jt}{2^{n+1}}}\|^2)$$

$$= \frac{C}{2^{n+1}} \xrightarrow[n \to \infty]{} 0 \qquad (**)$$

Hence $\{I_t^{(n)}\}$ converges as $n \to \infty$ as a sequence of random variables in $L^2(\Omega, dP^{(W)})$. The stochastic integral I_t of (1.10) as a random variable is defined as this limit, which is characterized by virtue of (*) by:

$$E^{(W)}\left(I_t^{\,2}\right) = E^{(W)}\left(\int_o^t ds \; \| f(\beta_s) \|^2\right) \qquad (1.11\text{ a})$$

and

$$E^{(W)}(I_t) = 0 \qquad (1.11\text{ b})$$

(because of the independence of forward increments in (1.19).

Note that $I_t^{(n)}$ of (1.10) is continuous in t, $P^{(W)}$ a.e. since β_t is so. It can be shown with some more work that the limit I_t remains $P^{(W)}$ a.e., continuous in t (the proof is given in Appendix 2, of this chapter).

In the above we assume that the integrand f was once differentiable with bounded derivative. The definition of I_t can now be extended, using (1.11), as a mean square integrable random variable for any $f \in L^2(\mathbb{R}^N, d^N x)$ such that the right hand side of (1.11a) converges. The right hand side of (1.11 a) has the explicit expression:

$$\int_o^t ds \int_{\mathbb{R}^N} dx \; \| f(x) \|^2 \; \frac{e^{-\frac{\|x\|^2}{2s}}}{(\sqrt{2\pi s})^N}$$

and this converges for example for any $L^2(\mathbb{R}^N)$ function f which is not singular at the origin. Polynomials would also do, etc.

The definition of the stochastic integral immediately goes through for more general integrands:

$$I_t = \int_o^t (f(\chi_s), d\beta_s) \qquad (1.12)$$

where $\chi_s = (\chi_s, \ldots, \chi_s^{(N)})$ are random variables (functionals of

Brownian motion) such that $\beta_t^i - \beta_s^i$, $t > s$ is independent of all $\mathcal{X}_s^j$ and such that

$$E^{(W)} \int_0^t ds \, \| f(\mathcal{X}_s) \|^2 < \infty \, . \tag{1.13}$$

These are the only properties that enter into the definition of the Ito integral.

Time change [3]

Consider the Ito Stochastic integral

$$I_t = \int_0^t (f(s), d\beta_s)$$

where $f(s)$ is now taken to be a deterministic function.

Provided

$$\int_0^t ds \, \| f(s) \|^2 < \infty$$

the previous Ito definition of I_t goes through. I_t is a one-dimensional Gaussian process. Moreover because of the independence from the past of Brownian forward increments (entering in the definition of I_t) we have,

$$E^{(W)} ((I_t - I_s) I_s) = 0 \tag{*}$$

Moreover

$$E^{(W)}\left(\left| I_t - I_s \right|^2 \right) = \int_s^t ds' \, \|f(s')\|^2 = \tau(t) - \tau(s) \qquad (**)$$

where

$$\tau(t) = \int_0^t ds' \, \|f(s')\|^2$$

is an increasing function of t. The fact that I_t is Gaussian, together with (*) and (**), suffice to characterise I_t as <u>1-dimensional Brownian motion with time change $t \longrightarrow \tau(t)$</u>, starting at the origin. I_t has transition probability (starting at origin)

$$p_t(0, dy) = p^{(W)}_{\tau(t)}(0, dy) = \frac{1}{\sqrt{2\pi\,\tau(t)}}\, e^{\frac{-|y|^2}{2\,\tau(t)}} \, dy \,. \qquad (***)$$

This will be used later.

§ 1.3 The Ito stochastic calculus in $\mathbb{R}^N$ [1, 2, 3]

In the previous section we defined the Ito stochastic integral

$$I_t = \int_0^t (f(\xi_s), \, d\beta_s)$$

under the condition that ξ_s is independent of $\beta_t - \beta_s$, $t > s$, and $E^{(W)}\left(\int_0^t ds \, \| f(\xi_s) \|^2 \right) < \infty$. This is often written in "differential" form:

$$d\, I_t = \sum_{i=1}^{N} f_i(\xi_s) \, d\beta^i_s \,.$$

The Ito stochastic calculus is based on the following properties of stochastic differentials stated in the "multiplication table":

	$d\beta_t^i$	dt
$d\beta_t^j$	$\delta^{ij}\, dt$	0
dt	0	0

(1.14)

which are true in probability $(P^{(W)})$. This will be derived presently. Because of the first entry, in taking differentials of functions of Brownian motion we have to go to the second order in the Taylor expansion. Thus, if $f(x_1,\dots,x_N)$ is a twice differentiable function on $\mathbb{R}^N$, with bounded derivatives,

$$df(\beta_t) = \sum_i \frac{\partial f(\beta_t)}{\partial x^i}\, d\beta_t^i + \frac{1}{2} \sum_{i,j} \frac{\partial^2 f}{\partial x^i \partial x^j}(\beta_t)\, d\beta_t^i\, d\beta_t^j$$

$$= \sum_i \frac{\partial f}{\partial x^i}(\beta_t)\, d\beta_t^i + \frac{1}{2} \sum_i \frac{\partial^2 f}{\partial x^{i\,2}}(\beta_t)\, dt$$

where we have used the Ito multiplication table.

Let

$$f' = \left(\frac{\partial f}{\partial x^1}, \dots, \frac{\partial f}{\partial x^N}\right) \quad \text{a vector in } \mathbb{R}^N$$

then we have, with Δ = Euclidean Laplacian in $\mathbb{R}^N$,

$$df(\beta_t) = (f'(\beta_t), d\beta_t) + \frac{1}{2}\,\Delta f\,(\beta_t)\,dt$$

or in integral form (1.15)

$$f(\beta_t) = f(\beta_o) + \int_o^t (f'(\beta_s), d\beta_s) + \int_o^t ds\,\frac{1}{2}\,\Delta f(\beta_s)$$

(1.15) is a special case of Ito's formula, and makes sense in $L^2(P^{(W)},\Omega)$.

We now state Ito's formula in general form.

For j=1, ..., m, let each $\sigma^j(x)$ a vector in $\mathbb{R}^N$, and $b^j(x)$ a scalar. x_s are random variables in $\mathbb{R}^N$ independent of $\beta_t - \beta_s$, $t > s$. Assume σ^j, b^j are such that

$$z_t^j = \int_o^t (\sigma^j(x_s), d\beta_s) + \int_o^t ds\; b^j(x_s)$$

or (1.16)

$$dz_t^j = (\sigma^j(x_t), d\beta_t) + b^j(x_t)\;dt$$

well defined (previous subsection) in $L^2(\Omega, P^{(W)})$.

Then if $f(z^1, \ldots, z^m)$ is a twice differentiable function with bounded derivatives in $\mathbb{R}^m$, then

$$df(z_t^1, \ldots, z_t^m) = \sum_{j=1}^m \frac{\partial f}{\partial z^j}(z_t^1, \ldots, z_t^m)dz_t^j + \frac{1}{2}\sum_{j,k=1}^m \frac{\partial^2 f}{\partial z^j \partial z^k}(z_t^1, \ldots, z_t^m)dz_t^j\,dz_t^k .$$

Using (1.16) and the Ito multiplication table (1.14)

$$dz_t^j\,dz_t^k = (\sigma^j(x_t), \sigma^k(x_t))\;dt$$

and hence

$$df(z_t^1,\dots,z_t^m) = \sum_{j=1}^m \frac{\partial f}{\partial z^j}(z_t^1,\dots,z_t^m)dz_t^j + \frac{1}{2}\sum_{j,k=1}^m \frac{\partial^2 f}{\partial z^j \partial z^k}(z_t^1,\dots,z_t^m) .$$

$$.(\sigma^j(x_t),\sigma^k x_t))dt . \qquad (1.17)$$

(1.16) -(1.17) gives the general case of Ito's formula.

A direct proof of 1.15, 1,17 would justify the Ito rules (1.14). We will prove (1.15) following [2], the proof of the general case 1.17 being similar.

Proof of (1.15)

We first make a preliminary observation revealing important properties of Brownian motion.

Let

$$V_{n,\alpha}(\beta) = \sum_{k=1}^{2^n} \left\| \beta_{\frac{kt}{2^n}} - \beta_{\frac{(k-1)t}{2^n}} \right\|^\alpha , \quad \alpha > 0 \qquad (1.18)$$

Then

$$E^{(W)}(V_{n,\alpha}(\beta)) = \sum_{k=1}^{2^n} E^{(W)}\left(\left\| \beta_{\frac{kt}{2^n}} - \beta_{\frac{(k-1)t}{2^n}} \right\| \right)^\alpha$$

$$= \sum_{k=1}^{2^n} \frac{1}{(2\pi t/2^n)^{\frac{N}{2}}} \int_{\mathbb{R}^N} d^N y \, \|y\|^\alpha \, e^{-\frac{y^2}{2\, t/2^n}}$$

$$= \left(\frac{t}{2^n}\right)^{\frac{\alpha}{2}} 2^n \, C_\alpha$$

where

$$C_\alpha = \int_{\mathbb{R}^N} d^N y \, \|y\|^\alpha \, e^{-\frac{\|y\|^2}{2}}$$

Hence

$$E^{(W)}(V_{n,\alpha}(\beta)) = C_\alpha\, t^{\alpha/2}\, 2^{n(1-\frac{\alpha}{2})} \xrightarrow[n\to\infty]{} \begin{cases} \infty, & \alpha < 2 \\ Nt, & \alpha = 2 \\ 0, & \alpha > 2 \end{cases}$$

With a little bit more work one can show that actually,

$$V_{n,\alpha}(\beta) \xrightarrow[p^{(W)}\,a.e]{} \begin{cases} \infty, & \alpha < 2 \\ Nt, & \alpha = 2 \\ 0, & \alpha > 2 \end{cases} \qquad (1.19)$$

<u>The proof of this assertion is relegated to Appendix 3 of Chapter I.</u>

The proof of (1.15) is now very simple. We assume that f is twice differentiable with bounded derivatives.

Now

$$f(\beta_t) - f(\beta_o) = \sum_{k=1}^{2^n} [\, f(\beta_{\frac{kt}{2^n}}) - f(\beta_{\frac{(k-1)t}{2^n}})\,]$$

$$= \sum_{k=1}^{2^n} [\, f(\beta_{\frac{kt}{2^n}}) - f(\beta_{\frac{(k-1)t}{2^n}}) - (\nabla f(\beta_{\frac{(k-1)t}{2^n}}), (\beta_{\frac{kt}{2^n}} - \beta_{\frac{(k-1)t}{2^n}})) -$$

$$-\frac{1}{2}\sum_{i,j}^{N} \frac{\partial^2 f}{\partial x^i \partial x^j}(\beta_{\frac{k-1)t}{2^n}})(\beta^i_{\frac{kt}{2^n}} - \beta^i_{\frac{(k-1)t}{2^n}})(\beta^j_{\frac{kt}{2^n}} - \beta^j_{\frac{(k-1)t}{2^n}})] \qquad \text{I}$$

$$+ \sum_{k=1}^{2^n} (\nabla f(\beta_{\frac{(k-1)t}{2^n}}), (\beta_{\frac{kt}{2^n}} - \beta_{\frac{(k-1)t}{2^n}})) \qquad \text{II}$$

$$+\frac{1}{2}\sum_{i,j=1}^{N} \frac{\partial^2 f}{\partial x^i\, \partial x^j}(\beta_{\frac{(k-1)t}{2^n}})(\beta^i_{\frac{kt}{2^n}} - \beta^i_{\frac{(k-1)t}{2^n}})(\beta^j_{\frac{kt}{2^n}} - \beta^j_{\frac{(k-1)t}{2^n}}) \qquad \text{III}$$

We now discuss the sums I, II, III separately. Because f is twice differentiable with bounded derivatives, we have

$$\left| f(x)-f(y)-(\nabla f(y), x-y) - \frac{1}{2} \frac{\partial^2 f(y)}{\partial x^i \partial x^j} (x^i - y^i)(x^j - y^j) \right|$$

$$\leq C \; \|x-y\|^3$$

Applying this to each term in the sum I, we have,

$$|I| \leq C \sum_{k=1}^{2^n} \left\| \beta_{\frac{kt}{2^n}} - \beta_{\frac{(k-1)t}{2^n}} \right\|^3 \underset{n \to \infty}{\longrightarrow} 0, \; P^{(W)} \text{ a.e} \qquad (1.21)$$

by (1.19) (case $\alpha > 2$).

On the other hand, by the very definition of the Ito integral

$$II \underset{n \to \infty}{\longrightarrow} \int_o^t (\nabla f(\beta_s), d\beta_s), \; P_W \text{ a.e.} \qquad (1.22)$$

Finally, by the analogue of case $(\alpha = 2)$ of (1.19), for components,

$$(\beta^i_{\frac{kt}{2^n}} - \beta^i_{\frac{(k-1)t}{2^n}})(\beta^j_{\frac{kt}{2^n}} - \beta^j_{\frac{(k-1)t}{2^n}}) \underset{n \to \infty}{\longrightarrow} \delta^{ij} \frac{t}{2^n}, \; P^{(W)} \text{ a.e.} \qquad (1.23)$$

Hence

$$III \underset{n \to \infty}{\longrightarrow} \frac{1}{2} \int_o^t ds \, (\Delta f)(\beta_s) \quad P^{(W)} \text{ a.e.}$$

Hence, taking $n \longrightarrow \infty$ in (1.20) we obtain the Ito formula (1.15).

End of proof.

The general case of the Ito formula, namely (1.17), can be proved by very similar methods. Finally we note that the boundedness on the p th derivatives $f^{(p)}(x), p \leq 2$ can be dropped. It suffices that $f^{(p)}, p \leq 2$, is in $L^2(\mathbb{R}^N)$ and $\mathbb{E}^{(W)} \int_0^t ds \, \|f^{(p)}(\chi_s)\|^2 < \infty$, for the validity of the Ito formula.

§ 1.4 Applications of the stochastic calculus

To get a feeling for the Ito calculus we first consider the following classical example:

Example 1 (Feynman -Kac formula)

Let V be a continuous non-negative function on $\mathbb{R}^N$.

Define the semi group T_t on $L^2(\mathbb{R}^N)$ by

$$(T_t f)(x) = E_x^{(W)} \left(f(\beta_t) \, e^{-\int_0^t ds \, V(\beta_s)} \right) \tag{1.24}$$

which is well defined. Restrict now to f twice differentiable with bounded derivatives. Then an easy application of the Ito calculus gives:

$$\frac{d}{dt}\bigg|_{t=0} T_t \, f(x) = \left(\frac{1}{2} \Delta - V(x)\right) f(x) \tag{1.25}$$

i.e. we can write $T_t = \exp - t H$

$$H = -\frac{1}{2} \Delta + V \tag{1.26}$$

Remark: (1.24) can be extended to a much wider class of potentials, V. See e.g [2] and references there in).

Hint: By the Ito calculus

$$d \left(e^{-\int_0^t ds \, V(\beta_s)} \right) = - \left(e^{-\int_0^t ds \, V(\beta_s)} \right) V(\beta_t) \, dt$$

$$d\, f(\beta_t) = (f'(\beta_t), d\beta_t) + \frac{1}{2}(\Delta f)(\beta_t)\, dt.$$

Now use the Ito formula (1.17) choosing, $z_t^{(1)} = e^{-\int_0^t ds\, V(\beta_s)}$

$z_t^{(2)} = f(\beta_t)$, and $F(z_t^1, z_t^2) = z_t^{(1)} z_t^{(2)}$ to obtain

$$d(T_t f)(x) = E_x^{(W)}\left(df(\beta_t)\, e^{-\int_0^t ds\, V(\beta_s)} - f(\beta_t)\, V(\beta_t)\, e^{-\int_0^t ds\, V(\beta_s)} dt\right).$$

Plugging in the expression for $df(\beta_t)$ and noting that $d\beta_t$ is independent of the past of β_t we obtain

$$d(T_t f)(x) = E_x^{(W)}\left(e^{-\int_0^t ds\, V(\beta_s)} \frac{1}{2}(\Delta f)(\beta_t)\, dt - V(\beta_t)\, e^{-\int_0^t ds\, V(\beta_s)} f(\beta_t)\, dt\right)$$

whence the result follows (given that β_t starts at x).

<u>The next example will have important applications in the theory of stochastic differential equations (Langevin equations) of Chapter II.</u>

<u>Example 2</u> : <u>Cameron - Martin - formula</u>

Let $b(x)$, $x \in \mathbb{R}^N$, be a bounded continuous vector valued function,

$$\sup_{x \in \mathbb{R}^N} \| b(x) \| \leq M. \tag{1.27}$$

(Weaker assumptions are discussed later)

Consider the random variable,

$$\xi_0^t(\beta) \equiv \int_0^t (b(\beta_s), d\beta_s) - \frac{1}{2} \int_0^t ds\, \| b(\beta_s) \|^2 \tag{1.28}$$

or

$$d\xi_o^t = (b(\beta_t), d\beta_t) - \frac{1}{2} \| b(\beta_t) \|^2 \, dt . \quad (1.29)$$

Note that because of (1.27)

$$E^{(W)} (e^{p\xi_o^t}) \leq e^{pMNt} , \; 1 \leq p < \infty \quad (1.30)$$

(The proof is given in appendix 1 to this chapter).

From (1.29) we have by the Ito calculus

$$d\xi_o^t d\xi_o^t = \| b(\beta_t) \|^2 \, dt \quad (1.31)$$

and

$$d(e^{\xi_o^t}) = e^{\xi_o^t} d\xi_o^t + \frac{1}{2} e^{\xi_o^t} d\xi_o^t d\xi_o^t$$

whence using (1.29), (1.31)

$$d(e^{\xi_o^t}) = e^{\xi_o^t}\Big((b(\beta_t), d\beta_t) - \frac{1}{2} \| b(\beta_t) \|^2 \, dt\Big)$$

$$+ \frac{1}{2} e^{\xi_o^t} \| b(\beta_t) \|^2 \, dt$$

and so:

$$d(e^{\xi_o^t}) = e^{\xi_o^t} (b(\beta_t), d\beta_t) \quad (1.32)$$

$$e^{\xi_o^t} = e^{\xi_o^s} + \int_s^t e^{\xi_o^{s'}} (b(\beta_{s'}), d\beta_{s'}) \quad (1.33)$$

and, in particular,

$$= 1 + \int_o^t e^{\xi_o^s} (b(\beta_s), d\beta_s) \quad (1.34)$$

Because of (1.27) and (1.30), the stochastic integral in (1.34) is well defined (mean square integrable) and its mean (expectation) vanishes. Hence:

$$E^{(W)}\left(e^{\xi_o^t} \right) = 1. \tag{1.35}$$

Because of (1.30) we can define the semi-group on $L^2(\mathbb{R}^N)$ (Cameron-Martin or "Langevin" semigroup) T_t by:

$$(T_t f)(x) = E_x^{(W)}\left(f(\beta_t)\, e^{\xi_o^t}\right) \tag{1.36}$$

with ξ_o^t defined by (1.28).

This semigroup is important in the description of time evolution of solutions of the Langevin stochastic differential equation (chapter II).

Let us compute its "differential" generator. For this purpose we restrict f to be a twice differentiable function which with its derivatives are bounded.

Apply the Ito calculus.

$$d\,(T_t f)\,(x) = E_x^{(W)}\left(d(f(\beta_t)\, e^{\xi_o^t})\right).$$

Now use,

$$df = (f'(\beta_t),\, d\beta_t) + \frac{1}{2}\,(\Delta f)\,(\beta_t)\,dt$$

together with (1.32)

$$d\left(e^{\xi_o^t} \right) = (b(\beta_t),\, d\,\beta_t)\; e^{\xi_o^t}$$

and the Ito calculus to obtain

$$d\,(f(\beta_t)\, e^{\xi_o^t}) = df(\beta_t).\; e^{\xi_o^t} + f(\beta_t)\, d\,(e^{\xi_o^t}) + e^{\xi_o^t}(b(\beta_t), f'(\beta_t))\,dt$$

$$= [(f'(\beta_t), d\beta_t) + \frac{1}{2} (\Delta f)(\beta_t)\, dt] e^{\xi_o^t}$$

$$+ f(\beta_t)(b(\beta_t), d\beta_t)\, e^{\xi_o^t} + e^{\xi_o^t} (b(\beta_t), f'(\beta_t))\, dt.$$

Hence

$$d\,(T_t f.)(x) = E_x^{(W)} \left(dt\, e^{\xi_o^t} \left(\frac{1}{2} (\Delta f)(\beta_t) + (b(\beta_t), f'(\beta_t)) \right)\right)$$

where we have used the independence of the Ito stochastic differentials from the past.

Hence

$$\left.\frac{d}{dt}\right|_{t=0} (T_t f)(x) = \left(\frac{1}{2} \Delta f(x) + \sum_i b^i(x) \frac{\partial f(x)}{\partial x^i} \right). \qquad (1.37)$$

Thus we have shown:

$$(e^{tL} f)(x) = E_x^{(W)} \left(f(\beta_t)\, e^{\int_o^t (b(\beta_s), d\beta_s) - \frac{1}{2} \int_o^t \| b(\beta_s) \|^2 ds} \right)$$

where

$$= E_x^{(W)} \left(f(\beta_t)\, e^{\xi_o^t(\beta)} \right) \qquad (1.38)$$

$$L = \frac{1}{2} \Delta + (b(x), \nabla\cdot)$$

∇ being the grad operator in $\mathbb{R}^N$.

The above is known as the Cameron-Martin formula; L may be called the Langevin operator, b is the "drift" vector.

<u>A Markov process.</u>

Let now $A \subset \mathbb{R}^N$, a (Borel) subset and χ_A the characteristic function of A.

$$\chi_A(x) = \begin{cases} 1, & x \in A \\ 0 & x \notin A. \end{cases}$$

Define,

$$p_t(x,A) = (e^{tL} \chi_A)(x) = E_x^{(W)}(\chi\{\beta_t \in A\}\ e^{\xi_o^t(\beta)}). \qquad (1.39)$$

Then $p_t(x,.)$ is a probability measure on $\mathbb{R}^N$, because of the above formula, and because $e^{tL}\ 1=1$ (since $1 = \chi_{\mathbb{R}^N}$) because of (1.35). Because e^{tL} is a semigroup, $p_t(x, dy)$ satisfies the Chapman-Kolmogoroff equation (1.3). Then, as in the case of Brownian motion explained earlier, by first defining joint probability distributions as in (1.5) using (1.39) and then verifying (1.8) (by an elementary calculation) we arrive at probability measures P_x on the path space $\Omega = C^o([0,\infty), \mathbb{R}^N)$, satisfying the Markov property. We will see later that such probability measures are associated with solutions of the Langevin equation.

Remark: Weaker assumptions on the drift.

In example (2) we assumed that the drift vector b was bounded and continuous. Continuity can be relaxed, and we may assume that b is bounded and measurable (replace (1.27) by

$$\|b\|_\infty = \text{ess.}\sup_{x \in \mathbb{R}^N} \|b(x)\| < M). \qquad (1.40)$$

All the previous considerations then go through.

Boundedness is however often too strong an assumption. Let us only assume b is only square integrable and

$$E_x^{(W)} \int_0^T ds \, \|b(\beta_s)\|^2 < \infty .$$

Then ξ_0^t of (1.28) is a well defined random variable. In that case, we can find a sequence $\{b_n\}$ of bounded continuous drifts converging to b, and $\xi_0^{t,n} \longrightarrow \xi_0^t$, almost surely. Then, by Fatou's lemma,

$$E^{(W)}(e^{\xi_0^t}) \leq \lim_{n \to \infty} (e^{\xi_0^{t,n}}) = 1, \text{ by the above.}$$

In this case also,

$$(T_t f.)(x) = E_x(f.(\beta_t)\, e^{\xi_0^t})$$

is well defined provided f is bounded. However the important property

$$T_t 1 = 1$$

is not guaranteed. We only have $T_t 1 < 1$. Additional considerations are now necessary (see later) to check that $T_t 1 = 1$.

Appendix 1 to Chapter 1.

We give the proof of (1.30) which was relegated here. Indeed, if we call I_t the stochastic integral in (1.28), then

$$I_t^{(n)} = \sum_{j=1}^{2^n} \left(\left(\beta_{\frac{jt}{2^n}} - \beta_{\frac{(j-1)t}{2^n}}\right),\ b\left(\beta_{\frac{(j-1)t}{2^n}}\right) \right) \xrightarrow[P^W \text{a.e}]{} I_t$$

as shown earlier.

Now

$$E^{(W)}(e^{p\xi_0^t}) \leq E^{(W)}(e^{pI_t})$$

and (by Fatou's lemma)

$$E^{(W)}(e^{pI_t}) \leq \lim_{n \to \infty} (e^{pI_t^{(n)}})$$

$$\leq \lim_{n \to \infty} E^{(W)} \prod_{j=1}^{2^n} e^{pM \left\| \beta_{\frac{jt}{2^n}} - \beta_{\frac{(j-1)t}{2^n}} \right\|} \quad \text{(have used 1.27)}$$

$$\leq \lim_{n \to \infty} \left(\prod_{j=1}^{2^n} E^{W} \left(e^{2^n pM \left\| \beta_{\frac{jt}{2^n}} - \beta_{\frac{(j-1)t}{2^n}} \right\|}\right)^{\frac{1}{2^n}} \right)$$

(we have used Holder's inequality)

and hence

$$E^{(W)}(e^{pI_t}) \leq \lim_{n \to \infty} \left(\prod_{j=1}^{2^n} e^{2^n pM \, E^{(W)} \left\| \beta_{\frac{jt}{2^n}} - \beta_{\frac{(j-1)t}{2^n}} \right\|^{2 \cdot \frac{1}{2^n}}} \right) .$$

Since
$$E^{(W)} \left\| \beta_{\frac{jt}{2^n}} - \beta_{\frac{(j-1)t}{2^n}} \right\|^2 = N \frac{t}{2^n}$$

we obtain straightaway

$$E^{(W)} (e^{pI_t}) \leq e^{pMNt}$$

which is (1.30)

Appendix 2: Proof of almost sure continuity of stochastic integral I_t

We take up after equation (1.11) and give the promised proof.

To this end we restrict t to the compact interval $[0,T]$ and show that the $L^2(P^W,\Omega)$ convergence of the sequence $\{I_t^{(n)}\}$ can be replaced by almost sure (i.e. $P^{(W)}$ a.e) uniform convergence. Infact, because $I_t^{(n)} - I_s^{(n)}$ is independent of β_s $(t>s)$, we have the inequality ("martingale inequality", see e.g [1,3]):

$$P^{(W)}\left\{ \sup_{0\le t\le T} \left| I_t^{(n)} - I_t^{(n-1)} \right| > \frac{1}{n^2} \right.$$

$$\le n^4 E^{(W)} \left| I_t^{(n)} - I_t^{(n-1)} \right|^2 \le C n^4 2^{-n} .$$

Hence

$$\sum_{n=1}^{\infty} P^{(W)} \{ \sup_{0\le t\le T} \left| I_t^{(n)} - I_t^{(n-1)} \right| > \frac{1}{n^2} \} < \infty$$

and hence for $n \ge n_o$ sufficiently large we have (by the Borel-Cantelli-lemma*):

$$\sup_{0\le t\le T} \left| I_t^{(n)} - I_t^{(n-1)} \right| \le \frac{1}{n^2} , \quad P_W \text{ a.e.}$$

Hence $\{ I_t^{(n)} \}$ converges uniformly almost surely, and thus the limit I_t is continuous, P_W a.e.

(*) Remark. We have used the Borel- Cantelli lemma which goes as follows. Let $\{E_n\}$ be an infinite sequence of measurable sets (events) in a probability space, P a probability measure. Define the set $E_{i.o} = \{ w \mid w$ belongs to an infinite number of $E_n\}$. The Borel Cantelli lemma states that if $\sum_n P(E_n) < \infty$, then $P(E_{i.o}) = 0$, (and so the complementary event $\tilde{E}_{io}$ takes place almost surely).

Proof: Let χ_E be characteristic function of event E. Then $\sum_n \chi_{E_n}(w) = \infty$, $w \in E_{i.o}$. But $\int (\sum_n \chi_{E_n}) dP = \sum_n P(E_n) < \infty$, so that $\sum_n \chi_{E_n} < \infty$, P a.e. The lemma follows.

Appendix 3 to chapter I (Proof of (1.19)).

In fact, the case $\alpha = 1$ was already shown when we checked that Brownian paths are almost surely of unbounded variation. Next take the case $\alpha = 2$. By the previous calculation,

$$E^{(W)}(V_{n,2}) = C_2 t, \quad C_2 = N$$

and it is easy to check using imdependence of forward Brownian increments with respect to the past that

$$E^{(W)}(|V_{n,2}|^2) = E^{(W)}(V_{n,4}) = C_4 t^2 2^{-n}.$$

Hence

$$E^{(W)}|V_{n,2} - C_2 t|^2 = E^{(W)}|V_{n,2} - E^{(W)}(V_{n,2})|^2$$

$$\leq E^{(W)}|V_{n,2}|^2 = E^{(W)}|V_{n,4}| = C_4 t^2 2^{-n}.$$

Hence, by the Tchehycheff inequality,

$$P^{(W)}(|V_{n,2} - C_2 t| > \frac{1}{n^2}) \leq n^4 E^{(W)}|V_{n,2} - C_2 t|^2 = C_4 t^2 n^4 2^{-n}.$$

Hence,

$$\sum_{n=1}^{\infty} P^{(W)} \left(\left| V_{n,2} - C_2 t \right| > \frac{1}{n^2} \right) < \infty$$

and so for $n \geq n_o$ large enough, by the Borel-Cantelli lemma, we have,

$$\left| V_{n,2} - C_2 t \right| < \frac{1}{n^2} \text{ , } P^{(W)} \text{ a.e}$$

Thus $V_{n,2} \xrightarrow[n \to \infty]{} Nt$, almost surely.

The last case $\alpha > 2$, can be proved similarly.

CHAPTER II

STOCHASTIC DIFFERENTIAL EQUATIONS IN $\mathbb{R}^N$

§ 2.1 Introduction: In the previous chapter we developed the Ito stochastic calculus in $\mathbb{R}^N$. In this chapter this calculus will be used in the study of a class of stochastic differential equations in $\mathbb{R}^N$. These equations may be thought of as finite dimensional approximations to those arising in stochastic field quantization. ('N' can be thought of as the "number of modes", an ultraviolet cutoff. Smoother UV cutoffs are easily introduced, in which case we remain in ∞ dimensions, but the relevant "soft" theory is easily modelled on that of this chapter. The hard part, namely renormalized stochastic differential equations in infinite dimensions, will be of paramount concern in subsequent chapters).

We consider the stochastic differential equation (S.D.E) in $\mathbb{R}^N$ (Langevin equation):

$$dx_t = d\beta_t + b(x_t)\, dt \tag{2.1}$$

$$x_o = x$$

which can also be written in the integral form

$$x_t = \beta_t + x + \int_o^t ds\ b(x_s)\, ds \tag{2.2}$$

where $b(x)$ is a vector valued function in $\mathbb{R}^N$ and β_t is standard Brownian motion in $\mathbb{R}^N$, starting at the origin.

One can also replace β_t by W_t, a Brownian motion with covariance

$$E^{(W)}(W_t^i \, W_s^j) = \min\ (t,s)\ K^{ij} \tag{2.3}$$

and K is a (constant) positive symmetric matrix.

We can write

$$W_t = \sigma \beta_t \tag{2.4}$$

$$K = \sigma \sigma^*$$

and replace (2.1) by:

$$dx_t = \sigma \beta_t + b(x_t)\, dt \tag{2.5}$$

b is called the drift vector, and K the (constant) diffusion matrix.
The case where σ (in 2.5) is variable (matrix valued function in $\mathbb{R}^N$) arises in the study of S.D.E on curved Riemannian manifolds, in which case it is a (local) N- bein (for the relevant theory see [3]).
We stick to the case σ is constant, and in fact stick with (2.1) because the necessary changes are trivial (for the purpose of this chapter). These changes will however be important in subsequent chapters, when we pass to infinite dimensions.

In this chapter we are concerned with solving (2.1) with very mild conditions on the drift. A relevant notion is that of the "weak" solution, to be explained later. However we shall proceed by adopting rather stringent conditions which will be progressively weakened.

§ 2.2 Bounded, Lipshitz continuous drift, strong solution.

We solve (2.1) (2.2) under the assumption

$$\| b(x) - b(y) \| \leq M \| x - y \| , \quad \forall\, x, y \in \mathbb{R}^N . \tag{2.5}$$

In this case the method of successive iterations converges.

Define:

$$x_t^{(n)} = \beta_t + \int_0^t b(x_s^{(n-1)})\, ds, \quad n \geq 1 \tag{2.6}$$

$$x_t^{(o)} = \beta_t .$$

Without loss of generality x_t starts at the origin.

Then,

$$x_t^{(n)} - x_t^{(n-1)} = \int_0^t (b(x_s^{(n-1)}) - b(x_s^{(n-2)}))\, ds$$

and hence, using (2.5),

$$\| x_t^{(n)} - x_t^{(n-1)} \| \leq M \int_0^t \| x_s^{(n-1)} - x_s^{(n-2)} \| \, ds$$

whence

$$\sup_{0 \leq t \leq T} \| x_t^{(n)} - x_t^{(n-1)} \| \leq M \int_0^T \| x_s^{(n-1)} - x_s^{(n-2)} \| \, ds$$

and, iterating on this,

$$\sup_{0 \leq t \leq T} \| x_t^{(n)} - x_t^{(n-1)} \| \leq M^n \int_0^T ds_1 \int_0^{s_1} ds_2 \ldots \int_0^{s_{n-1}} ds_n \, \| \beta_{s_n} \| .$$

Hence

$$E^{(W)}\left(\sup_{0\le t\le T}\|x_t^{(n)}-x_t^{(n-1)}\|\right)\le M^n\int_0^T ds_1\int_0^{s_1} ds_2\ldots\int_0^{s_{n-1}} ds_n\, E^{(W)}\|\beta_{s_n}\|^2$$

Using,

$$E^{(W)}\|\beta_{s_n}\|^2 = N s_n ,$$

we obtain

$$E^{(W)}\left(\sup_{0\le t\le T}\|x_t^{(n)}-x_t^{(n-1)}\|\right)\le NM^n\frac{T^n}{n!} . \quad (2.7)$$

Finally, using Tchebycheff's inequality, and (2.7)

$$P^{(W)}\left(\sup_{0\le t\le T}\|x_t^{(n)}-x_t^{(n-1)}\| > \frac{1}{2^n}\right)\le 2^n NM^n\frac{T^n}{n!} . \quad (2.8)$$

Now

$$\sum_{n=1}^{\infty} P^{(W)}\left(\sup_{0\le t\le T}\|x_t^{(n)}-x_t^{(n-1)}\| > \frac{1}{2^n}\right) < \infty .$$

Hence by Borel-Cantelli, for $n \ge n_o$ suff. large

$$\sup_{0\le t\le T}\|x_t^{(n)}-x_t^{(n-1)}\|\le\frac{1}{2^n} , \quad P^{(W)} \text{ a.e.}$$

Thus the sequence $\{x_t^{(n)}\}$ converges uniformly in $[0,T]$ almost surely $(P^{(W)})$, and the limit x_t is a continuous process satisfying (2.2). By construction x_t is a measurable function $x_t(\beta)$ of Brownian paths β_s, $0 \le s \le t$.

It is easy to check that the solution x_t is unique, and hence, by a standard argument [1], the process is Markovian. Namely, if we define the operator T_t on bounded measurable functions in $\mathbb{R}^N$ by

$$(T_t f)(x) = E_x^{(W)} (f(x_t(\beta))) \tag{2.9}$$

then T_t satisfies the semigroup property.

Let f be a twice differentiable function with bounded derivatives. Since x_t satisfies (2.1), we obtain by the Ito calculus (chapter I),

$$df(x_t) = (\nabla f(x_t), dx_t) + \frac{1}{2} \frac{\partial^2 f}{\partial x^i \partial x^j} (x_t)\, dx_t^i\, dx_t^j$$

$$= (\nabla f(x_t), d\beta_t) + (\tfrac{1}{2}\Delta f + (b, \nabla f)(x_t))\, dt . \tag{2.10}$$

Whence from (2.9), using the fact that $d\beta_t$ is independent of x_t,

$$d(T_t f)(x) = E_x^{(W)} (\tfrac{1}{2}\Delta f(x_t) + (b, \nabla f)(x_t))\, dt.$$

Hence

$$\frac{d}{dt}\bigg|_{t=0} (T_t f)(x) = Lf(x)$$

$$L = \frac{1}{2}\Delta + \sum_i b^i(x) \frac{\partial}{\partial x^i} \tag{2.11}$$

L is the infinitisimal generator on the above subspace of functions and coincides with that in (1.38) (§ 1.4, Ex 2). Hence we have the Cameron- Martin formula (chapter 1) giving an explicit representation for (2.9):

$$(e^{tL} f)(x) = E_x^{(W)} (f(x_t(\beta)))$$

$$= E_x^{(W)} (f(\beta_t)\, e^{\int_0^t (b(\beta_s), d\beta_s) - \frac{1}{2}\int_0^t \|b(\beta_s)\|^2 ds}) \tag{2.12}$$

As shown earlier (§ 1.4, Ex 2), we also have

$$e^{tL} 1 = 1$$

which checks with the validity of (2.12), since $P_x^{(W)}$ is probability measure.

§ 2.3 Smooth, non-bounded drifts. We now turn to the case when the drift b in (2.1 - 2.2) is smooth (C^∞) but unbounded, so that condition (2.5) is not satisfied globally. But the drift being smooth, (2.5) is satisfied locally.

Let

$$K_N = \{x \mid \|x\| \leq R_N\}, \quad R_N < R_{N+1} < \cdots$$

with $R_N \longrightarrow \infty$, $N \longrightarrow \infty$. Then

$$K_N \subset K_{N+1} \subset \cdots$$

is an increasing sequence of compacts. The drift being C^∞, it is easy to show that for every N, we can find a smooth b_N such that

(i) $b_N = b$ on K_N

(ii) support $b_N \subset K_{N+1}$.

As a consequence for each N, b_N satisfies (2.5).

Hence, by § 2.2, for each N the SDE

$$dx_t = d\beta_t + b_N(x_t)\, dt, \quad x_o = x \tag{2.13}$$

has a unique solution $x_t^{(N)}$, defined for all $t > 0$, which is a continuous Markov process.

Let τ_N be the first exit time of $x_t^{(N)}$ from K_N. Namely

$$\tau_N = \inf. \{ t \geq 0 \mid \| x_t^{(N)} \| > R_N \} \tag{2.14}$$

τ_N is a random variable, and $\{ \tau_N \}$ is an increasing sequence of exit times ("stopping times).

Because $b_N = b_{N+1}$ on K_N, it can be shown* that

$$x_t^{(N)} = x_t^{(N+1)} \quad , t < \tau_N .$$

(By this we mean, $x_t^{(N)}(\beta) = x_t^{(N+1)}(\beta)$

for $t < \tau_N(\beta) = \inf. \{ t \geq 0 \mid x_t^{(N)}(\beta) > R_N \}$.

Let

$$\tau \equiv \sup_N \tau_N . \tag{2.15}$$

Define a Markov process x_t, $t < \tau$ by requiring

$$x_t = x_t^{(N)} \quad \text{for} \quad t < \tau_N . \tag{2.16}$$

We say that x_t is non-explosive if

---

* This is because each $x_t^{(N)}$ satisfies the "strong Markov property" [1] which in turn follows from the fact that $e^{tL^{(N)}}$ transforms bounded continuous functions into bounded continuous functions.

$$P^{(W)}\{\tau=\infty\}=1 \qquad (2.17)$$

When x_t is non-explosive, it is defined for all times. For a non-explosive process, the representation (2.12) remains valid with L given by (2.11). Moreover,

$$e^{tL}\, 1 = 1.$$

We now give a useful criterion, due to Khasminski [4], for the absence of explosion based on the technique of "Liapounov functions".

Criterion for the absence of explosion :

Let x_t be the drift process of this section. Let V be a non-negative, twice differentiable function on $\mathbb{R}^N$ such that

(i) $V_R = \inf\limits_{\|x\| > R} V(x) \longrightarrow \infty \quad R \longrightarrow \infty$

(ii) there exists $c > 0$ such that (2.18)

$$LV \leq cV$$

Then the process x_t is non- explosive.

The above criterion is very useful, and some examples of its use will be found later. The proof, which is very simple and involves the use of the Ito calculus, is given in Appendix II A.

§ 2.4 Weak solutions: In the two previous subsections the stochastic differential equation (2.1) was studied under very restrictive conditions on the drift vector b(x), viz b(x) is either globally Lipshitz continuous (§ 2.2) or smooth (§ 2.3). (In the latter case under the assumption of no-explosion). In both cases equation (2.1) has a unique solution x_t which is a continuous Markov process with differential generator,

$$L = \frac{1}{2} \Delta + (b(x), \nabla \cdot)$$

Moreover x_t is a measurable function of β_s, $0 \leq s \leq t$. Such a solution is called a strong solution.

In this subsection we will relax considerably the conditions on the drift $b(x)$. We will merely assume that b is measurable and satisfies, for any $T > 0$,

$$E^{(W)} \int_0^T ds \, \| b(\beta_s) \|^{2(1+\epsilon)} < \infty \qquad (2.19)$$

for some $\epsilon > 0$, and

$$E^{(W)} \left(e^{2(1+\frac{1}{\epsilon}) \xi_0^T (\beta)} \right) < \infty \qquad (2.20)$$

where $\xi_0^t(\beta)$ was defined in (1.28). These two conditions are often satisfied in practice, even when the drift b has no continuity properties, and is not bounded.

In the present situation we will no longer obtain in general a strong solution. Instead we will obtain a weak solution. By this we mean the following:

Rewrite (2.2) as:

$$x_t - \int_0^t ds \, b(x_s) \, ds = \beta_t^{(x)} \qquad (2.21)$$

where $\beta_t^{(x)}$ is Brownian motion starting at x. Then a <u>weak solution</u> of the non-linear equation (2.21) is a Markovian* family of measures P_x on $\Omega_T = C^o([0,T), \mathbb{R}^N)$ such that the joint probability distributions under P_x of the random variable

$$\tilde{Z}_t = x_t - \int_o^t ds\ b(x_s)\, ds \qquad (2.22)$$

coincide with the known distributions of β_t^x, (in (2.22) above x_t is a path in Ω).

The key to the construction of the weak solution is the Cameron-Martin formula (1.38) of Example 2, § 1.4. This was proven under condition (1.27) and more generally under (1.40), (see the Remark), assuming that b is measurable and bounded. In this case, as explained after (1.38), we can construct a Markovian family of measures P_x on Ω. We will now relax the boundedness condition, and replace it by (2.19 - 2.20). Moreover we will <u>define</u> a semigroup T_t by (1.38), namely:

$$(T_t f)(x) = E_x^{(W)}(f(\beta_t)\, e^{\xi_o^t(\beta)}) \qquad (2.23)$$

with

$$\xi_o^t(\beta) = \int_o^t (b(\beta_s), d\beta_s) - \frac{1}{2}\int_o^t ds\ \|b(\beta_s)\|^2 . \qquad (2.24)$$

Because

---

* in other words $(T_t f)(x) = E^{(P_x)}(f(x_t))$, with x_t paths in Ω, defines a semigroup on bounded measurath functions.

$$E^{(W)}\left(\int_0^t ds\ \|b(\beta_s)\|^2\right) < \infty$$

ξ_0^t is a random variable. As explained in the Remark at the end of § 1.4, we have

$$E^{(W)}\left(e^{\xi_0^t}\right) \leq 1$$

so that (2.23) is well defined if f is a bounded measurable function However condition (2.20) guarantees a stronger result, namely

$$E^{(W)}\left(e^{\xi_0^t}\right) = 1 \tag{2.25}$$

<u>The proof is relegated to Appendix II B</u>

We can now construct a Markov process, starting from the transition probabilities (1.39) obtained from (2.23) <u>as explained after (1.38)</u>, § 1.4. Because of (2.25), these transition probabilities are countably additive probability measures, so that the Kolmogoroff construction can be applied. The resulting markovian family of measures P_x are supported on $\Omega = C^0([0,\infty), \mathbb{R}^N)$, as can bechecked by verifying* the Kolmogoroff criterion (1.8) (see footnote after 1.8). (see footnote after 1.8).

It remains to verify that (P_x, Ω_T) gives the desired <u>weak solution</u>, for any time interval $[0,T]$, in the sense explained earlier. Going back to (2.21) and (2.22) we have to verify that the P_x transition probability of $\tilde{Z}_t$ is identical to the Wiener transition probability $p_t(x, dy)$.

* This uses (2.23), the stability bound (2.20), and the fact that β_t satisfies (1.8). We omit the details.

Let f be a bounded measurable function. By definition,

$$\left.\begin{aligned} & E^{(\tilde{P}_x)}(f(\tilde{Z}_t)) = E_x^{(W)}(f(Z_t)\, e^{\xi_o^t}) \\ & \text{where} \\ & Z_t = \beta_t - \int_o^t ds\; b(\beta_s) \end{aligned}\right\} \tag{2.26}$$

Define on $\Omega_T = C^o([0,T], \mathbb{R}^N)$, the probability measure $\tilde{P}^{(W)}$ by

$$d\tilde{P}^{(W)} = e^{\xi_o^T}\, dP^{(W)} \tag{2.27}$$

Let $\tilde{E}^{(W)}$ stand for expectation with respect to $\tilde{P}^{(W)}$.

Define also,

$$\xi_s^t = \int_s^t (b(\beta_{s'}), d\beta_{s'}) - \frac{1}{2}\int_s^t ds' \left\| b(\beta_{s'}) \right\|^2 \tag{2.28 a}$$

Then following the steps leading to (1.34), but now starting from ξ_s^t we obtain by Ito's formula:

$$e^{\xi_s^t} = 1 + \int_s^t e^{\xi_s^{s'}} (b(\beta_{s'}), d\beta_{s'}) \,. \tag{2.28 b}$$

Because of (2.19), (2.20) the stochastic integral in (2.26b) is mean-square integrable*, and thus a standard Ito integral. As such it is independent of β_u, $u \leq s$. We thus have

* Take the expectation of the square of it, and then use Holder's inequality.

$$E^{(W)}\left(e^{\xi_s^t} \middle| \mathcal{F}_s\right) = 1 \qquad (2.28\text{ c})$$

where $\mathcal{F}_s$ is the least σ-subalgebra of events generated by $\beta_u, 0 \leq u \leq s$ and $E^{(W)}(.\,|\mathcal{F}_s)$ denotes conditional expectation*.

We then obtain,

$$\tilde{E}^{(W)}(f(Z_t)) = E^{(W)}\left(f(Z_t)\, e^{\xi_o^T}\right)$$

$$=E^{(W)}\left(f(Z_t) e^{\xi_o^t} e^{\xi_t^T}\right) = E^{(W)}\left(f(Z_t) e^{\xi_o^t} E^{(W)}\left(e^{\xi_t^T} \middle| \mathcal{F}_t\right)\right)$$

$$=E^{(W)}\left(f(Z_t)\, e^{\xi_o^t}\right) \qquad (*)$$

(we have used (2.28c) and standard properties of conditional expectations* as projectors). From (2.26) and (*), we have

$$E^{(P_x)}(f(\tilde{Z}_t)) = \tilde{E}_x^{(W)}(f.(Z_t)). \qquad (2.29)$$

We now use Girsanov's theorem [1], which states that for $0 \leq t < T$

$$Z_t = \beta_t - \int_o^t ds\; b(\beta_s)$$

is Brownian motion with respect to the probability measure $\tilde{P}^{(W)}$ (given by (2.24)) with the same covariance as that of β_t. Hence from (2.29) we have that the P_x transition probability of $\tilde{Z}_t$ is just that

---

* If $\mathcal{F}$ is the σ-algebra of events for $0 \leq t < \infty$, then $\mathcal{F}_s \subset \mathcal{F}$ and $L^2(\Omega, \mathcal{F}_s, P^W)$ is a closed subspace of $L^2(\Omega, \mathcal{F}, P^{(W)})$. Then recall that conditional expectation $E(.\,|\mathcal{F}_s)$ is just the projector onto this closed subspace.

of Brownian motion β_t. This concludes the proof that we have obtained a weak solution.

<u>Exercise</u> :

Define the semigroup $\tilde{T}_t$ by (2.26) :

$$(\tilde{T}_t f.)(x) = E_x^{(W)} (f(Z_t) e^{\xi_o^t})$$

and take f a twice differential function of compact support. Now use the Ito calculus to check that the differential generator of $\tilde{T}_t$ is $\frac{1}{2}\Delta$, which is that of Brownian motion.

§ 2.5 <u>Invariant Measures and Ergodicity</u> <u>A</u> Let (Ω, P_x) be a weak solution of the stochastic differential equation in $\mathbb{R}^N$ as in § 2.4:

$$dx_t = d\beta_t + b(x_t)\, dt \qquad (2.30)$$

$$x_o = x$$

where the measurable drift vector field b(x) satisfies (2.19 - 2.20). (P_x, Ω) is a continuous Markov process in $\mathbb{R}^N$ and the associated semigroup e^{tL} on bounded measurable functions:

$$(e^{tL} f)(x) = E_x^{(W)} (f(\beta_t) e^{\xi_o^t}) \qquad (2.31)$$

has (differential) generator,

$$L = \frac{1}{2}\Delta + (b(x), \nabla .) \qquad (2.32)$$

on $C_b^2(\mathbb{R}^N)$.

Suppose now that f is positive i.e. it is not strictly vanishing and $f \geq 0$, a.e. in $\mathbb{R}^N$ (a.e. with reference to Lebegue measure). Then since $e^{\xi_o^t} > 0$ a.e., (2.31) shows that

$$e^{tL} f > 0 , \text{ a.e} \tag{2.33}$$

We say that e^{tL} is <u>positivity improving.</u>

Let $p_t(x, dy)$ be the transition probability. Then it can be shown (see Appendix II C) that $p_t(x, dy)$ admits a density with respect to Lebesgue measure:

$$p_t(x , dy) = p_t(x, y) \, dy \tag{2.34}$$

where $p_t(x, y)$ is jointly measurable in x, y, non-negative and ofcourse the total mass is one.

Condition (2.33) now $\Rightarrow$

$$p_t(x, y) > 0, \quad \text{a.e.} \tag{2.35}$$

B. <u>Invariant measure</u>

We say that the semigroup e^{tL} admits an <u>invariant (probability) measure</u> μ, if for every Borel set $A \subseteq \mathbb{R}^N$,

$$\int_{\mathbb{R}^N} p_t (x, A) \, d\mu(x) = \mu(A). \tag{2.36}$$

It can be shown (Appendix II C) that μ admits a measurable density with respect to Lebesgue measure, i.e.

$$d\mu(x) = \rho(x) \, dx \tag{2.37}$$

with $\rho \geq 0$, measurable and

$$\int_{\mathbb{R}^N} \rho(x)\, dx = 1 . \tag{2.38}$$

From (2.36) and (2.37) we deduce

$$\int_{\mathbb{R}^N} p_t(x,y)\, \rho(x)\, dx = \rho(y), \quad \text{a.e.} \tag{2.39}$$

C Uniqueness

The positivity improving property (2.33) implies that whenever there is an invariant measure μ, it is unique. The proof is a' la Perron - Frobenius. For suppose that μ_i, i=1,2 are two invariant measures, and ρ_i, i=1,2 the respective densities. Define

$$f = \rho_1 - \rho_2 \tag{2.40}$$

which satisfies

$$\int_{\mathbb{R}^N} f(x)\, dx = 0 . \tag{2.41}$$

The positivity improving property will be shown to imply that f is of fixed sign, in which case (2.41) implies that $f = 0$, a.e and we are done. To see that f is of fixed sign, write (2.39) in operator notation:

$$(T_t^* \rho)(y) \equiv \int_{\mathbb{R}^N} p_t(x,y)\, \rho(x)\, dx = \rho(y)$$

or

$$T_t^* \rho = \rho \tag{2.42}$$

and (2.40) implies that

$$T_t^* f = f . \tag{2.43}$$

Now

$$\| f \|^2 = (f, f)_{L^2(\mathbb{R}^N, dx)} = (T_t^* f, f) \leq (T_t^* |f|, |f|) \leq \| f \|^2$$

where we have used that $p_t(x,y)$ is positive and T_t^* has operator norm less than or equal to one.

$$(T_t^* f, f) = (T_t^* |f|, |f|) . \tag{2.44}$$

Decompose

$$f = f_+ - f_-$$

where $f_+ = f$, whenever f is positive, otherwise zero and $f_- = -f$, whenever f is negative, otherwise zero. Hence $|f| = f_+ + f_-$. From (2.44) we have

$$(T_t^* f_+, f_-) + (T_t^* f_-, f_+) = 0 . \tag{2.45}$$

Since the kernel of T_t^* is strictly positive, (2.45) implies that either $f_+ = 0$ or $f_- = 0$. Suppose $f_+ = 0$ then $f = f_-$ of fixed sign and (2.41) $\Longrightarrow$ $f_- = 0$. Hence the invariant measure μ is unique.

D Fokker - Planck operator

The condition (2.36) for an invariant measure can be written also as:

$$\int_{\mathbb{R}^N} d\mu(x)(e^{tL}f)(x) = \int_{\mathbb{R}^N} d\mu(x)\; f(x) \qquad (2.46)$$

where f is a bounded measurable function.

Using the Ito calculus (as in Ex 2 of § 1.4) we have, for f a test function (C^∞ function of compact support) and $\beta_o = x$,

$$f(\beta_t)\, e^{\xi_o^t} = f(x) + \int_o^t ds\; e^{\xi_o^t} Lf(\beta_s) + \int_o^t e^{\xi_o^t} (\nabla f(\beta_s) + f(\beta_s) b(\beta_s), d\beta_s)$$

with L given by (2.32). The Ito calculus is justified because the last term on the r.h.s. is a well defined Ito stochastic integral because of the conditions (2.19 - 2.20) on the drift vector field b(x). Taking Wiener expectations as in (2.31), and recalling that the stochastic integral has zero mean, we obtain

$$(e^{tL}f)(x) = f(x) + \int_o^t ds\; e^{sL}(Lf)(x) \qquad (2.47)$$

(which is formally obvious). Integrating the above with respect to the invariant measure μ, we obtain:

$$\int_{\mathbb{R}^N} d\mu(x)\; Lf(x) = 0 \qquad (2.48)$$

or

$$\int_{\mathbb{R}^N} dx \;\rho(x)\; Lf(x) = 0 \tag{2.49}$$

since μ admits a probability density ρ with respect to dx. Note that the drift being only assumed to be measurable ρ is only measurable. However, since f is a test function, we can write (2.49) in the <u>distributional sense</u>

$$L^* \rho = 0 \tag{2.50}$$

where

$$L^* = \frac{1}{2}\,\Delta - \frac{\partial}{\partial x^i}\,(b^i(x)\,.\,) \tag{2.51}$$

is the formal adjoint of L in $L^2(dx)$. L^* is known as the <u>Fokker-Planck operator</u>, and ρ the <u>equlibrium density</u>.

(2.47) can be written in terms of the transition probability as:

$$\int p_t(x, dy)\, f(y) = f(x) + \int_0^t ds \int p_s\,(x, dy)\,(L_{(y)} f\,)(y)$$

where f is a test function. Hence

$$\int \frac{\partial p_t}{\partial t}\,(x, dy)\;\; f(y) = \int p_t(x, dy)\,(L_{(y)} f\,)(y).$$

and we have in the <u>distributional sense</u>, for $t > 0$,

$$\frac{\partial p_t}{\partial t}\,(x, dy) = L^*_{(y)}\; p_t\,(x, dy) \tag{2.52}$$

(2.52) is the Fokker -Planck equation. As stated earlier, $p_t(x, dy)$ admits a density (transition probability density) $p_t(x, y)$ with respect to Lebesgue measure. Hence (2.57) reads for $t > 0$,

$$\frac{\partial p_t}{\partial t}(x, y) = L_y^* p_t(x, y) \tag{2.53}$$

where the r.h.s. is in the sense of distributions.

Densities of invariant measures are called equilibrium densities

E Ergodicity

Whenever a Markovian (i.e. positivity preserving) semigroup e^{tL} admits an (unique) invariant measure μ, we have ergodicity (see e.g [5]). For f a bounded measurable function,

$$\lim_{T \to \infty} \frac{1}{T} \int_0^T dt\; e^{tL} f(x) = \int_{\mathbb{R}^N} d\mu(y)\; f(y), \quad \text{a.e. in } x. \tag{2.46}$$

Since the l.h.s. admits a probabilistic interpretation we get a "constructive" procedure for obtaining μ.

If in addition, for $f, g \in L^2(d\mu)$

$$(f, e^{tL} g)_{L^2(d\mu)} \xrightarrow[t \to \infty]{} \left(\int f d\mu\right)\left(\int g d\mu\right) \tag{2.47}$$

we say that e^{tL} is mixing. This is the case whenever e^{tL} has a mass gap.

Finally, we say that an invariant measure μ is reversible if e^{tL} is selfadjoint in $L^2(d\mu)$.

F Boundedness properties of semigroup

Note that whenever we have an invariant measure μ, then

$$\| e^{tL} f \|_{L^p(d\mu)} \leq \| f \|_{L^p(d\mu)}, \quad 1 \leq p \leq \infty \tag{2.48}$$

and the map $[0, \infty) \longrightarrow L^p(d\mu)$ given by

$$t \longrightarrow e^{tL} f, \text{ for every fixed } f \in L^p(d\mu) \text{ is}$$

continuous (in $L^p(d\mu)$), i.e. strongly continuous.

Infact, if $f \in L^1(d\mu)$

$$\| e^{tL} f \|_{L^1(d\mu)} = \int_{\mathbb{R}^N} d\mu(x) \left| \int p_t(x, dy) f(y) \right| \leq \int d\mu \int p_t(x, dy) \, | f(y) |$$

(by positivity preserving)

$$\leq \int d\mu \, e^{tL} \, |f|(x) = \int d\mu \, |f|(x)$$

$$= \| f \|_{L^1(d\mu)}$$

where we have used that μ is invariant.

If $f \in L^\infty(\mathbb{R}^N)$, we have from (2.31) and $E_x^{(W)}(e^{\xi_0^t}) = 1$, (2.21), that

$$\| e^{tL} f \|_{L^\infty} \leq \| f \|_{L^\infty} .$$

Finally if $1 < p < \infty$, then

$$\| e^{tL} f \|^p_{L^p(d\mu)} = \int d\mu(x) \left| \int p_t(x, dy) f(y) \right|^p$$

and $\left| \int p_t(x, dy) f(y) \right| = \left| \int p_t(x, dy) \ 1 . \ f(y) \right|$

$$\leq \Big(\underbrace{\int p_t(x,dy)\, 1}_{1}\Big)^{1/q} \Big(\int p_t(x,dy)\,|f(y)|^p\Big)^{1/p}, \quad \frac{1}{p}+\frac{1}{q}=1$$

$$\leq \left(e^{tL}|f|^p(x)\right)^{1/p}$$

whence, using μ is invariant, we have

$$\|e^{tL}f\|^p_{L^p(d\mu)} \leq \int d\mu(x)\,|f|^p(x)\,.$$

We omit the proof of strong continuity of e^{tL}.

<u>Criterion for the existence of invariant measures</u>

Given $R > 0$, let K_R be the closed ball of radius R in $\mathbb{R}^N$ and $\tilde{K}_R$ its complement. Then for e^{tL} to admit an invariant measure it is necessary and sufficient that

$$\lim_{R\to\infty}\ \liminf_{T\to\infty}\ \frac{1}{T}\int_0^T dt\ p_t(x,\tilde{K}_R) = 0 \tag{2.49}$$

where $p_t(x,.)$ is the transition probability of the semigroup. Since e^{tL} is positivity improving, we have uniqueness of invariant measures, so that (2.49) also is the condition for ergodicity. The proof of (2.49) is standard and can be found in [4].

Suppose now that the drift $b(x)$ not only satisfies (2.19 -2.20), but is also C^∞. Then we fall into the frame-work of §2.3, and the weak solution is also a strong non-explosive solution. Based on (2.49) we have now the following test for the existence of an invariant measure:

Hasminski invariant measure test, [4]

Let $V(x)$ be twice differentiable and (i) $V(x) \geq 0$.

(ii) $\sup_{\|x\| > R} LV(x) = -A_R$, where for R sufficiently large

$A_R > 0$, and $A_R \longrightarrow \infty$, as $R \longrightarrow \infty$.

(iii) $\sup_{x \in \mathbb{R}^N} LV(x) < c$, for some C. Then e^{tL} admits an invariant measure.

Appendix II A

Proof of the Khasminski non-explosion test, [4].

We start after (2.18). Set

$$W(t,x) = V(x)e^{-ct}$$

Then, by (2.18)

$$\left(\frac{\partial}{\partial t} + L\right) W(t,x) = (LV - cV)\, e^{-ct} \leq 0, \qquad (*)$$

Let

$$\tau_N = \text{first exit time from } K_N$$

$$\tau_N(t) = \min(\tau_N, t)$$

(recall construction of the process x_t in § 2.3). Then $x_{\tau_N(t)}$, the stopped process, satisfies the integral equation:

$$x_{\tau_N(t)} = x + \beta_{\tau_N(t)} + \int_0^{\tau_N(t)} ds\ b(x_s)$$

and Ito's formula applied to $W(\tau_N(t), x_{\tau_N(t)})$

gives

$$W(\tau_N(t), x_{\tau_N(t)}) - W(0,x)$$

$$= \int_0^{\tau_N(t)} (\text{grad } W(s, x_s), d\beta_s)$$

$$+ \int_0^{\tau_N(t)} ds \left(\frac{\partial}{\partial s} + L\right) W(s, x_s)$$

(for stochastic integrals with stopping times as end points see [1]).

Taking the Wiener expectation of the previous formula, and recalling that the stochastic integral has vanishing expectation, we get

$$E^{(W)}(W(\tau_N(t), x_{\tau_N(t)})) - W(0,x) = E^{(W)}\left(\int_0^{\tau_N(t)} ds \left(\frac{\partial}{\partial s} + L\right) W(s, x_s)\right)$$

$$\leq 0 \qquad (**)$$

by (*).

Hence,

$$E^{(W)}(V(x_{\tau_N(t)})) = E^{(W)}(W(\tau_N(t), x_{\tau_N(t)})\, e^{c\,\tau_N(t)})$$

$$\leq E^{(W)}(V(x)\, e^{c\,\tau_N(t)}), \text{ (have used (**))}$$

Hence,

$$E^{(W)}(V(x_{\tau_N(t)})) \leq e^{ct}\ V(x) \quad , \text{ by definition of } \tau_N(t)$$

$$(***)$$

Now use,

$$E^{(W)}(V(x_{\tau_N(t)})) \geq (\inf_{x \notin K_N} V(x))\ P^{(W)}\ \{\|x_{\tau_N(t)}\| > R_N\}$$

and (***) to get

$$P^{(W)}\{\|x_{\tau_N}(t)\| > R_N\} \leq \frac{e^{ct}\,V(x)}{\inf\limits_{x \notin K_N}(V(x))} \xrightarrow[N \to \infty]{} 0 \qquad (****)$$

because of (2.18 (i)).

Since, $P\{t > \tau_N\} = P\{\|x_{\tau_N}(t)\| > R_N\}$

we have therefore

$$\lim_{N \to \infty} P\{\tau_N < t\} = 0$$

and hence the explosion time $\tau = \infty$, almost surely.

Appendix II B

Proof of (2.25)

Approximate b by a sequence $\{b_n\}$ of bounded, measurable functions, $b_n \longrightarrow b$, a.e. Define $\xi_o^{t,n}$ by (2.24), replacing b by b_n. Then $\xi_o^{t,n} \xrightarrow[n \to \infty]{} \xi_o^t$ in $L^2(p^W, \Omega)$. Now

$$e^{\xi_o^t} = e^{\xi_o^{t,n}} + (e^{\xi_o^t} - e^{\xi_o^{t,n}})$$

and

$$E^{(W)}(e^{\xi_o^t}) = 1 + E^{(W)}(e^{\xi_o^t} - e^{\xi_o^{t,n}}) \qquad (*)$$

since

$$E^{(W)}(e^{\xi_o^{t,n}}) = 1$$

as b_n is bounded by (1.35) and Remark following (1.38).

Now use:

$$\left| e^{\xi_o^t} - e^{\xi_o^{t,n}} \right| < \left| \xi_o^t - \xi_o^{t,n} \right| \left(e^{\xi_o^t} + e^{\xi_o^{t,n}} \right).$$

Taking the expectation and using the Schwarz inequality,

$$E^{(W)}\left(\left| e^{\xi_o^t} - e^{\xi_o^{t,n}} \right|\right) \leq \left(E^{(W)} \left| \xi_o^t - \xi_o^{t,n} \right|^2\right)^{\frac{1}{2}} \left(\left(E^{(W)}\left(e^{2\xi_o^t}\right)\right)^{\frac{1}{2}} + \left(E^{(W)}\left(e^{2\xi_o^{t,n}}\right)\right)^{\frac{1}{2}} \right). \qquad (**)$$

Now use (2.20), $E^{(W)}\left(e^{2\xi_o^{t,n}}\right) \leq C_t$, and $\xi_o^{t,n} \underset{n \to \infty}{\longrightarrow} \xi_o^t$ in $L^2(\Omega, dP^{(W)})$

to deduce

$$e^{\xi_o^{t,n}} \underset{n \to \infty}{\longrightarrow} e^{\xi_o^t}$$

in $L^1(\Omega, dp^W)$. From (*) and (***), (2.21) follows.

End of proof.

Appendix II C

In this appendix we give an elementary proof (*) that transition probabilities and invariant measures associated with weak solutions of stochastic differential equations in $\mathbb{R}^N$ under conditions (2.19 - 2.20) are absolutely continuous with respect to Lebesgue measure in $\mathbb{R}^N$.

The transition probability $p_t(x, dy)$ satisfies for $t > 0$ the Fokker- Planck equation (2.52).

$$(L_y^* - \frac{\partial}{\partial t})\ p_t(x, dy) = 0$$

in the sense of distributions, Suppressing the x dependence, we have, in the sense of distributions,

$$(\frac{1}{2}\ \Delta_y - \frac{\partial}{\partial t})\ p_t(dy) = \sum_i \frac{\partial}{\partial y^i}\ (b^i(y)\ p_t(dy)) \qquad (*)$$

Note that $b^i(y)$ is in $L^1\ (p_t(dy))$, so that the above formula makes sense. In fact

$$\int_{\mathbb{R}^N} p_t\ (dy)\ |\ b^i(y)| = E^{(W)}(\ |b^i(\beta_t)|\ e^{\xi_t})$$

$$\leq \left(E^{(W)} |\ b^i(\beta_t)|^2\right)^{\frac{1}{2}} (E^{(W)}(e^{2\xi_t}))^{\frac{1}{2}}$$

$$= (\int p_t^{(W)}(dy)\ |b^i(y)|^2)^{\frac{1}{2}} (E^{(W)}(e^{2\xi_t}))^{\frac{1}{2}}$$

$$< \infty$$

because of (2.19 - 2.20).

Define

$$M_t = \frac{1}{2}\ \Delta_y - \frac{\partial}{\partial t}$$

and its formal adjoint in $L^2(\mathbb{R}^N \times \mathbb{R},\ dy \otimes dt)$

$$M_t^* = \frac{1}{2}\ \Delta_y + \frac{\partial}{\partial t}$$

Let $f_s(x)$ be a test function, C^∞ function in s, x

(x) based on a suggestion of M.S. Narasimhan.

of compact support in (s,x). Then from (*)

$$\int_0^T ds < p_s , M_s^* f_s > = -\sum_i \int_0^T ds < p_s , b^i \frac{\partial f_s}{\partial y^i} >$$

or
$$\int_0^T ds \int_{\mathbb{R}^N} p_s(dy) \, (M_s^* f_s)(y) = \int_0^T ds \int_{\mathbb{R}^N} p_s(dy) \, b^i(y) \frac{\partial f_s(y)}{\partial y^i} \qquad (**)$$

Let $\epsilon(x,s)$ be a C^∞ function of compact support in $\mathbb{R}^N \times [0,T]$, such that $\epsilon = 1$ in a neighbourhood of $x=0$, $s=t$ with $0<t<T$. Let $p_t^{(W)}(x)$ be the Wiener probability density (fundamental solution of the heat equation $M_t f = 0$). Then

$$M_s^* \, p_{t-s}^{(W)}(x-y) = \delta(x-y) \, \delta(t-s)$$

for $s \leq t < T$, and $p_{t-s}^{(W)} = 0$, for $s > t$.

Now choose

$$f_s(y) = \int_{\mathbb{R}^N} \epsilon(y-y',s) \, p_{t-s}^{(W)}(y-y') \, g(y') \, dy' \qquad (***)$$

where g is a test function. Then

$$M_s^* f(y) = g(y) \, \delta(t-s)$$

$$+ \int F(y-y',s) \, g(y') \, dy'$$

where F is a compactly supported function in $L^1(ds \otimes dy)$. (We use the fact that $p_s^{(W)}(y)$ and its first spatial derivative are in $L^1_{loc}(ds \otimes dy)$).

Hence

$$\int_0^T ds \int_{\mathbb{R}^N} p_s(dy)\,(M_s^* f)(y) = \int p_s(dy) g(y) + \int_{\mathbb{R}^N} dy\, \tilde{F}(y) g(y) \qquad (****)$$

where $\tilde{F}$ is in $L^1(dy)$.

Similarly, from (***) we have

$$\frac{\partial f_s(y)}{\partial y^i} = \int_{\mathbb{R}^N} G_i(y-y', s)\, g(y')dy' \qquad (+)$$

where G_i is an $L^1(dy \otimes ds,\ \mathbb{R}^N \otimes [0,T])$. Plugging in (****) and (+) in (**) we have

$$\int_{\mathbb{R}^N} p_t(dy)\, g(y) = \int_{\mathbb{R}^N} dy\, \tilde{G}(y)\, g(y)$$

where $\tilde{G} \in L^1(dy)$. This shows that $p_t(dy)$ is absolutely continuous.

Very similarly one can show that if μ is an invariant measure so that

$$L^* \mu = 0$$

in the sense of distributions, then μ is absolutely continuous with respect to Lebesgue measure on $\mathbb{R}^N$.

CHAPTER III

ORNSTEIN-UHLENBECK PROCESSES IN $\mathbb{R}^N$ AND PERTURBATIONS

This chapter is in preparation for stochastic field theory

§ 3.1 Ornstein-Uhlenbeck process in $\mathbb{R}^N$: Let $\{e_i, i=1,\ldots,N\}$ be an orthonormal basis in $\mathbb{R}^N$. If $x = x_1 e_1 + \ldots + x_n e_n$ is in $\mathbb{R}^N$, then $x = (x^1,\ldots,x^N)$ also denotes a point in $\mathbb{R}^N$. Let $\epsilon \geq 0$, and λ_i, $i = 1,\ldots,N$, >0 be real numbers. Let $C_N = \mathrm{diag}(\lambda_1, \ldots, \lambda_N)$ be the diagonal matrix with respect to the basis $\{e_i\}$. Define the Wiener process in $\mathbb{R}^N$:

$$W_t^{(N)} = C_N^{\frac{1-\epsilon}{2}} \beta_t \tag{3.1}$$

where β_t is standard normalized Brownian motion in $\mathbb{R}^N$. Consider the process $x_t^{(N)} \in \mathbb{R}^N$ given as the unique solution of the (Langevin) equation:

$$\begin{aligned} dx_t^{(N)} &= dW_t^{(N)} - \frac{1}{2} C_N^{-\epsilon} x_t^{(W)} dt \\ x_o^{(N)} &= x \end{aligned} \tag{3.2}$$

which in components in the basis $\{e_i\}$ reads:

$$\begin{aligned} dx_t^{(N),i} &= \lambda_i^{(1-\epsilon)/2} d\beta_t^i - \frac{1}{2} \cdot \lambda_i^{-\epsilon} x_t^i \, dt \\ x_o^{(N),i} &= x^i \end{aligned} \tag{3.3}$$

$i=1,\ldots,N$; no sum on i

For different i, we have independent processes.

The equation (3, 2) being linear we have the solution:

$$x_t^{(N)} = e^{-\frac{t}{2} C_N^{-\epsilon}} x + \int_0^t e^{-\frac{1}{2}(t-s) C_N^{-\epsilon}} dW_t^{(N)} \qquad (3.4)$$

which in components reads

$$x_t^{(N),i} = e^{-\frac{t}{2} \lambda_i^{-\epsilon}} x^i + \int_0^t e^{-\frac{1}{2}(t-s) \lambda_i^{-\epsilon}} \lambda_i^{\frac{(1-\epsilon)}{2}} d\beta_s^i \qquad (3.5)$$

$$i = 1, \ldots, N$$

where on the r.h.s. we have a well defined Ito stochastic integral. Often one chooses $C_N = I$ (or equivalently $\epsilon = 1$ in the above context) but we hold on to the more general form for the purposes of the following chapters.

(3.4 - 3.5) defines a continuous Gaussian Markov process $x_t^{(N)}(W)$ as a measurable function of the Wiener process $W_s^{(N)}, s \leq t$. It will be called an Ornstein-Uhlenbeck process (O. U process) and its infinite dimensional generalization (chapter V) gives the stochastic free field. It is of much more importance than Brownian motion itself, which will play only an auxiliary role. The O. U process is ergodic whereas Brownian motion (in a non-compact space) is not.

Properties

From the remark on time change (§ 1.2, after equation 1.13) we have that the stochastic integral in (3.5) is a random variable with the same distributions as Brownian motion on $\mathbb{R}^1$ with time change:

$$t \longrightarrow \tau_i(t) = \lambda_i\,(1 - e^{-\lambda_i^{-\epsilon} t}) \qquad (3.6)$$

so that the component process $x_t^{(i)}$ has transition probability on $\mathbb{R}^1$

$$p_t^{(i)}(x^i, dy^i) = p^{(W)}_{\tau_i(t)}\left(e^{-\frac{t}{2}\lambda^{-\epsilon}} x^i, dy^i\right) \qquad (3.7)$$

where

$$p_t^{(W)}(x^i, dy^i) = \frac{1}{\sqrt{2\pi t}}\, e^{-\frac{|x^i - y^i|^2}{2t}}\, dy^i$$

is the Brownian motion transition probability on $\mathbb{R}^1$.

Hence the O. U. process $x_t^{(N)}(W)$ in $\mathbb{R}^N$ has transition probability on $\mathbb{R}^N$ (<u>Mehler formula</u>):

$$p_t(x, dy) = \bigotimes_{i=1}^{N} p^{(W)}_{\tau_i(t)}\left(e^{-\frac{t}{2}\lambda_i^{-\epsilon}} x^i, dy^i\right) \qquad (3.8)$$

The transition probability density of (3.8) is known as the <u>Mehler kernel</u>.

Note that as $t \longrightarrow \infty$, $\tau_i(t) \longrightarrow \lambda_i$, and in the sense of measures (i.e. weakly),

$$p_t(x, dy) \longrightarrow \left(\prod_{i=1}^{N} \left(\frac{1}{\sqrt{2\pi\lambda_i}}\right)\right) e^{-\sum_{i=1}^{N} \frac{|y_i|^2}{2\lambda_i}}\, d^N y \equiv \mu_{C_N}(dy) \qquad (3.9)$$

μ_{C_N} is a Gaussian probability measure on $\mathbb{R}^N$ of mean 0 and covariance C_N, i.e.

$$\int_{\mathbb{R}^N} d\mu_{C_N}\, e^{i(x,e)} = e^{-\frac{1}{2}(e,\, C_N e)}$$

for any vector e in $\mathbb{R}^N$. (3.9) shows that μ_{C_N} is an invariant measure and, being unique, $x_t^{(N)}(W)$ is ergodic. Check that (3.8) can be written for every Borel set $B \subset \mathbb{R}^N$:

$$p_t(x,B) = \mu_{\tilde{C}_{N,t}}\left(B - e^{-\frac{t}{2} C_N^{-\epsilon}} x\right) \tag{3.10}$$

where

$$\tilde{C}_{N,t} = (1 - e^{-t C_N^{-\epsilon}}) C_N \tag{3.11}$$

and $\mu_{\tilde{C}_{N,t}}$ is the Gaussian measure of mean 0 and covariance $\tilde{C}_{N,t}$.

<u>Ornstein - Uhlenbeck semigroup</u>

The O. U. semigroup $e^{tL_o^{(N)}}$ is defined by:

$$(e^{tL_o^{(N)}} f)(x) = E_x^{(W)} (f(x_t^{(N)}(W))) = \int p_t(x, dy)\; f(y) \tag{3.12}$$

where $p_t(x,dy)$ is given by the Mehler formula (3.8) or (3.10).

Since μ_{C_N} is an invariant measure we have, by virtue of § 2.5 D, the following boundedness properties:

$e^{tL_o^{(N)}}$ is a strongly continuous contractive semigroup on $L^p(d\mu_{C_N}, \mathbb{R}^N)$, $1 \leq p \leq \infty$. By § 2.5 A, or explicitly from the strict positivity of the Mehler kernel, $e^{tL_o^{(N)}}$ is positivity improving. Moreover $e^{tL^{(N)}} 1 = 1$.

We have a further boundedness property, <u>hypercontractivity:</u>

For $t > T$, sufficiently large,

$$\left\| e^{tL_o^{(N)}} f \right\|_{L^4(d\mu_{c_N})} \leq \left\| f \right\|_{L^2(d\mu_{c_N})} \tag{3.13}$$

This can be checked by explicitly using the Mehler kernel (see, [6]).

Finally, using the Mehler formula, one checks that $e^{tL_o^{(N)}}$ is symmetric in $L^2(d\mu_{c_N})$. By virtue of the earlier boundedness and continuity property of $e^{tL_o^{(N)}}$, $L_o^{(N)}$ the infinitesimal generator is a selfadjoint operator on $L^2(d\mu_{c_N})$. On the dense subspace of C^∞ functions of compact support, $L_o^{(N)}$ has the explicit expression

$$L_o^{(N)} = \frac{1}{2} \sum_{i=1}^{N} \lambda_i^{(1-\epsilon)} \frac{\partial^2}{\partial {x^i}^2} - \frac{1}{2} \sum_{i=1}^{N} \lambda_i^{-\epsilon} x^i \frac{\partial}{\partial x^i} \tag{3.14}$$

For $\epsilon = 0$, define

$$-\mathbb{N} = L_o^{(N)} \Big|_{\epsilon = o} \tag{3.15}$$

$$-\mathbb{N} = \frac{1}{2} \sum_{i=1}^{N} \lambda_i \frac{\partial^2}{\partial {x^i}^2} - \frac{1}{2} \sum_{i=1}^{N} x^i \frac{\partial}{\partial x^i} = \sum_{i=1}^{N} -\mathbb{N}_i$$

$-\mathbb{N}$ is the Hermite operator on $\mathbb{R}^N$, and $e^{-t\mathbb{N}}$ is called the <u>Hermite semigroup</u>. $\mathbb{N}$ coincides with the <u>number operator</u> for N harmonic oscillators of frequencies $\lambda_1, \dots, \lambda_N$, acting on "Fock space" $= L^2(d\mu_{c_N})$. For $\epsilon = 1$ we get the <u>Hamiltonian,</u> on $L^2(d\mu_{c_N})$. (3.14) interpolates between these cases, but μ_{c_N} is always the invariant measure).

The Hermite polynomials

$$H_n(x_i, \lambda_i) = \frac{(-\lambda_i)^n}{\sqrt{n!}} \, e^{\frac{x_i^2}{2\lambda_i}} \left(\frac{\partial}{\partial x^i}\right)^n e^{-\frac{x_i^2}{2\lambda_i}} \tag{3.16}$$

are eigen functions of $\mathbb{N}_i$:

$$\mathbb{N}_i \; H_n(x^i, \lambda^i) = \frac{n}{2} H_n(x^i, \lambda_i) \tag{3.17}$$

They give an orthonormal basis of $L^2\left(\frac{1}{\sqrt{2\pi\lambda_i}} \, e^{\frac{x_i^2}{2\lambda_i}} \, dx^i\right)$

Let $\underline{\lambda} = (\lambda_1, \ldots, \lambda_N)$. Define

$$H^{(n)}_{p_1 \ldots p_N}(x, \underline{\lambda}) = \prod_{i=1}^{N} H_{p_i}(x_i, \lambda_i), \quad \sum_{i=1}^{N} p_i = n \tag{3.18}$$

with p_i non-negative integers.

For $n \neq m$, $H^{(n)}_{p_1 \ldots p_N}$ and $H^{(m)}_{q_1 \ldots q_N}$ are orthogonal in $L^2(d\mu_{c_N}, \mathbb{R}^N)$. They provide an orthonormal basis of $L^2(d\mu_{c_N})$.

We denote by $\mathcal{H}_n$ the subspace spanned by the $\{H^{(n)}_{p_1 \ldots p_N}\}$ for fixed n, and $\overline{\mathcal{H}}_n$ its closure in $L^2(d\mu_{c_N})$.

Then

$$L^2(d\mu_{c_N}) = \bigoplus_{n=0}^{\infty} \overline{\mathcal{H}}_n \, . \tag{3.19}$$

Let P be a homogeneous polynomial of degree n. Let π_n be the orthogonal projector of $L^2(d\mu_{c_N})$ onto $\mathcal{H}_n$. Define

$$:P: \; = \sqrt{n!} \;\; \pi_n \;\; P \tag{3.20}$$

:P: is called a <u>Wick polynomial.</u> Wick ordering thus depends on P <u>and</u> the Gaussian measure. To be explicit, we can write $:P:_{\mu_{c_N}}$.

Let u_i, i=1, ..., n, be fixed vectors in $\mathbb{R}^N$. Then $(x,u_1)(x,u_2)\ldots(x,u_n)$ is a homogeneous polynomial in $x_1,\ldots,x_N$ of degree n. Verify, using the Hermite recursion relation, that the Wick polynomial $:(x,u_1)(x,u_2)\ldots(x,u_n):$ satisfies the recursion relation

$$\begin{aligned}
&(x,u_1) : (x,u_2)\ldots(x,u_n): \\
&\quad = \; :(x,u_1)(x,u_2)\ldots(x,u_n): \\
&\quad\quad + \sum_{j=2}^{n} (u_1, C_N u_j) \; :(x,u_1)\ldots(x,u_{j-1})(x,u_{j+1})\ldots(x,u_n): \\
&\quad =: (x,u_1)(x,u_2)\ldots(x,u_n): \\
&\quad\quad + \sum_{j=2}^{n} < (x,u_1)(x,u_j) >_{\mu_{c_N}} : (x,u_1)\ldots(x,y_{j-1})(x,u_{j+1})\ldots(x,u_n):
\end{aligned} \tag{3.21}$$

Moreover,

$$\frac{\partial}{\partial x^i} : (x,u)^n : \; = n : (x,u)^{n-1} : \; u_i \tag{3.22}$$

<u>§ 3.2 Perturbation of the O. U. process</u>

Let b(x) be a measurable vector in $\mathbb{R}^N$ such that $b \in L^p(d\mu_{c_N})$, $1 \leq p < \infty$ and satisfying the analogue of (2.19), (2.20) with the O. U. process $x_t^{(W)}$ substituted for Brownian motion namely:

$$E^{(W)}\int_0^T ds \left\| C_N^{-\frac{(1-\epsilon)}{2}} b(x_s(w))\right\|^{2(1+\delta)} < \infty \quad \text{(i)}$$

$$E^{(W)}\left(e^{2\left(1+\frac{1}{\delta}\right)\xi_0^T}\right) < \infty \quad \text{(ii)} \tag{3.23}$$

for any $\delta > 0$, where

$$\xi_0^t \equiv \int_0^t (b(x_s(W)), C_N^{\frac{-(1-\epsilon)}{2}} d\beta_s) - \frac{1}{2}\int_0^t \left\| C_N^{-\frac{(1-\epsilon)}{2}} b(x_s(W))\right\|^2 ds \tag{3.24}$$

and $W_s = C_N^{\frac{1}{2}(1-\epsilon)} \beta_s$ as in (3.1). ξ_0^t is a well defined random variable, because of (3.23)i)).

Define a semigroup* e^{tL} by:

$$(e^{tL} f)(x) = E_x^{(W)}(f(x_t(W))\, e^{\xi_0^t}) \tag{3.25}$$

Clearly, by virtue of the above conditions,

$$\| e^{tL} f\|_{L^\infty} \le \|f\|_{L^\infty} . \tag{3.26}$$

Approximate b by a sequence $\{b_j\}$ of bounded, measurable functions. Then, as in Example 2 § 1.4, show, using the Ito calculus:

* The suffix N (referring to $\mathbb{R}^N$) has been dropped.

$$E^{(W)}\left(e^{\xi_o^{t,j}}\right) = 1 .$$

Then a repeat of Appendix II B, using 3.23 (ii) gives,

$$E^{(W)}\left(e^{\xi_o^t}\right) = 1 \tag{3.27}$$

Hence

$$e^{tL}\, 1 = 1 \tag{3.28}$$

Conditions (3.23) justify the use of the Ito calculus directly on (3.25). Let f be a C^{∞} function of compact support. Starting from (3.25), (3.24) and (3.2), use the Ito calculus as in Ex 2, § 1.4, to show that on such f,

$$L = L_o + (b(x), \nabla) \tag{3.29}$$

where L_o is given by (3.14). (In other words (3.25) is the Cameron-Martin formula for perturbation of the O.U. differential generator L_o, instead of the Brownian motion differential generator $\frac{1}{2}\Delta$, by the drift vector field b).

As in Ex 2, § 1.4, starting from (3.25), (3.27) we can construct a continuous Markov process in $\mathbb{R}^N$. Namely, define for every $B \subset \mathbb{R}^N$, a Borel set,

$$p_t(x, B) = (e^{tL}\chi_B)(x) = E_x^{(W)}\left(\chi_{x_t \in B}\; e^{\xi_t}\right) \tag{3.30}$$

Then because of (3.27) $p_t(x, dy)$ is a probability measure satisfying the Chapman - Kolmogoroff equation. Then the Kolmogoroff construction gives a Markovian family of probability measures indexed by $\mathbb{R}^N$, (P_x, Ω), (using Kolmogoroff's theorem one verifies the continuity of sample paths).

We now consider the Stochastic differential equation

$$d\hat{x}_t = dW_t - (\tfrac{1}{2} C_N^{-\epsilon} x_t - b(\hat{x}_t))\, dt$$

$$\hat{x}_o = x \tag{3.31}$$

which can be written as an integral equation

$$\hat{x}_t = x_t + \int_o^t ds \; d^{-\frac{1}{2}(t-s) C_N^{-\epsilon}} b(\hat{x}_s) \tag{3.32}$$

where $x_t = x_t(W)$ is the O.U. process of § 3.1, starting at x.

(P_x, Ω) solves (3.32) in the weak sense, (§ 2.4). Namely, taking $\hat{x}_t$ as a path in Ω, the random variable

$$\tilde{Z}_t \equiv \hat{x}_t - \int_o^t ds \; e^{-\frac{1}{2}(t-s) C_N^{-\epsilon}} b(\hat{x}_s) \tag{3.33}$$

has P_x distributions coinciding with the known distributions of the O.U. process x_t.

The proof of this statement is very similar to that in § 2.4 except that we are perturbing with reference to the O.U. process. It is sufficient to check it at the level of transition probabilities. By definition of P_x,

$$E^{(P_x)}(f(\tilde{Z}_t)) = E_x^{(W)}\left(f(Z_t)\, e^{\xi_o^t}\right) \tag{3.34}$$

$$Z_t = x_t - \int_0^t ds\; e^{-\frac{1}{2}(t-s) C_N^{-\epsilon}} b(x_s) \qquad (3.35)$$

Define on $\Omega_T = C^o([0,T], \mathbb{R}^N)$, the probability measure $\tilde{P}^{(W)}$:

$$d\tilde{p}^{(W)} = e^{\xi_o^T} d\,p^{(W)}$$

Then show, similarly to (2.25) - (2.27),

$$E^{(W)}\left(e^{\xi_s^t} \middle| \mathcal{F}_s\right) = 1 \qquad (3.36)$$

where

$$\xi_s^t = \int_s^t \left(C_N^{-(1-\epsilon)} b(x_s), dW_s\right) - \frac{1}{2}\int_s^t ds\, \left\| C_N^{-\frac{1-\epsilon}{2}} b(x_s)\right\|^2 \qquad (3.37)$$

so that we again have (2.28), namely

$$\tilde{E}^{(W)}(f(Z_t)) = E^{(W)}\left(f(Z_t)\, e^{\xi_o^t}\right) \qquad (3.38)$$

and hence, from (3.34, 3.35) and (3.38)

$$E^{(P_x)}(f(\tilde{Z}_t)) = \tilde{E}^{(W)}(f(Z_t)) \qquad (3.39)$$

It remains to show that Z_t is an O.U. process with reference to $\tilde{P}^{(W)}$.

Show that (Girsanov's theorem): , [1],

$$\tilde{W}_t \equiv W_t - \int_0^t ds\; b(x_s(W)) \qquad (3.40)$$

is a Wiener process in Ω_T with respect to $\tilde{P}^{(W)}$, with the same covariance as W_t. Then Z_t is clearly the unique solution of

$$d Z_t = d \tilde{W}_t - \frac{1}{2} C^{-\epsilon} Z_t \, dt \tag{3.41}$$

and it follows that under $\tilde{P}^{(W)}$, Z_t is an O.U. process, and we are done.

§ 3.3 Gradient vector fields as perturbing drifts (this section is written with a view to generalization to infinite dimensions).

We consider the special case of § 3.2 when the perturbing drift is a gradient vector field which we write as:

$$b(x) = -\frac{1}{2} C_N^{1-\epsilon} \nabla V(x)$$

or

$$b_i(x) = -\frac{1}{2} \lambda_i^{1-\epsilon} \frac{\partial V(x)}{\partial x^i} \tag{3.42}$$

in the orthonormal basis of the beginning of § 3.2.

On the "potential" function $V(x)$, assumed to be measurable, we impose further:

$$V, \nabla_i V \text{ and } L_o V \in L^p(d\mu_c), \; 1 \leq p < \infty \tag{3.43 a}$$

$$e^{L_o V}, \; e^{-V} \in L^1(d\mu_c). \tag{3.43 b}$$

(Derivatives above are in the distributional sense).

A Existence of Weak solution

These conditions guarantee, as we shall see, conditions (3.23) for almost all starting points and hence we have a weak solution by § 3.2. (We have not assumed smoothness, but put ourselves in a situation which will generalize in an infinite dimensional context).

Consider the O.U process $x_t(W)$, $x_o = x$ of (3.4) as a function of x and W. It gives a (jointly) measurable map

$$\mathbb{R}^N \times \Omega \longrightarrow \mathbb{R}^N$$

$$(x, w) \longrightarrow x_t(x; W)$$

where we have indicated explictly the dependence on the starting point. The measurability is with respect to the product measure

$$dQ = d\mu_c \otimes dP^{(W)} \quad \text{on } \mathbb{R}^N \times \Omega .$$

Let F stand for any of the functions, $V, \nabla_i V, L_o V$ of (3.43 a). Then $F(x_t(x;W))$ is jointly measurable function

$$\mathbb{R}^N \times \Omega \longrightarrow \mathbb{R}$$

Suppose

$$\int d\mu_c \; E^{(W)} \int_o^T ds \; \| F\, x_t(x;W) \|^p < \infty \qquad (3.44)$$

$$1 \leq p < \infty$$

then by Fubini's theorem,

$$E^{(W)} \int_o^T ds \; \| F(x_t(x; W)) \|^p < \infty , \qquad \mu_c \text{ a.e.} \qquad (3.45)$$

$$1 \leq p < \infty$$

As for checking (3.44), by Fatou's lemma

$$\ell\text{hs of } 3.44 \leq \underline{\lim}_{n \to \infty} \int d\mu_c \; E_x^{(W)} \left(\left(\frac{T}{n}\right) \sum_{j=1}^{n} \| F(x_{\frac{jt}{n}}) \|^p \right)$$

$$= \lim_{n \to \infty} \frac{T}{n} \sum_{j=1}^{n} \int d\mu_c \; e^{\frac{jt}{n} L_o} \| F(x) \|^p = \lim_{n \to \infty} \frac{T}{n} \sum_{j=1}^{n} \int d\mu_c \| F(x) \|^p$$

since μ_c is an invariant measure

$$= T \| F \|^p_{L^p(d\mu_c)} < \infty , \text{ by } (3.43 \text{ a})$$

Hence (3.45) is satisfied. Choosing $F = b(x)$ of (3.42) we see that condition (3.23 (i)) is satisfied.

Next we proceed to checking (3.23(ii)).

Approximate V by a sequence of bounded C^∞ functions $\{V_n\}$ converging to V in $L^p(d\mu_c)$. Then for each such V_n, by Ito's formula,

$$V_n(x_t) - V_n(x) = \int_o^t (\nabla V_n(x_s), C^{\frac{1}{2}(1-\epsilon)} d\beta_s) + \int_o^t ds \; L_o V_n(x_s) .$$

Because of (3.45), each term in the above formula converges in $L^2(\Omega, dP^{(W)})$ as $n \to \infty$. Hence, taking $n \to \infty$

$$V(x_t) - V(x) - \int_o^t ds \; L_o V(x_s) = \int_o^t (\nabla V(x_s), dW_s) \qquad (3.46)$$

which is valid in $L^2(\Omega, p^W)$, μ_c-a.e. in the starting point x.

Substituting the expression (3.42) in (3.24), and using (3.46) we get

$$\xi_o^t = \frac{1}{2} V(x) - \frac{1}{2} V(x_t) + \frac{1}{2} \int_o^t ds \; L_o V(x_s) - \frac{1}{8} \int_o^t ds \left\| C^{\frac{1-\epsilon}{2}} \nabla V(x_s) \right\|^2$$

(3,47)

Hence

$$E_x^{(W)}(e^{p\xi_o^t}) = e^{\frac{p}{2} V(x)} E_x^{(W)}\left(e^{-\frac{p}{2} V(x_t) + \frac{p}{2} \int_o^t ds \; L_o V(x_s) - \frac{p}{8} \int_o^t ds \|C^{\frac{1-\epsilon}{2}} \nabla V(x_s)\|^2}\right)$$

$$\leq e^{p/2 \; V(x)} E_x^{(W)}\left(e^{-\frac{p}{2} V(x_t)} e^{\frac{p}{2} \int_o^t ds \; L_o V(x_s)}\right) \quad (3.48)$$

$$\leq e^{\frac{p}{2} V(x)} \left(E_x^{(W)}(e^{-p V(x_t)})\right)^{\frac{1}{2}} \left(E_x^{(W)}(e^{p \int_o^t L_o V(x_s)})\right)^{\frac{1}{2}}$$

(3.49)

By the same argument as leading from (3.44) to (3.45), it is sufficient to check

$$\int d\mu_c \, E_x^{(W)}(e^{-p V(x_t)}) < \infty \qquad \text{(i)}$$

(3.50)

$$\int d\mu_c \, E_x^{(W)}(e^{p \int_o^t L_o V(x_s)}) < \infty \qquad \text{(ii)}$$

As for (3.50) (i)

$$\int d\mu_c \, E_x^{(W)}(e^{-p V(x_t)}) = \int d\mu_c \, e^{tL_o}(e^{-pV})(x)$$

$$= \int d\mu_c \, e^{-p\,V(x)} \text{ , since } \mu_c \text{ is invariant.}$$

$$< \infty \text{ , by (3.43 b).}$$

As for (3.50) (ii), using Fatou's lemma

$$\int d\mu_c E_x^{(W)} \left(e^{p\int_0^t L_o V(x_s)}\right) \leq \lim_{n \to \infty} \int d\mu_c \, E_x^{(W)} \left(\prod_{j=1}^{n} e^{\frac{pt}{n} L_o V(x_{\frac{jt}{n}})} \right)$$

$$\leq \lim_{n \to \infty} \prod_{j=1}^{n} \left(\int d\mu_c \, E_x^{(W)} \left(e^{pt\, L_o V(x_{\frac{jt}{n}})} \right) \right)^{\frac{1}{n}} \text{ , by Holder's}$$

inequality

$$= \lim_{n \to \infty} \prod_{j=1}^{n} \left(\int d\mu_c \, e^{\frac{jt}{n} L_o} \left(e^{pt L_o V(x)} \right) \right)^{\frac{1}{n}}$$

$$= \lim_{n \to \infty} \prod_{j=1}^{n} \left(\int d\mu_c \, e^{pt\, L_o V(x)} \right)^{\frac{1}{n}}$$

$$= \int d\mu_c \, e^{pt L_o V(x)} < \infty \text{ , by (3.43 b),} \tag{3.51}$$

Hence 3.50 (i), (ii) hold and from (3.49), and argument following (3.44) we have

$$E_x^{(W)} \left(e^{p\, \xi_o^t}\right) < \infty , \quad \mu_c \text{ a.e.} \tag{3.52}$$

Hence we have shown that conditions (3.43 a-b) guarantee that conditions

3.23 (i), (ii) are satisfied for almost all starting points and hence by § 3.2 we have a weak solution of (3.31) with b given by (3.42). Note that we have shown that a weak solution exists for almost all starting points.

B Ergodicity

Because of (3.43 a-b)

$$Z = \int_{\mathbb{R}^N} d\mu_c \, e^{-V} < \infty \tag{3.53}$$

and we can define a probability measure μ on $\mathbb{R}^N$, by

$$d\mu = \frac{1}{Z} e^{-V} d\mu_c \tag{3.54}$$

Our first observation is that the semi-group e^{tL}, associated with the weak solution, first restricted to bounded measurable functions,

$$(e^{tL} f)(x) = E_x^{(W)} \left(f(x_t) \, e^{\xi_0^t}\right) \tag{3.55}$$

is symmetric in $L^2(d\mu)$.

Recall from the previous subsection that e^{tL} is a contraction on L^∞, so that on L^∞, (3.55) is well defined. Let $f, g \in L^\infty$

$$(f, e^{tL} g)_{L^2(d\mu)} = \int d\mu_c(x) \, e^{-V(x)} f(x) \, E_x^{(W)}(f(x_t) \, e^{\xi_0^t})$$

$$= \int d\mu_c(x) \, e^{-\frac{1}{2} V(x)} f(x) \, E_x^{(W)}\left(e^{-\frac{1}{2} V(x_t)} f(x_t) e^{-\int_0^t \vartheta(x_s) \, ds}\right)$$

(3.56)

where we have used (3.47) and defined

$$\vartheta(x) \equiv -\frac{1}{2} L_o V(x) + \frac{1}{8} \| C^{\frac{1-\epsilon}{2}} \nabla V(x) \|^2 \tag{3.57}$$

Define also,

$$\tilde{f}(x) = e^{-\frac{1}{2} V(x)} f(x) \tag{3.58}$$

and

$$e^{t\tilde{L}} \tilde{f}(x) = E_x^{(W)}(\tilde{f}(x_t) e^{-\int_o^t \vartheta(x_s) ds}) \tag{3.59}$$

Hence from (3.56), (3.58, (3.59)

$$(f, e^{tL} g)_{L^2(d\mu)} = (\tilde{f}, e^{t\tilde{L}} \tilde{g})_{L^2(d\mu_c)} = (\tilde{g}, e^{t\tilde{L}} \tilde{f})_{L^2(d\mu_c)}$$

$$= (g, e^{tL} f)_{L^2(d\mu)} \tag{3.60}$$

where we have used that $e^{t\tilde{L}}$ is symmetric in $L^2(d\mu_c)$.
Hence e^{tL} is symmetric in $L^2(d\mu)$, restricted to bounded measurable functions.

From § 3.2 we also have

$$e^{tL} 1 = 1 \tag{3.61}$$

since $E_x^{(W)}(e^{\xi_o^t}) = 1$

Hence, using symmetry in $L^2(d\mu)$, and (3.61)

$$\int d\mu\, e^{tL} f = (1, e^{tL} f)_{L^2(d\mu)} = (e^{tL} 1, f)_{L^2(d\mu)} = (1,f)_{L^2(d\mu)}$$

$$= \int d\mu\, f. \qquad (3.62)$$

Hence μ is an invariant measure, and since e^{tL} is positivity improving (by the argument of § 2.5 A), μ is the unique invariant measure (§ 2.5 C). Hence e^{tL} is an ergodic semigroup on L^∞, (§ 2.5 E), and by (§2.5F), e^{tL} is a strongly continuous contraction on $L^p(d\mu)$, $1 \le p \le \infty$. In particular e^{tL} is a strongly continuous self adjoint contraction on $L^2(d\mu)$ and its infinitesimal generator L is a self-adjoint operator.

CHAPTER IV

THE FINITE VOLUME $(\phi^4)_2$ EUCLIDEAN MEASURE, [7,8,9]

In Chapter III we have considered the O.U. process in $\mathbb{R}^N$ and perturbations through drift vector fields. We saw that under suitable conditions one can obtain continuous Markov process as weak solutions of non-linear stochastic differential equations of the Langevin type. Moreover for appropriate "gradient vector fields", these Markov processes are symmetric with unique invariant measure μ. Thus one way to obtain μ is to take ergodic limits of the Markov process.

In this and following chapters we will apply this method in order to obtain Euclidean field theory measures as limiting distributions of Markov processes. The method may be called stochastic quantization [11]. As mentioned in the introduction, this has a long history in statistical mechanics. In quantum field theory, this method was shown to work rigorously (non-perturbatively) for the continuum $P(\varphi)_2$ model in [10], with the ultraviolet cutoff removed already at the level of the Markov process, i.e. the Markov process is renormalized. The following chapters are devoted to explaining the content of [10]. Recovering the $P(\varphi)_2$ measure this way can be considered to be a non-trivial (non perturbative) test case of stochastic quantization.

In order to explain all this, in the present chapter we record some important facts about the $P(\varphi)_2$ Euclidean field theory measure (in finite space-time volume), [7-9].

4.1 Gaussian probability measures on $\mathcal{D}'(\Lambda)$.

Let $\Lambda = (-L, L)^2 \subset \mathbb{R}^2$, be a bounded open set. $C_o^\infty(\Lambda)$ consists of C^∞ functions with compact support in Λ. Let $K \subset \Lambda$ be a compact set. Then $\mathcal{D}_K(\Lambda)$ consists of C^∞ functions with support in K and convergence is with respect to seminorms $\|\cdot\|_{K,s}$

$$\|f\|_{K,s} = \sup_{x \in K} \left| D^s f(x) \right| \qquad (4.1)$$

$s=0, 1, 2, \ldots$. Then recall that the test function space $\mathcal{D}(\Lambda)$ is

$$\mathcal{D}(\Lambda) = \bigcup_{K \subset \Lambda} \mathcal{D}_K(\Lambda) \qquad (4.2)$$

considered as a strict inductive limit, [5]. If f is a test function it is necessarily C^∞ with support in some compact $K \subset \Lambda$; a sequence $\{f_p\}$, $p = 1, 2, \ldots$, of test functions converges, if they all have support in some compact K, and they converge in every seminorm $\|\cdot\|_{K,s}$, $s=0, 1, 2, \ldots$. $\mathcal{D}'(\Lambda)$, the topological dual i.e. the space of continuous linear functionals on $\mathcal{D}(\Lambda)$, is the space of distributions.

Let C be a bounded, selfadjoint, positive operator on $L^2(\Lambda)$. By Minlos' theorem [7,8] we have a Borel* probability measure $\mu_c^{(o)}$ on $\mathcal{D}'(\Lambda)$ which is Gaussian of mean 0 and covariance C. In other words, for $f \in \mathcal{D}(\Lambda)$.

---

* The Borel field Σ_Λ is the σ - completion of the algebra of cylinder sets.

$$\int_{\mathcal{D}'(\Lambda)} d\mu_c^{(o)}(\varphi)\, e^{i\varphi(f)} = e^{-\frac{1}{2}(f,\, Cf)_{L^2(\Lambda)}} \tag{4.3}$$

We shall choose

$$C = (-\Delta_D + 1)^{-1} \tag{4.4}$$

where Δ_D is the Dirichlet Laplacian on $L^2(\Lambda)$ (i.e. Dirichlet boundary condition on $\partial\Lambda$). (We omit the suffix D on C). C is the Dirichlet covariance. It is also a compact operator on $L^2(\Lambda)$. Let $\{\lambda_n\}$, each $\lambda_n > 0$, be the eigen values and $\{e_n\}$ the corresponding orthonormal eigen functions (which are C^∞). We have $\lambda_n \longrightarrow 0$, as $n \longrightarrow \infty$. Note that $\sum_{n=1}^{\infty} \lambda_n$ diverges, but

$$\sum_{n=1}^{\infty} \lambda_n^{1+\delta} < \infty \tag{4.5}$$

for any $\delta > 0$. $\mu_c^{(o)}$ is the Dirichlet Euclidean free field measure. Dirichlet boundary conditions will be in use in this and succeeding chapters. Thus the suffix D is omitted in C. However inorder to simplify estimates, sometimes we will use free boundary conditions in intermediate steps together with domination arguments. The free covariance C_F is the restriction of $(-\Delta+1)^{-1}$, on $L^2(\mathbb{R}^2)$, to $L^2(\Lambda)$. $\mu_{c_F}^{(o)}$ is the corresponding Gaussian measure on $\mathcal{D}'(\Lambda)$. We have the following FACTS [chapter VII, [8]].

1) Restricted to any compact $K \subset \Lambda$, $\mu_c^{(o)}$ and $\mu_{c_F}^{(o)}$ are equivalent measures and $d\mu_c^{(o)}/d\mu_{c_F}^{o}$ is in $L^p(\mathcal{D}', d\mu_{c_F}^{o})$, some $p > 1$

2) For $f \in L^2(\Lambda)$

$$(f, Cf)_{L^2(\Lambda)} \leq (f, C_F f)_{L^2(\Lambda)} \tag{4.6}$$

3) For integral kernels, with $x \neq y \in \Lambda$

$$C(x,y) \leq C_F(x-y) \tag{4.7}$$

4) $$\delta C(x) = \lim_{y \to x} \delta C(x,y) = \lim_{y \to x} [C_F(x-y) - C(x,y)] < \infty \tag{4.8}$$

for all $x \in \Lambda$.

Define

$$B = C_F - C \tag{4.9}$$

Then because of 2)

$$(f, Bf)_{L^2(\Lambda)} \geq 0 \tag{4.10}$$

and B is a positive, selfadjoint, bounded operator on $L^2(\Lambda)$. Hence it can be considered as the covariance of another Gaussian measure $\mu_B^{(o)}$ on $\mathcal{D}'(\Lambda)$. Now consider the product measure $\mu_B \otimes \mu_c$ on $\mathcal{D}'(\Lambda) \times \mathcal{D}'(\Lambda)$. Let F be a measurable function on $\mathcal{D}'(\Lambda)$. We define a measurable function $F(.,.)$ on $\mathcal{D}'(\Lambda) \times \mathcal{D}'(\Lambda)$, by

$$F(\varphi', \varphi) = F(\varphi' + \varphi) \tag{4.11}$$

Then

$$\int_{\mathcal{D}'(\Lambda)\times\mathcal{D}'(\Lambda)} d\mu_B(\varphi')\, d\mu_c(\varphi)\, F(\varphi',\varphi) = \int_{\mathcal{D}'(\Lambda)} d\mu_{c_F}(\varphi)\, F(\varphi) \tag{4.12}$$

§ 4.2 Wick powers and the interacting $(\varphi^4)_2$ Euclidean (Dirichlet) measure.

We can define Wick powers $:\varphi^n:_c(f)$ in the standard way, (C is the Dirichlet covariance). Use $\{(e_n, \lambda_n)\}$, where λ_n, e_n are eigen values / eigen functions of C, to define the sequence of random variables $\{\varphi^{(N)}(f)\}$:

$$\varphi^{(N)}(f) = \sum_{n=1}^{N} \varphi_n\, (e_n, f) \tag{4.13}$$

where the φ_n are independent Gaussian random variables of mean 0 and covariance λ_n. Note that $\varphi^{(N)}(f)$ are cylinder functions and

$$\int d\mu_c\, \varphi^{(N)}(f)\, \varphi^{(N)}(g) = (f, C_{(N)}\, g) \tag{4.14}$$

and $C_{(N)}$ has integral kernel

$$C_{(N)}(x,y) = \sum_{n=1}^{N} \lambda_n\, e_n(x)\, e_n(y) \tag{4.15}$$

As $N \longrightarrow \infty$ $C_{(N)}$ converges in norm to C.

We have

$$\varphi^{(N)}(f) \xrightarrow[N\to\infty]{} \varphi(f) \text{ in } L^p(d\mu_c, \mathcal{D}'(\Lambda)) \quad 1 \le p < \infty \tag{4.16}$$

Let

$$H_n(y;\lambda) = \frac{(-\lambda)^n}{n!} e^{\frac{y^2}{2\lambda}} \left(\frac{d}{dy}\right)^n e^{-\frac{y^2}{2\lambda}} \qquad (4.17)$$

be the $n^{\underline{th}}$ Hermite polynomial in 1 variable (see 3.16).

Then the Wick power $:(\varphi^{(N)}(x))^n:_c$ is defined by substitution:

$$:(\varphi^{(N)}(x))^n:_c = \sqrt{n!}\, H_n(\varphi^{(N)}(x)\,;\, C_{(N)}(x,x)) \qquad (4.18)$$

and

$$:(\varphi^{(N)})^n:_c(f) = \int_\Lambda :(\varphi^{(N)}(x))^n:_c f(x) d^2x \qquad (4.19)$$

$:(\varphi^{(N)})^n:_c(f)$ is a well defined random variable in

$L^p(d\mu_c, \mathcal{D}'(\Lambda))$, $1 \le p < \infty$

Note that for $N > M$,

$$\| :(\varphi^{(N)})^n:_c(f) - :(\varphi^{(N)})^n:_c(f) \|^2_{L^2(d\mu_c)}$$

$$= (f, C^n_{(N)} * f)_{L^2(\Lambda)} - (f, C^n_{(M)} * f)_{L^2(\Lambda)} \qquad (4.20)$$

where $*$ is convolution product with $(C^n_M * f)(x) = \int dy\, [C_M(x,y)]^n f(y)$.

Note that $C_N \longrightarrow C$, as $N \longrightarrow \infty$ and $(f, C^n * f)_{L^2(\Lambda)} < \infty$.

Hence taking $N, M \longrightarrow \infty$, the r.h.s. of 4.20 $\longrightarrow 0$. Hence the sequence $:(\varphi^{(N)})^n:_c$ converges as $N \longrightarrow \infty$ in $L^2(d\mu_c)$. In a similar way one deduces convergence in $L^p(d\mu_c)$, $1 \leq p < \infty$. The limit is denoted $:\varphi^n:_c$. (Without Wick ordering φ^n is not μ_c measurable because of ultraviolet divergences)

Domination

From (4.9 - 4.12), we can write the Gaussian random variable φ' with mean 0 and covariance C_F as the sum of independent random variables:

$$\varphi' = \varphi + \varphi_B \tag{4.21}$$

where φ is Gaussian with mean 0 , covariance C and φ_B is Gaussian with mean 0 and covariance B. Then we have:

$$:(\varphi'(f))^n:_{C_F} = \sum_{m=o}^{n} \binom{n}{m} :(\varphi(f))^m:_c :(\varphi_B(f))^{n-m}:_B \tag{4.22}$$

From (4.21) and (4.22), we have

$$:\varphi^n:_c(f) = \int d\mu_B(\varphi_B) :(\varphi+\varphi_B)^n:_{C_F}(f) \tag{4.23}$$

Hence, using that μ_B is a probability measure,

$$\left| :\varphi^n:_c(f)\right|^p \leq \int d\mu_B(\varphi_B) \left| :(\varphi+\varphi_B)^n:_{C_F}(f)\right|^p$$

and

$$\left\| :\varphi^n:_c(f)\right\|^p_{L^p(d\mu_c)} \leq \int d\mu_B(\varphi_B)\, d\mu_c(\varphi_c) \left| :(\varphi+\varphi_B)^n:_{C_F}(f)\right|^p$$

$$= \int d\mu_{C_F}(\varphi') \left| :(\varphi')^n:_{C_F}(f)\right|^p, \text{ by } (4.12)$$

$$\| : \varphi^n :_c (f) \|_{L^p(d\mu_c)} \leq \| :\varphi^n :_{c_F} (f) \|_{L^p(d\mu_{c_F})} \tag{4.24}$$

2) Choose

$$V(\varphi) = \frac{\lambda}{4} \int_\Lambda :\varphi^4:_c(x)\, d^2x\ , \ \lambda > 0, \text{ real.} \tag{4.25}$$

$$\text{Then} \quad V \in L^p(d\mu_c), \quad 1 \leq p < \infty \tag{4.26}$$

Moreover,

$$e^{-V} \in L^1(d\mu_c) \tag{4.27}$$

The last statement again follows from a domination argument, and Nelson's estimate, [7-9],

$$e^{-V_F} \in L^1(d\mu_{c_F}) \tag{4.28}$$

where

$$V_F(\varphi) = \frac{\lambda}{4} \int_\Lambda :\varphi^4:_{c_F}(x)\, d^2x, \quad \lambda > 0 \tag{4.29}$$

We will not prove Nelson's estimate here, since a more complicated estimate (of which 4.28 is a special case), will be proved later (chapter VI). The proof of (4.27) follows , [8],:

Using (4.23), (4.25) can be written

$$V(\varphi) = \int d\mu_B(\varphi_B)\ V_F(\varphi + \varphi_B) \tag{4.30}$$

and by Jensen's inequality

$$e^{-V(\varphi)} \leq \int d\mu_B(\varphi_B)\, e^{-V_F(\varphi+\varphi_B)} \tag{4.31}$$

whence

$$\int d\mu_c(\varphi)\, e^{-V(\varphi)} \leq \int d\mu_c(\varphi)\, d\mu_B(\varphi_B)\, e^{-V_F(\varphi+\varphi_B)}$$

$$= \int d\mu_{c_F}(\varphi')\, e^{-V_F(\varphi')} \quad \text{, by} \quad (4.12)$$

$$< \infty \quad \text{, by } (4.28). \tag{4.32}$$

The interacting $(\varphi^4)_2$ Dirichlet - Euclidean measure μ on $\mathcal{D}'(\Lambda)$ is defined by

$$d\mu = \frac{1}{Z}\, d\mu_c^{(o)}(\varphi)\, e^{-V(\varphi)} \tag{4.33}$$

where

$$Z = \int_{\mathcal{D}'(\Lambda)} d\mu_c^{(o)}(\varphi)\, e^{-V(\varphi)} < \infty \text{, by} \quad (4.27)$$

In the next two chapters we will show how μ can be rigorously recovered via "stochastic quantization". We will use the method of chapter III generalized to ∞ dimensions.

<u>Remark</u> We remind the reader of the following well-known obvious points. 1) The integer 'N' in intermediate steps is an ultraviolet (UV) cutoff. 2) As is well-known $\varphi^n, n \geq 2$

is not μ_c measurable, because of UV divergence indeed

$$\int d\mu_c \, (\varphi(x))^2 = C(x,x) = \infty$$

Since C is not of trace class.

3) $:\varphi^n:_c(f)$ which is a random variable and in $L^p(d\mu_c)$, $1 \leq p < \infty$ is the renormalized product. Thus $V_F(\varphi)$ (in 4.29) and $V(\varphi)$ (in 4.25) are in $L^p(d\mu_c)$, $1 \leq p < \infty$, but they are not bounded below (as infinite subtractions were involved) Hence Nelson's a priori estimate which proves that the renormalized interacting measure exists is non-trivial !

CHAPTER V

THE STOCHASTIC FREE FIELD, [10], SOME WICK POWERS AND AN ITO FORMULA

§ 5.1 The stochastic free Dirichlet field in two space-time dimensions is an infinite dimensional generalization of the O. U. process of § 3.1. It is an O. U. process with values in $\mathcal{D}'(\Lambda)$ with the Gaussian measure $\mu_c^{(o)}$ of chapter IV as its unique invariant measure. It will satisfy a Langevin equation in $\mathcal{D}'(\Lambda)$.

We first introduce a Wiener process W_t with values in $\mathcal{D}'(\Lambda)$ and covariance:

$$E^{(W)}(W_t(f)W_s(g)) = \min(t,s)\,(f, C^{(1-\epsilon)} g)_{L^2(\Lambda)} \tag{5.1}$$

Here $E^{(W)}$ is integration with respect to the Wiener measure $P^{(W)}$ on $\Omega = C^o([0,\infty), \mathcal{D}'(\Lambda))$. The Wiener paths are assumed to start at the origin. $P^{(W)}$ can be characterized as follows. First define

$$C_t = t\,C^{1-\epsilon}, \quad t > 0 \text{ and } 0 \leq \epsilon \leq 1 \quad (\epsilon \text{ is held fixed}) \tag{5.2}$$

and $\mu_{c_t}^{(o)}$ the corresponding Gaussian measure on $\mathcal{D}'(\Lambda)$ of mean 0 and covariance C_t. Then the Wiener process W_t has transition probability ($W_t + \varphi$ is the Wiener process starting at φ).

$$P^{(W)}(W_t + \varphi \in B) = p_t^{(W)}(\varphi, B) = \mu_{c_t}^{(o)}(B - \varphi) \tag{5.3}$$

where B is a Borel set in $\mathcal{D}'(\Lambda)$.

Once (5.3) is given, $P^{(W)}$ is completely characterized by, for $t > s$:

(i)

$$P^{(W)}(W_t - W_s \in B) = p_{t-s}^{(W)}(B) \tag{5.4}$$

and

(ii)

$$P^{(W)}(W_s \in A,\ W_t - W_s \in B) = p_s(0, A)\, p_{t-s}^{(W)}(0, B) \tag{5.5}$$

Then $P^{(W)}$ is a Gaussian Borel probability measure on $\Omega = C^o([0,\infty), \mathcal{D}'(\Lambda))$.

As noted in chapter III we have an eigen basis $\{(\lambda_n, e_n)\}$ of the Dirichlet covariance C, and $\{e_n\}$ is an orthonormal basis of $L^2(\Lambda)$. Consider the sequence of random variables

$$W_t^{(N)} = \sum_{n=1}^{N} \lambda_n^{\frac{1-\epsilon}{2}} \beta_t^{(n)} e_n \tag{5.6}$$

where $\beta_t^{(n)}$ are independent normalized Brownian motions. Then $W_t^{(N)}(f) \longrightarrow W_t(f)$ in the mean square convergence sense, We have

$$\lambda_n^{\frac{1-\epsilon}{2}} \beta_t^{(n)} = (W_t, e_n) \tag{5.7}$$

and the convergence is in $L^2(\Omega, d\,P^{(W)})$.

We now pass to the O. U. process which is of fundamental importance. This process $\varphi_t \in \mathcal{D}'(\Lambda)$ is obtained as the solution of the Langevin equation:

$$d\varphi_t = dW_t - \frac{1}{2} C^{-\epsilon} \varphi_t \, dt$$

$$\varphi_o = \varphi . \tag{5.8}$$

The solution φ_t is:

$$\varphi_t = e^{-\frac{t}{2} C^{-\epsilon}} \varphi + \int_o^t e^{-\frac{1}{2}(t-s) C^{-\epsilon}} dW_s \tag{5.9}$$

which gives φ_t as an explicit function $\varphi_t(\varphi;W)$ (with values in $\mathcal{D}'(\Lambda)$) of the starting point φ, and the Wiener process W_s, $0 \leq s \leq t$.

Note that the stochastic integral with values in $\mathcal{D}'(\Lambda)$.

$$I_t = \int_o^t e^{-\frac{1}{2}(t-s) C^{-\epsilon}} dW_s \tag{5.10}$$

is well defined in the Ito sense in $L^2(\Omega, dP^{(W)})$

To see this first introduce the finite dimensional approximation (5.6) $W_t^{(N)}$ to W_t in (5.10) and call the resulting object $I_t^{(N)}$:

$$I_t^{(N)}(f) = \sum_{n=1}^{N} \left(\int_o^t ds \; e^{-\frac{1}{2}(t-s) \lambda_n^{-\epsilon}} \lambda_n^{\frac{1-\epsilon}{2}} d\beta_s^{(n)} \right) (e_n, f) \tag{5.11}$$

(5.11) is well defined as a sum of 1- dimensional Ito stochastic integrals. Then, for any integer $p \geq 0$,

$$E^{(W)}(|I_t^{(N+p)}(f)-I_t^{(N)}(f)|^2) = \sum_{n=N+1}^{N+p} (f,e_n)\lambda_n(1-e^{-t\lambda_n^{-\epsilon}})(e_n,f) \tag{5.12}$$

Now the sum

$$\sum_{n=1}^{\infty} (f,e_n)\lambda_n(1-e^{-t\lambda_n^{-\epsilon}})(e_n,f) = (f, C(1-e^{-tC^{-\epsilon}})f)_{L^2(\Lambda)} < \infty .$$

Hence as $N \longrightarrow \infty$, the r.h.s. of (5.12) $\longrightarrow 0$. Hence the sequence $I_t^{(N)}$ converges in $L^2(\Omega, d\,P^{(W)})$, and the limit defines (5.10) as an Ito stochastic integral. By the method of Appendix 2 of chapter 1, we obtain that I_t is almost surely continuous. Note that the solution (5.9) gives a map

$$\mathcal{D}'(\Lambda) \times \Omega \longrightarrow \mathcal{D}'(\Lambda)$$

$$(\varphi, W) \longrightarrow \varphi_t(\varphi; W) \tag{5.13}$$

which is jointly measurable.

If we introduce the eigen basis $\{(e_n, \lambda_n)\}$ of C, $\{e_n\}$ supplying an orthonormal basis of $L^2(\Lambda)$, then φ_t can be written as a sum of random variables

$$\varphi_t = \sum_{n=1}^{\infty} \varphi_t^{(n)} e_n \tag{5.14}$$

where the $\varphi_t^{(n)}$ are the independent O.U. processes of § 3.1, equation (3.3) (where we identify φ_t^n with x_t^n). Here $\varphi_t^{(n)} = (\varphi_t, e_n)$ and the sum (5.14) converges in $L^2(\Omega, d\,P^W)$.

From the Mehler formula for the transition probability of each component process derived in § 3.1, we obtain the Mehler formula for the transition probability

$$p_t(\varphi, B) = P^{(W)}(\varphi_t(\varphi; W) \in B) \tag{5.15}$$

of φ_t. Define

$$\tilde{C}_t = (1 - e^{-tC^{-\epsilon}})\, C \tag{5.16}$$

Then

$$p_t(\varphi, B) = \mu^{(o)}_{\tilde{C}_t}(B - e^{-\frac{t}{2} C^{-\epsilon}} \varphi) \tag{5.17}$$

which is the Mehler formula. (Here $\mu^{(o)}_{\tilde{C}_t}$ is the Gaussian measure on $\mathcal{D}'(\Lambda)$ of mean 0 and covariance $\tilde{C}_t$).

As $t \longrightarrow \infty$, we have from (5.16, 5.17)

$$p_t(\varphi, .) \underset{t \to \infty}{\longrightarrow} \mu^{(o)}_C \tag{5.18}$$

in the sense of weak convergence of measures. As a consequence the O.U. semigroup e^{tL_o}

$$(e^{tL_o} f)(\varphi) = \int p_t(\varphi, d\varphi') f(\varphi') = E^{(W)}(f(\varphi_t(\varphi; W))) \tag{5.19}$$

which is a contraction on $L^{\infty}(\mathcal{D}'(\Lambda), \mu^{(o)}_c)$ is ergodic and mixing (§ 2.5 E) with unique invariant measure $\mu^{(o)}_c$. Furthermore (§ 2.5 F), e^{tL_o} is a strongly continuous contractive semigroup on $L^p(\mathcal{D}'(\Lambda), d\mu_c^{(o)})$ $1 \leq p \leq \infty$

Furthermore, as in § 3.1, e^{tL_o} is selfadjoint on $L^2(d\mu_c)$, and

$$e^{tL_o} 1 = 1 \tag{5.20}$$

We also have hypercontractivity, which is a regularity property of the O.U. semigroup: for $t > T$ sufficiently large [7-9]

$$\| e^{tL_o} f \|_{L^4(d\mu_c^{(o)})} \leq \| f \|_{L^2(d\mu_c^{(o)})} \tag{5.21}$$

Finally, from (5.18)

$$\lim_{t \to \infty} E_\varphi (f(\varphi_t)) = \int d\mu_c(\varphi) f(\varphi) \tag{5.22}$$

(for integrable functions), so we have "stochastically quantized" the free field.

§ 5.2 Wick ordering

For every starting point φ, and $t > 0$, the O.U transition probability gives a Gaussian measure $p_t(\varphi, .)$ given by (5.17).

Fix $\epsilon > 0$. Then for every $t > 0$, $e^{-tC^{-\epsilon}}$ is a traceclass operator in $L^2(\Lambda)$. This and (5.16) implies, [7,8], that $\mu_{c_t}^{(o)}$ and $\mu_c^{(o)}$ are equivalent Gaussian measures. Also note that for the difference of integral kernels

$$\delta \tilde{C}_t(x,y) = \tilde{C}_t(x,y) - C(x,y) = (e^{-tC^{-\epsilon}} C)(x,y) \tag{5.23}$$

satisfies (for $\epsilon > 0$)

$$\lim_{y \to x} \delta \tilde{C}_t (x,y) \equiv \delta \tilde{C}_t (x,x) < \infty. \qquad (5.24)$$

Next note that $p_t(\varphi,.)$ is obtained from $\mu^{(o)}_{\tilde{C}_t}$ by translation by the vector $e^{-\frac{t}{2}C^{-\epsilon}}\varphi$, where $\varphi \in \mathcal{D}'(\Lambda)$.

<u>Fix $\epsilon > 0$.</u> Then for every $t > 0$, convolution with $e^{-\frac{t}{2}C^{-\epsilon}}(x,y)$ is infinitely regularizing, and the vector in question is a smooth function. It follows [7,8] that $p_t(\varphi,.)$ and $\mu^{(o)}_{\tilde{C}_t}$ are equivalent Gaussian measures on $\mathcal{D}'(\Lambda)$. combining all this we have : <u>for fixed $\epsilon > 0$, for every $t > 0$ and starting point φ, $p_t(\varphi,.)$ and $\mu^{(o)}_C$ are equivalent</u> Gaussian measures on $\mathcal{D}'(\Lambda)$

It can be checked that the Radon-Nikolym derivative

$$f_{t,\varphi}(\varphi') = \frac{p_t(\varphi, d\varphi')}{\mu_C(d\varphi')} \qquad (5.25)$$

is in $L^2(d\mu_C)$. By explicit computation using the Mehler formula,

$$\begin{aligned} a(t,\varphi)^2 &\equiv \int |f_{t,\varphi}(\varphi')|^2 \, d\mu_C(\varphi') \\ &= [\det(I - e^{-2tC^{-\epsilon}})]^{-\frac{1}{2}} \exp\{(\varphi, C^{-1} e^{-tC^{-\epsilon}} (1+e^{-tC^{-\epsilon}})\varphi)_{L^2(\Lambda)}\} \\ &< \infty \end{aligned} \qquad (5.26)$$

since, for $t, \epsilon > 0$, $e^{-tC^{-\epsilon}}$ is of trace class and maps distributions into smooth functions.

The above properties of $p_t(\varphi, d\varphi')$ which hold when $\epsilon > 0$ are <u>crucial</u>. <u>For this reason we fix $\epsilon > 0$ in this and</u>

succeeding chapters.

Suppose now $h \in L^{2p}(d\mu_c)$. Then

$$E^{(W)}(|h(\varphi_t(\varphi; W))|^p) = \int p_t(\varphi, d\varphi')|h(\varphi')|^p$$

$$= \int \mathcal{P}_{t,\varphi}(\varphi')|h(\varphi')|^p \quad d\mu_c(\varphi')$$

$$\leq a(t, \varphi) \quad \|h\|^p_{L^{2p}(d\mu_c)} \tag{5.27}$$

As a consequence, provided $0 < \epsilon \leq 1$

$$h \in L^{2p}(d\mu_c), \ 1 \leq p < \infty \Longrightarrow h(\varphi_t) \in L^p(dP^{(W)}, \Omega) \quad 1 \leq p < \infty \tag{5.28}$$

where $\varphi_t(\varphi; W)$ is the O. U. process.

Note that the sample paths of φ_t are almost surely continuous. Using this and (5.28), where we assume $h \in L^{2p}(d\mu_c)$ for every p, $1 \leq p < \infty$, we have that

$$\int_0^t ds \quad h(\varphi_s) \tag{5.29}$$

belongs to $L^p(\Omega, dp^{(W)})$ for every p, $1 \leq p < \infty$.

§ 5.3 : An Ito formula

We shall now derive an Ito formula, valid in $L^2(\Omega, dp^W)$ which will be in use in the next chapter when we obtain the stochastic quantization of the interacting $(\varphi^4)_2$ theory. So far we chose $0 < \epsilon \leq 1$, for reasons explained above. We shall now be led to restrict the range of values of ϵ even further.

Let us now restrict ϵ to the range $0 < \epsilon < \frac{1}{2}$. In this case each of the functions

$$V(\varphi) = \frac{\lambda}{4} \int_{\Lambda} d^2x \, :\varphi^4:_c (x)$$

$$H(\varphi) = \left(:\varphi^3:_c , C^{1-\epsilon} :\varphi^3:_c \right)_{L^2(\Lambda)} \tag{5.30}$$

$$G(\varphi) = \lambda : (\varphi^3 , C^{-\epsilon} \varphi):_c$$

belong to $L^p(d\mu_c)$ for every p such that $1 \leq p < \infty$. (This is easily checked by Feynman graph computations. $0 < \epsilon < 1$ would have sufficed for V, H. The stronger condition $0 < \epsilon < \frac{1}{2}$ is necessary for G). As a consequence, by (5.29), (5.28)

$$\int_0^t ds \quad h(\varphi_s) \quad , h = V, H, G \tag{5.31}$$

belong to $L^p(\Omega, dp^{(W)})$, $1 \leq p < \infty$.

As explained earlier in this section, the infinitesimal generator L_o of the O.U. semigroup e^{tL_o} is selfadjoint on $L^2(d\mu_c)$. The domain of L_o contains as a dense subspace the space of twice differentiable cylindrical functions which with their derivatives are in $L^2(d\mu_c)$. On this subspace L_o is realized as a second order elliptic differential operator.

Introducing the orthonormal basis $\{e_n\}$ of $L^2(\Lambda)$ where $\{\lambda_n, e_n\}$ is an eigen basis of the covariance operator C, define

$$\varphi^{(N)} = \sum_{n=1}^{n} \varphi_n e_n \tag{5.32}$$

If h is a Wick power e.g. any of the functions in (5.30), then $h(\varphi^{(N)})$ belongs to the above subspace and $h(\varphi^{(N)}) \longrightarrow h(\varphi)$ in $L^2(d\mu_c)$.

Moreover,

$$L_o h(\varphi^{(N)}) = \frac{1}{2} \sum_{n=1}^{N} \left(\lambda_n^{(1-\epsilon)} \frac{\partial^2}{\partial \varphi_n^2} - \lambda_n^{-\epsilon} \varphi_n \frac{\partial}{\partial \varphi_n} \right) h(\varphi^{(N)}) \tag{5.33}$$

We recognize on the r.h.s. of (5.33) the finite dimensional Hermite operator (3.14) of § 3.1.

Now choose h = V, the function defined in (5.30). Then by straight-forward calculation:

$$L_o V(\varphi^{(N)}) = -\frac{1}{2} G(\varphi^{(N)}) \tag{5.34}$$

where G is given in (5.30).

By finite dimensional Ito calculus (of earlier chapters) applied to the function $V(\varphi^{(N)})$, with

$$\varphi_t^{(N)} = \sum_{n=1}^{N} \varphi_t^{(n)} e_n = \sum_{n=1}^{N} (\varphi_t, e_n)\, e_n$$

the finite dimensional approximation to the O. U. process $\varphi_t(\varphi_t^{(N)}(W) \longrightarrow \varphi_t(W)$ in $L^2(dP^W, \Omega))$ starting at $\varphi^{(N)}$

we obtain

$$V(\varphi_t^{(N)}(W)) = V(\varphi^{(N)}) + \lambda \int_0^t (:(\varphi_s^{(N)}(W))^3:, dW_s)_{L^2(\Lambda)}$$

$$+ \int_0^t ds \; L_o \; V(\varphi_s^{(N)}(W))$$

or

$$V(\varphi_t^{(N)}(W)) = V(\varphi^{(N)}) + \int_0^t \lambda \int_0^t (:(\varphi_s^{(N)}(W))^3:, dW_s) - \frac{1}{2}\int_0^t ds\, G(\varphi_s^{(N)}(W)) \tag{5.35}$$

where we have used (5.34).

If we call

$$I_t^{(N)} = \int_0^t (:(\varphi_s^{(N)}(W))^3:, d\,W_s).$$

Then, for any $p > 0$

$$E^{(W)}(|I_t^{(N+p)} - I_t^{(N)}|^2)$$

$$= E^{(W)} \int_0^t ds \left\| C^{\frac{1-\epsilon}{2}} :(\varphi_s^{(N+p)}(W))^3: - C^{\frac{1-\epsilon}{2}} :(\varphi_s^{(N)}(W))^3: \right\|^2_{L^2(\Lambda)}$$

$$\longrightarrow 0, \text{ as } N \longrightarrow \infty \tag{5.36}$$

Since

$$E^{(W)} \int_0^t ds \; (:\varphi_s^3:, C^{1-\epsilon} :\varphi_s^3:)_{L^2(\Lambda)} < \infty$$

by (5.30 - 5.31). Hence $I_t^{(N)} \xrightarrow[N \to \infty]{} I_t$ in $L^2(\Omega, dP^W)$. By (5.30), (5.31) all other terms in (5.35) also converge in $L^2(\Omega, dP^W)$. Hence in (5.35) taking $N \to \infty$ in $L^2(\Omega, dP^W)$ <u>we obtain the Ito formula</u>.

$$V(\varphi_t(W)) = V(\varphi) + \lambda \int_0^t (:(\varphi_s(W))^3:, dW_s) - \frac{1}{2}\int_0^t ds\; G(\varphi_s(W)) \quad (5.38)$$

<u>valid in $L^2(\Omega, dP^W)$. This formula will be used in the next chapter</u>.

<u>Remark</u> Note that § 5.1, § 5.2 are actually valid in any number of dimensions. This is no longer true for § 5.3 which is valid in two dimensions. However for dimensions bigger than 2, Wick ordering does not suffice for defining renormalized field products, as is well-known.

CHAPTER VI

THE STOCHASTIC $(\varphi^4)_2$ MODEL, [10]

§ 6.1 In the previous chapter (§ 5.1), we constructed a stochastic (Dirichlet) free field as an Ornstein-Uhlenbeck process in $\mathcal{D}'(\Lambda)$ with the free (Dirichlet) Euclidean field measure $\mu_c^{(o)}$ on $\mathcal{D}'(\Lambda)$ as its unique invariant measure. This was a generalization of the finite dimensional case treated in § 3.1. In the present chapter we will construct a stochastic (Dirichlet) $(\varphi^4)_2$ model. It is a continuous ergodic Markov process in $\mathcal{D}'(\Lambda)$ with the interacting (Dirichlet) $(\varphi^4)_2$ Euclidean field theory measure μ of (4.33) as its unique invariant measure. The model is specified by demanding that the Markov process is a diffusion and satisfies, in the sense of weak solutions, a nonlinear stochastic differential equation obtained by perturbing the Langevin equation (5.8) by a suitable drift term. It is a generalization to infinite dimensions of § 3.2, and more particularly of § 3.3.

The main point is to choose the drift carefully (it is a generalization of (3.42)) so that (i) the differential generator L is symmetric in $L^2(d\mu)$ and (ii) the perturbation is not too singular. (i) ensures that the equilibrium measure μ remains unchanged and (ii) ensures that the drift vector field lives in a suitable dense subspace of $\mathcal{D}'(\Lambda)$, so that the fundamental estimates of the type (3.43 a),b)) go through. In this case § 3.3 A,B generalize immediately. However for the stability estimates (especially the first of 3.43 b)) to go through, the interval in which the free parameter ϵ lies must be quite small, infact $0 < \epsilon < \frac{1}{10}$. We shall explain as we go along how the interval in which ϵ lies gets narrowed down to $(0, \frac{1}{10})$. It is only in this range that we have obtained the desired weak solution. (We emphasize that the stability estimates are inherently non-perturbative; if the non-linear drift is μ_c - Wick ordered*, then as follows from § 5.2, the perturbation series "solution" of the non-linear Langevin equation will have no ultraviolet divergences in the much bigger interval $0 < \epsilon \leq 1$). For full details on this chapter see [10].

§ 6.2 Recall that L_o, the differential generator of the O.U. process of chapter V is symmetric in $L^2(d\mu_c^{(o)})$, and given formally by:

$$L_o = \frac{1}{2}\iint_{\Lambda\times\Lambda} d^2x\, d^2y\, C^{1-\epsilon}(x,y)\frac{\delta^2}{\delta\varphi(x)\delta\varphi(y)} - \frac{1}{2}\iint_{\Lambda\times\Lambda} d^2x\, d^2y\, C^{-\epsilon}(x,y)\varphi(y)\frac{\delta}{\delta\varphi(x)} \tag{6.1}$$

If we add a drift vector field b on $\mathcal{D}'(\Lambda)$ as a perturbation and demand

$$L = L_o + \int_{\Lambda} d^2x\, b(\varphi)(x)\frac{\delta}{\delta\varphi(x)} \tag{6.2}$$

to be formally symmetric in $L^2(d\mu)$, then we must choose

* In other words the vacuum and mass counter terms are "borrowed" from the equilibrium theory; this renormalization does not affect the temporal homogeneity of the process.

$$b(\varphi) = -\frac{1}{2} C^{1-\epsilon} \frac{\delta V}{\delta \varphi} \text{ (in operator notation)} \quad (6.3)$$

$$= -\frac{\lambda}{2} C^{1-\epsilon} :\varphi^3:_c \quad (6.4)$$

where V is given in (5.30). (Notice that (6.3) is just the generalization of 3.42 (§3.3) to infinite dimensions). The corresponding stochastic differential equation (non-linear Langevin equation with b as perturbing drift) to be considered is:

$$\hat{\varphi}_t = dW_t - \frac{1}{2} (C^{-\epsilon} \hat{\varphi}_t + \lambda C^{1-\epsilon} :\hat{\varphi}_t^3:_c) dt$$

$$\hat{\varphi}_o = \varphi \quad (6.5)$$

which can also be written as an integral equation:

$$\hat{\varphi}_t = \varphi_t - \frac{\lambda}{2} \int_0^t ds\; e^{-\frac{1}{2}(t-s) C^{-\epsilon}} C^{1-\epsilon} :\hat{\varphi}_s^3:_c \quad (6.6)$$

where φ_t is the O.U. process of §5.1, 5.2.

Note that for $0 < \epsilon \leq 1$, the drift $b(\varphi)$ given by (6.4) is such that for every testfunction f, $b(\varphi)(f)$ belongs to $L^p(d\mu_c, \mathcal{D}'(\Lambda))$, $1 \leq p < \infty$. Hence by § 5.2, $b(\varphi_t)(f)$ belongs to $L^p(dP^W, \Omega)$, $1 \leq p < \infty$ where $\Omega = C^o([0,\infty), \mathcal{D}'(\Lambda))$ is our path space, and φ_t is the O.U. process. In particular, in this range $0 < \epsilon \leq 1$, equations (6.5) and (6.6) make sense in $L^2(dP^W, \Omega)$, (and the perturbation expansion of $\hat{\varphi}_t$ has no ultraviolet divergences). However, as mentioned earlier, to obtain the non-perturbative weak solution more stringent bounds on ϵ will have to be given. To obtain the weak solution of 6.5 - 6.6 we adopt the strategy of § 3.2 and § 3.3. We begin by defining a semigroup e^{tL} on $L^{\infty}(d\mu_c)$ by:

$$(e^{tL}f)(\varphi) = E^{(W)}(f(\varphi_t(\varphi;W))\ e^{\xi_o^t}) \tag{6.7}$$

where

$$\xi_o^t = \int_o^t (C^{-(1-\epsilon)} b(\varphi_s(\varphi;W)), dW_s)_{L^2(\Lambda)} - \frac{1}{2}\int_o^t ds \left\| C^{-\frac{(1-\epsilon)}{2}} b(\varphi_s(\varphi;W))\right\|^2_{L^2(\Lambda)}$$

or

$$\xi_o^t = -\frac{\lambda}{2}\int_o^t (:(\varphi_s(\varphi,W))^3:_c, dW_s) - \frac{\lambda^2}{8}\int_o^t ds \left\| C^{\frac{1-\epsilon}{2}} :(\varphi_s(\varphi,W))^3:_c \right\|^2_{L^2(\Lambda)}$$

(6.8)

where $\varphi_s(\varphi, W)$ is the O.U. process of (5.9).

For (6.7) to be well defined, it is necessary that ξ_o^t is a random variable. Using the isometry of the stochastic integral ($(L^2(\Omega, dP^W)$ limit of Ito integrals as in § 5.3), it is sufficient to check that the last term in (6.8) is a random variable. But this term can be identified with

$$-\frac{\lambda^2}{8}\int_o^t ds\ H(\varphi_s(\varphi;W)) \tag{6.9}$$

where H is given in (5.30) of § 5.3. From the analysis therein (§ 5.3, 5.2), we know that $H \in L^p(d\mu_c)$, $1 \leq p < \infty$, <u>provided</u> <u>$0 < \varepsilon < 1$</u>, and hence (6.9) belongs to $L^p(\Omega, dP^W)$, $1 \leq p < \infty$, (thus we have already narrowed down the range from $0 < \varepsilon \leq 1$). In this range, $0 < \varepsilon < 1$, ξ_o^t <u>is a random variable and</u>

we furthermore claim:

$$E^{(W)}\left(e^{\xi_o^t}\right) \le 1 \qquad (6.10)$$

From (6.10) it follows that e^{tL}, as given by 6.7, is a contraction on $L^{\infty}(d\mu_c)$

$$\| e^{tL} f \|_{L^{\infty}(d\mu_c)} \le \| f \|_{L^{\infty}(d\mu_c)} \qquad (6.11)$$

To prove (6.10), approximate φ_s by the sequence

$$\varphi_s^{(N)} = \sum_{n=1}^{N} \varphi_{s,n}\, e_n = \sum_{n=1}^{N} (\varphi_s, e_n)\, e_n$$

(see 5.14), and recall that $\varphi_s^{(N)}(f) \xrightarrow[N\to\infty]{} \varphi_s(f)$ in $L^2(\Omega, d^W)$.

$\varphi_s^{(N)}$ is the finite dimensional O.U. process. Define $\xi_o^{t,N}$ by substituting $\varphi_s^{(N)}$ for φ_s in (6.8). Then $\xi_o^{t,N} \xrightarrow[N\to\infty]{} \xi_o^t$ in $L^2(\Omega, d^W)$. But for each fixed N,

$$E^{(W)}\left(e^{\xi_o^{t,N}}\right) = 1 \qquad (6.12).$$

To see this note that for $N < \infty$, Wick constants are finite. So we can undo the Wick ordering in $:(\varphi^{(N)})^3:$, and the latter is now a C^{∞} function on $\mathbb{R}^N$. By chapter 2, we have a strong solution of the finite dimensional approximation to (6.5) (obtained by substituting $\hat{\varphi}_t^{(N)}$ for $\hat{\varphi}_t$, and $W_t^{(N)}$ for W_t in 6.5) defined upto an explosion time τ_{∞}. That $\tau_{\infty} = \infty$ can be shown by a simple application of the Hasminski test. Thus the strong solution of the finite dimensional system is defined for all times, Hence

$$e^{tL(N)}\ 1 = 1 \tag{6.13}$$

and by the Cameron- Martin formula (6.12) follows. (6.12) and Fatou's lemma $\Longrightarrow$ (6.10), and the latter $\Longrightarrow$ (6.11).

We now claim that for $0 < \epsilon < \frac{1}{10}$,

$$E^{(W)}(e^{\xi_o^t}) = 1\ , \quad \mu_c \quad \text{a.e.} \tag{6.14}$$

and, from § 3.2, this will ensure the existence of a weak solution. We know from § 3.2, that to obtain (6.14), it is sufficient to prove the stability bound,

$$E^{(W)}(e^{p\,\xi_o^t}) < \infty\ , \quad \text{for } 2 \leq p < \infty\ , \quad \mu_c \ \text{a.e.} \tag{6.15}$$

(6.15), together with (6.12), imply (6.14) by repeating the argument of Appendix II B, chapter II. Thus it boils down to proving (6.15), and for this we will have narrow down the interval in which ϵ lies to $0 < \epsilon < \frac{1}{10}$. Why this is so will be seen presently.

To prove (6.15) (in the range $0 < \epsilon < \frac{1}{10}$) we use the method of § 3.3. Recall that this involves the use of an Ito formula (3.46) which, in our case here, is nothing but (5.38) of chapter V (the validity of this in $L^2(\Omega, dP^W)$ only requires $0 < \epsilon < \frac{1}{2}$).

From the Ito formula (5.38) we obtain:

$$\xi_o^t = \frac{1}{2}\, V(\varphi) - \frac{1}{2}\, V(\varphi_t(\varphi;W)) - \frac{1}{4}\int_o^t ds\ G(\varphi_s(\varphi;W))$$

$$- \frac{\lambda^2}{8}\int_o^t ds \left\| C^{\frac{1-\epsilon}{2}} :\varphi_s(\varphi,W)^3:_c \right\|^2_{L^2(\Lambda)} \tag{6.16}$$

with V, G given by (5.30).

Hence,

$$E^{(W)}(e^{p\xi_o^t}) \leq e^{\frac{p}{2}V(\varphi)} E^{(W)}(e^{-\frac{p}{2}V(\varphi_t(\varphi;W)) - \frac{p}{4}\int_o^t ds\, G(\varphi_s(\varphi;W))})$$

$$\leq e^{\frac{p}{2}V(\varphi)} (E^{(W)}(e^{-p\,V(\varphi_t(\varphi;W))}))^{\frac{1}{2}} (E^{(W)}(e^{-\frac{p}{2}\int_o^t ds\, G(\varphi_s(\ ;W))}))^{\frac{1}{2}} \tag{6.17}$$

Now $\varphi_t(\varphi;W)$, given by (5.9), is $\mu_c \otimes P^{(W)}$ measurable map:

$$\mathcal{D}'(\Lambda) \times \Omega \longrightarrow \mathcal{D}'(\Lambda) \tag{6.18}$$

and using § 5.2 (5.28) and §5.3, $V(\varphi_t(\varphi;W))$ and $G(\varphi_t(\varphi;W))$ are $\mu_c \otimes P^{(W)}$ measurable maps

$$\mathcal{D}'(\Lambda) \times \Omega \longrightarrow \mathbb{R}. \tag{6.19}$$

Hence if we can prove

$$\left.\begin{aligned} &\int d\mu_c\, E^{(W)}(e^{-p\,V(\varphi_t(\varphi;W))}) < \infty \quad \text{(i)} \\ &\int d\mu_c\, E^{(W)}(e^{-\frac{p}{2}\int_o^t ds\, G(\varphi_s(\varphi;W))}) < \infty \quad \text{(ii)} \end{aligned}\right\} \tag{6.20}$$

Then (by Fubini's theorem)

$$\left.\begin{aligned} &E^{(W)}(e^{-p\,V(\varphi_t(\varphi;W))}) < \infty \\ &E^{(W)}(e^{-\frac{p}{2}\int_o^t ds\, G(\varphi_s(\varphi;W))}) < \infty \end{aligned}\right\} \quad \mu_c \text{ a.e} \tag{6.21}$$

which $\Longrightarrow$ from (6.17)

$$E^{(W)}(e^{p\xi_o^t}) < \infty, \quad \mu_c \text{ a.e.}$$

Thus it is sufficient to prove (6. 20) which is the same as the conditions (3. 50). We have

$$\int d\mu_c \; E^{(W)}(e^{-p\,V(\varphi_t(\varphi;W))}) = \int d\mu_c \; e^{tL_o} e^{-p\,V}(\varphi)$$

$$= \int d\mu_c \; e^{-pV(\varphi)} \quad \text{since } \mu_c \text{ is invariant under } e^{tL_o}$$

$$< \infty, \text{ by (4. 27)}$$

As for (6. 20) (ii), following exactly the steps leading to (3. 51), we have:

$$\int d\mu_c \; E^{(W)}\left(e^{-\frac{p}{2}\int_o^t ds\; G(\varphi_s(\varphi;W))}\right)$$

$$\leq \int d\mu_c \; e^{-p\frac{t}{2}\,G(\varphi)} \tag{6. 22}$$

We now have the fundamental stability bound given by

<u>Proposition</u> For $0 < \epsilon < \frac{1}{10}$,

$$\int d\mu_c \, e^{-p\frac{t}{2}\,G(\varphi)} < \infty, \; 1 \leq p < \infty \tag{6. 23}$$

where G is given by (5. 30), namely

$$G(\varphi) = \lambda : (\varphi^3, C^{-\epsilon}\varphi):_c \; , \lambda > 0 .$$

<u>Proof:</u> Given in the following section § 6. 3. <u>It is in the proof of (6. 23) that above condition on ϵ comes in.</u>

Given the above proposition, we have completed the proof of (6.14) Let $\mathcal{F}_s$ be the smallest σ-algebra with respect to which $W_{t'}, 0 \leq t' \leq s$ is measurable. Then similarly to (6.14) we also have for $s \leq t$

$$E^{(W)}\left(e^{\xi_s^t} \,\middle|\, \mathcal{F}_s\right) = 1 \quad , \mu_c \quad \text{a.e.} \tag{6.24}$$

Then, exactly as in § 3.2, we have a weak solution of (6.5) or (6.6). In other words, if for every $T > 0$ $\Omega_T = C^o([0,T], \mathcal{D}'(\Lambda))$. Then we have a Markovian family of Borel probability measures $\hat{P}_\varphi$ such that for $0 \leq t \leq T$ the process

$$\hat{Z}_t \equiv \hat{\varphi}_t + \frac{\lambda}{2} \int_0^t ds\; e^{-\frac{(t-s)}{2} C^{-\epsilon}} C^{1-\epsilon} :\hat{\varphi}_s^3:_C \tag{6.25}$$

(with $\hat{\varphi}_s$ a path in Ω) has the same $\hat{P}_\varphi$ joint probability distributions as the known distributions of the O.U. process φ_t of (5.9).

Remark: The measure $\hat{P}_\varphi$ is of course that given by the Kolmogoroff construction starting from the transition probabilities $\hat{p}_t(\varphi,.)$ given by (6,7):

$$\hat{p}_t(\varphi, B) = E^{(W)}\left(\chi\{\varphi_t(\varphi;W) \in B\}\, e^{\xi_o^t}\right) \tag{6.26}$$

where B is a Borel set in $\mathcal{D}'(\Lambda)$. (The countably additivity of Wiener measure together with (6.14) assures us that $\hat{p}_t(\varphi,.)$ are countably additive probability measures). The fact that $\hat{P}_{\varphi}$ is supported on $C^{o}([0,\infty), \mathcal{D}'(\Lambda))$ follows from an application of Kolmogoroff's theorem [see chapter I] since we have the estimate [10]:

For $0 \leq s \leq t \leq T$,

$$E^{(\ )}(|\hat{\varphi}_t(f) - \hat{\varphi}_s(f)|^4) \leq C\ |t-s|^2 (C_T)^{t-s} \tag{6.27}$$

where C, C_T are constants (depending on the test function f).

Let us now return to the semigroup e^{tL} given by (6.7) associated to the weak solution of (6.5). Observe that (6.14) implies:

$$e^{tL}\ 1 = 1 \tag{6.28}$$

Moreover e^{tL} is symmetric in $L^2(d\mu)$, exactly as in § 3.3, (eqn 3.55 - eqn. 3.60) since:

$$(f, e^{tL} g)_{L^2(d\mu)} = (\tilde{f}, e^{t\tilde{L}} \tilde{g})_{L^2(d\mu_c)} \tag{6.29}$$

where

(i) $\tilde{f} = \dfrac{1}{Z^{\frac{1}{2}}}\, e^{-\frac{1}{2}V} f$

This gives a unitary map $L^2(d\mu) \longrightarrow L^2(d\mu_c)$

and

(ii) $(e^{t\tilde{L}}\tilde{g})(\varphi) = E^{(W)}(\tilde{g}(\varphi_t(\varphi; W))\, e^{-\int_0^t ds\, \vartheta(\varphi_s(\varphi; W))})$

$$\mathcal{V}(\varphi) = \frac{\lambda}{4} :(\varphi^3, C^{-\epsilon}\varphi):_{L^2(\Lambda)} + \frac{\lambda^2}{8}(:\varphi^3:, C^{1-\epsilon}:\varphi^3:)_{L^2(\Lambda)}$$

(proposition giving (6, 23) assures stability)

(To obtain (6. 28) start from definition of μ, and 6.7, and use the Ito formula (5, 38)).

From the symmetry of $e^{t\tilde{L}}$ in $L^2(d\mu_c)$ follows the symmetry of e^{tL} in $L^2(d\mu)$. This fact together with (6. 28) implies for every f a bounded measurable function on $\mathcal{D}'(\Lambda)$:

$$\int_{\mathcal{D}'(\Lambda)} d\mu(\varphi)\, e^{tL} f(\varphi) = \int_{\mathcal{D}'(\Lambda)} d\mu(\varphi)\, f(\varphi) \tag{6. 30}$$

Hence μ is an invariant measure. Moreover e^{tL} is positivity improving (by the same argument as leading to (2. 33). Hence by the Perron- Frobenius argument, the invariant measure μ is unique. It follows that e^{tL} is an ergodic semigroup. Note that by virtue of § 2.5 F (which is independent of dimension), e^{tL} is a strongly continuous contraction on $L^p(d\mu)$, $1 \leq p \leq \infty$. Moreover on $L^2(d\mu)$, e^{tL} is selfadjoint and its infinitesimal generator L is a non-positive self adjoint operator (Langevin operator).

In our case however we have an even stronger result, namely e^{tL} is mixing:

$$\lim_{t\to\infty} (f, e^{tL} g)_{L^2(d\mu)} = \left(\int d\mu\, f\right)\left(\int d\mu\, g\right) \tag{6. 31}$$

As a consequence, for any bounded measurable function F on $\mathcal{D}'(\Lambda)$

$$\lim_{t \to \infty} E_{\varphi}^{(\hat{P})}(F(\hat{\varphi}_t)) = \int d\mu(\varphi)\, F(\varphi), \quad \mu \text{ a.e.} \tag{6.32}$$

which fulfills the program of stochastic quantization.

Remark: (6.31) follows from the fact that e^{tL} has not only a unique ground state (1) but also a mass gap. The latter fact follows from the representation 6.29, (ii), and a) e^{tL_0} is a hypercontractive semigroup with a mass gap, together with b) $e^{-\vartheta}$ is in $L^1(d\mu_c)$ which follows from proposition giving (6.23). This result is standard in constructive field theory [6] and the proof is omitted.

§ 6.3 In this section we will prove the proposition leading to (6.23)

Define

$$G \equiv :(\varphi^3, C^{-\epsilon}\varphi):_C. \tag{6.33}$$

Here C will be the covariance with free boundary conditions. In fact by the domination argument of § 4.2 leading to 4.32 the Proposition below implies a similar estimate for our field with Dirichlet boundary conditions in Λ.

We will prove , [10],

Proposition: For any $\lambda > 0$ and for ϵ restricted to the range $0 < \epsilon < \frac{1}{10}$

$$e^{-\lambda G} \in L^1(d\mu_c). \tag{6.34}$$

Note that $G \in L^p(d\mu_c), 1 \leq p < \infty$, for $\epsilon < \frac{1}{2}$. The restriction $\epsilon > 0$ was imposed to ensure that the transition probabilities of the O.U.

process are absolutely continuous with respect to μ_c. The upper bound $\epsilon < \frac{1}{10}$ turns out to be sufficient for (6.34) to hold.

The proof of the above proposition and (6.34) is based on a series of lemmata [10].

Lemma 1: For $0 < \epsilon < 1$,

$$(\varphi, C^{3-\epsilon}\varphi) \geq \int_\Lambda d^2x\, \varphi^4(x) \tag{6.35}$$

Proof: We let $\|\cdot\|_p$ denote the $L^p(\Lambda)$ norm.

Define

$$a_\epsilon = \int_0^\infty ds \frac{s^{-1+2\epsilon}}{1+s^2} > 0,$$

which converges for $0 < \epsilon < 1$.

We have the representation, converging for $0 < \epsilon < 1$,

$$(-\Delta+1)^\epsilon = a_\epsilon^{-1} \int_0^\infty ds\, s^{-1+2\epsilon}(1 - s^2(-\Delta+1+s^2)^{-1}).$$

Hence

$$(\varphi^3, (-\Delta+1)^\epsilon \varphi) = a_\epsilon^{-1} \int_0^\infty ds\, s^{-1+2\epsilon}(\|\varphi\|_4^4 - s^2(\varphi^3, C_{s^2+1}\varphi))$$

$$\geq a_\epsilon^{-1} \int_0^\infty ds\, s^{-1+2\epsilon}(\|\varphi\|_4^4 - s^2|(\varphi^3, C_{s^2+1}\varphi)|) \tag{6.36}$$

where

$$C_{s^2+1} = (-\Delta+1+s^2)^{-1}$$

Using Holder's inequality,

$$|(\varphi^3, C_{s^2+1}\varphi)| \leq \|\varphi\|_4^3 \, \|C_{s^2+1}\varphi\|_4 \tag{6.37}$$

Now use Young's convolution inequality

$$\|f * \varphi\|_p \leq \|f\|_q \|\varphi\|_r, \quad \frac{1}{p} = \frac{1}{q} + \frac{1}{r} - 1$$

with the choice f = integral kernel of C_{s^2+1}, $p = r = 4$, $q = 1$

Note that

$$\|f\|_1 \leq \frac{1}{s^2+1}$$

Hence

$$\|C_{s^2+1}\varphi\|_4 \leq \frac{1}{s^2+1}\|\varphi\|_4 \tag{6.38}$$

From (6.37), (6.38), we have

$$|(\varphi^3, C_{s^2+1}\varphi)| \leq \frac{1}{s^2+1}\|\varphi\|_4^4 \tag{6.39}$$

From (6.36), (6.39)

$$(\varphi^3, (-\Delta+1)^{\epsilon}\varphi) \geq a_\epsilon^{-1} \int_0^\infty ds\, s^{-1+2\epsilon}\left(1 - \frac{s^2}{s^2+1}\right)\|\varphi\|_4^4 = \|\varphi\|_4^4 .$$

Next we turn to Lemma 2 .

Define UV cutoff fields $\varphi_{\mathcal{H}}(x)$ by

$$\varphi_{\mathcal{H}}(x) = \int_{K \leq \mathcal{H}} \frac{d^2 K}{(2\pi)^2} \, e^{iK} \, \tilde{\varphi}(K).$$

Define

$$G_\varkappa = :(\varphi_\varkappa^3 , C^{-\epsilon} \varphi_\varkappa):_C = :(\varphi_\varkappa^3 , (-\Delta+1)^\epsilon \varphi_\varkappa):_C . \quad (6.40)$$

Then for $0 < \epsilon < \frac{1}{2}$

$$G_\varkappa \xrightarrow[\varkappa \to \infty]{} G, \text{ in } L^p(d\mu_c), \ 1 \leqq p < \infty.$$

Undoing the Wick ordering

$$G_\varkappa = (\varphi_\varkappa^3 , (-\Delta+1)^\epsilon \varphi_\varkappa) - 3(C_\varkappa(0)(\varphi_\varkappa , (-\Delta+1)^\epsilon \varphi_\varkappa)$$

$$+ C_\varkappa^{1-\epsilon}(0)(\varphi_\varkappa , \varphi_\varkappa)) + 3|\Lambda| C_\varkappa(0) C_\varkappa^{1-\epsilon}(0) , \quad (6.41)$$

where $C_\varkappa(0) = C_\varkappa(0,0)$ and $C_\varkappa(x,y)$ is the integral kernel of $C_\varkappa$ the covariance of $\varphi_\varkappa$, $C_\varkappa^{1-\epsilon}(0) = C_\varkappa^{1-\epsilon}(0,0)$, and $C_\varkappa^{1-\epsilon}(x,y)$ is the integral kernel of $C_\varkappa^{1-\epsilon}$.

Now use primitive positivity, i.e. Lemma 1, and

$$(\varphi_\varkappa , (-\Delta+1)^\epsilon \varphi_\varkappa) \leqq (\varkappa^2+1)^\epsilon \ (\varphi_\varkappa , \varphi_\varkappa)$$

to obtain from (6.41)

$$G_\varkappa \geqq \int_\Lambda d^2x \, ((\varphi_\varkappa(x))^4 - 3(C_\varkappa(0)(\varkappa^2+1)^\epsilon + C_\varkappa^{1-\epsilon}(0))(\varphi_\varkappa(x))^2$$

$$+ 3C_\varkappa(0) C_\varkappa^{1-\epsilon}(0)) \quad (6.42)$$

Now use

$$C_\varkappa(0) \sim O(\ln\varkappa), \ C_\varkappa^{1-\epsilon}(0) \sim O(\varkappa^{2\epsilon} \ln\varkappa)$$

to obtain immediately:

Lemma 2

$$G_\varkappa \geqq - \text{const } 0(\varkappa^{4\epsilon} (\ln\varkappa)^2) \qquad (6.43)$$

Define

$$\tilde{G}_\varkappa \equiv G - G_\varkappa \qquad (6.44)$$

Then it is straightforward to verify via Feynman graph calculations

Lemma 3

$$\int d\mu_c \left|\tilde{G}_\varkappa\right|^{2j} \leqq (j!)^4 b^j ((\ln\varkappa)^m \varkappa^{-2+4\epsilon})^j \qquad (6.45)$$

for any j, some $m > 0$. b is a constant independent of j and

The proof of the proposition at the beginning of the appendix now follows from Lemmata 2 and 3 by Nelson's argument [9].

Namely, we have

$$\mu_c \{G \leqq -\text{const}(\varkappa^{4\epsilon} (\ln\varkappa)^2 - 1\} \leqq \mu_c \{|\tilde{G}_\varkappa|^{2j} \geqq 1\} \leqq \int d\mu_c |\tilde{G}|^{2j}$$

$$\leqq (j!)^4 b^j ((\ln\varkappa)^m \varkappa^{-2+4\epsilon})^j \qquad (6.46)$$

Using Stirling's approximation for $j!$ and an optimal $\varkappa$-dependent choice of j, we have

$$\mu_c \{G \leqq - \text{const } (\varkappa^{4\epsilon} (\ln\varkappa)^2) - 1\} \leqq e^{-\text{const } \varkappa^{\frac{2-4\epsilon}{4}} (\ln\varkappa)^{\frac{-m}{4}}} \qquad (6.47)$$

This estimate, together with Lemma 2, assures us that for $\epsilon < \frac{1}{10}$, $e^{-\lambda G} \in L^1 (d\mu_c)$, $\lambda > 0$, and the proposition has been proved.

REFERENCES

1. A. Friedman : Stochastic differential equations, Vol. I. New York: Academic Press 1975.

2. B. Simon: Functional integration and Quantum Physics. New York: Academic Press 1975.

3. H. Mc Kean: Stochastic integrals. New York: Academic Press 1969.

4. R. Z. Hasminski: Stochastic Stability of differential equations. Sijthoff and Noordhoff 1980 (Netherlands)

5. K. Yosida: Functional analysis. Berlin, Heidelberg, New York: Springer 1966.

6. J. Glimm and A. Jaffe: Quantum field models: (Les Houches (1970) in Quantum field theory and Statistical Mechanics, C. De Witt and R. Stora (eds.) New York: Gordon and Breach 1971.

7. J. Glimm and A, Jaffe: Quantum physics. Berlin, Heidelberg, New York: Springer 1981.

8. B. Simon: The $P(\varphi)_2$ Euclidean (quantum) field theory. Princeton, NJ: Princeton University Press. 1974.

9 E, Nelson: In: Constructive Quantum field theory. Lecture Notes in Physics. Vol. 25, G.Velo and A. Wightman (eds). Berlin, Heidelberg and New York: Springer 1973.

10. G. Jona-Lasinio and P.K. Mitter: Comm. Math. Phys. 101, 409-436, (1985).

11. G. Parisi and Y - S Wu: Sci. Sin 24, 483 (1981)

12. W. Faris and G-Jona Lasinio: J. Phys A 15, 3025 (1982).

13. T.M. Liggett: Interacting Particle Systems. Berlin, Heidelberg, New York: Springer, 1985.
This book has an exhaustive bibliography containing in particular references to the many papers of R. Holley and D. Stroock following the original work of R. Glauber on stochastic I sing systems. In addition, note the following papers on stochastic Heiseberg modds:

W. Faris: J. Funct. Anal. 32, 342 (1979); Trans. Am. Math. Soc. 261, 579 (1980).

W. Wick: Comm. Math. Phys. 81, 361 (1981).

14. D. Zwanziger: Nucl. Phys B 192, 259 (1981) L. Beaulieu and D. Zwanziger: Nucl. Phys B 193, 163 (1981)
E. Floratos, J. Iliopoulos: Nucl. Phys. B 214, 392 (1983)
E. Floratos, J. Iliopoulos and D. Zwanziger: Nucl. Phys B 241 221 (1984).
Z. Bern, M.B. Halpern, L. Sadun, C. Taubes: LBL-19900, UCB-PTH- 85/39.

15 E. Seiler: 1984 Schladming Lectures. Max Planck Institut, Munich, Preprint. In Erhard Seiler's lectures additional references to the physics literature on stochastic quantization will be found.

ALGEBRAIC STRUCTURE OF CHIRAL ANOMALIES

by

R. STORA
LAPP, Annecy-le-Vieux, France

I. INTRODUCTION

Chiral anomalies are objects discovered as such about fifteen years ago[1]. Preceding the discovery of dilatation anomalies, they constitute one of the major conquests of the period of the seventies through which the analysis of perturbative expansions essentially came to a final acceptable stage.

They were recognized as important, not only in connection with our views on symmetry breaking, but also, somewhat later, as major obstructions to the consistency of the perturbative treatment of fully quantized gauge theories.

Over the last five years or so their mathematical structure has been substantially better understood, but there are indications that more work may be needed in view, for instance, of the idea, vigorously defended by L. Faddeev[2], among others, that an anomalous gauge theory may very well turn out to possess a consistent interpretation, at the non perturbative level, of course. The example of the chiral Schwinger model analyzed by R. Jackiw and R. Rajaraman[3] does indeed force us to keep in mind that, so far, anomalies have only been shown to spoil the perturbative regime.

To come back to the present, developments have taken place in several complementary directions: in view of the often severe computational difficulties met in direct Feynman graph perturbative computations of anomalies[4] several lines of attack have been devised to analyze their structure.

At the most primitive level, the algebraic analysis of the Wess Zumino consistency conditions allows to reduce the computations to those of a finite number of numerical coefficients. This method often requires sharpening the algebraic formulation of the symmetry affected by anomalies. On the other hand, neither does it explain the role of chirality, nor does it explain the arithmetic regularity of the various coefficients involved, which has proved crucial in the analysis of anomaly cancellation mechanisms.

These more detailed, very important structural aspects can be reached by application of sophisticated versions of index theory which have however so far failed to incorporate ab initio the crucial concept of locality[5].

It is to be hoped that, in the long range, the whole subject will find itself fully contained within a "local" index theory, but, at the moment, the algebraic theory and the index theory appeal to different cohomology theories and actually both approaches have provided results which the other one is not able to produce[6].

Since the coverage of the subject has fortunately been shared between Paul Ginsparg[7] and myself, I will mostly describe here the algebraic aspects of chiral anomalies, exercising however due care about the topological delicacies involved here and there. I will most of the time illustrate the structure and methods in the context of gauge anomalies and will eventually make contact with results obtained from index theory. I will then go into two sorts of generalizations: on the one hand, generalizing the algebraic set up yields e.g. gravitational and mixed gauge anomalies, supersymmetric gauge anomalies, anomalies in supergravity theories (σ model anomalies will be treated in P. Ginsparg's lectures); on the other hand most constructions applied to the first - and eventually second - cohomologies which characterize anomalies easily extend to higher cohomologies. Although the latter have not so far received firmly founded physical interpretations, they have appeared in the topological analysis and definitely belong to the theoretical framework.

Section II is devoted to a description of the general set up as it applies to gauge anomalies. It owes much to an article by J. Mañes, B. Zumino[8] and myself, whose writing has now come to an end (referred as MSZ in the body of these notes).

Section III deals with a number of algebraic set ups which characterize more general types of anomalies: gravitational and mixed gauge anomalies, supersymmetric gauge anomalies, anomalies in supergravity theories. It also includes brief remarks on σ models and a reminder on the full BRST algebra of quantized gauge theories.

A mathematical appendix is devoted to a description of the general cohomological constructions which underly the whole analysis.

II. GAUGE ANOMALIES

(Perturbative current algebra anomalies[9]):

Let $S(\phi)$ be a classical action involving "matter" fields ϕ, transforming linearly under an internal compact Lie symmetry group G, of the renormalizable type, $\Gamma(\phi)$ the corresponding vertex functional

$$\Gamma(\phi) = S(\phi) + \sum_{n=1}^{\infty} \hbar^n \Gamma^{(n)}(\phi) \tag{1}$$

If S is invariant under the action of G it is in particular invariant under the action of Lie G. This is expressed by a Ward identity

$$W_{cl.}(\omega)\, S(\phi) = 0 \qquad \omega \in \text{Lie } G \tag{2}$$

where $W_{cl.}(\omega)$ is a functional differential operator linear in ω. This set up covers the situation where the internal symmetry G is softly broken. One can then show that the renormalized perturbative series representing $\Gamma(\phi)$ can be defined in such a way that

$$W(\omega)\, \Gamma(\phi) = 0 \tag{3}$$

where $W(\omega)$ fulfills the commutation relations

$$[W(\omega), W(\omega')] = W([\omega,\omega']) \qquad \omega,\omega' \in \text{Lie } G \tag{4}$$

Let $\mathcal{G}$ be the gauge group associated with G (maps from space time to G in the simplest case). Then it is easy to extend $S(\phi)$ into $S(\phi,a)$, where

a is an external classical gauge field transforming under $\mathcal{G}$ in the well known way, and $W_{cl.}(\omega)$ into $\mathcal{W}_{cl.}(\underline{\omega})$, $\underline{\omega} \in \mathrm{Lie}\,\mathcal{G}$, in such a way that

$$\mathcal{W}_{cl.}(\underline{\omega})\, S(\phi,a) \;=\; 0 \tag{5}$$

This is easily carried out by "minimal coupling".

The question is then whether $\Gamma(\phi)$ can be extended into $\Gamma(\phi,a)$ and $\mathcal{W}_{cl.}(\underline{\omega})$ into $\mathcal{W}(\underline{\omega})$ in such a way that

$$\begin{aligned} &\mathcal{W}(\underline{\omega})\, \Gamma(\phi,a) \overset{?}{=} 0 \\ &[\mathcal{W}(\underline{\omega}), \mathcal{W}(\underline{\omega}')] = \mathcal{W}([\underline{\omega},\underline{\omega}']) \end{aligned} \qquad \underline{\omega},\underline{\omega}' \in \mathrm{Lie}\,\mathcal{G} \tag{6}$$

It is found that this is in general not the case, precisely when the matter fields contain chiral spinors. Rather, one has an anomalous Ward identity

$$\mathcal{W}(\underline{\omega})\, \Gamma(\phi,a) \;=\; \int_M \mathfrak{a}(\underline{\omega},a) \tag{7}$$

where $\mathfrak{a}(\underline{\omega},a)$ is a differential form linear in $\underline{\omega}$, a local polynomial in a and its derivatives, and $\int_M$ denotes space time integration. This describes the situation for $d = 4$-dimensional perturbatively renormalizable theories. The locality property of $\mathfrak{a}$ is due to the locality properties of perturbative expansions; the fact it has canonical dimension 4 comes from power counting. $\mathfrak{a}$ is partly characterized by the Wess Zumino consistency conditions which follow from the commutation relations Eq.(6):

$$\mathcal{W}(\underline{\omega}) \int \mathfrak{a}(\underline{\omega}',a) - \mathcal{W}(\underline{\omega}') \int \mathfrak{a}(\underline{\omega},a) - \int \mathfrak{a}([\underline{\omega},\underline{\omega}']) \;=\; 0 \;. \tag{8}$$

By the algebraic Poincaré lemma[10)] applied to the local functional $\mathfrak{a}(\underline{\omega},a)$, one has

$$\mathcal{W}(\underline{\omega})\, \mathfrak{a}(\underline{\omega}',a) - \mathcal{W}(\underline{\omega}')\, \mathfrak{a}(\underline{\omega},a) - \mathfrak{a}([\underline{\omega},\underline{\omega}']) \;=\; 0 \mod d \tag{9}$$

where "mod d" means "modulo" the exterior differential of a form local in $\underline{\omega},\underline{\omega}'$,a, and their derivatives.

The Wess Zumino consistency condition characterizes $\int \mathfrak{a}(\underline{\omega},a)$ as an element of $H^1(\text{Lie}\ \mathcal{G},\ \Gamma^{loc}(a))$ (cf. Appendix) where $\Gamma^{loc}(a)$ denotes the space of local functionals of a, because a change of renormalization prescriptions, which does not introduce a dependence on ϕ and alters $\Gamma(\phi,a)$ by a local counterterm $\Gamma^{loc}(a)$, alters $\mathfrak{a}(\underline{\omega},a)$ by an amount $\mathcal{W}(\underline{\omega})\ \Gamma^{loc}(a)$. In the case of $d = 4$ renormalizable theories the elimination of the linearly transforming "matter" fields requires a bit more than the Wess Zumino consistency conditions when G contains U(1) factors; once $\Gamma^{loc}(a,\phi)$ is reduced to $\Gamma^{loc}(a)$ there only remains to compute $H^1(\text{Lie}\ \mathcal{G},\ \Gamma^{loc}(a))$. The result, which will be described later turns out to be completely expressible in terms of the differential form $\underline{a}$, and its exterior derivative, not separately on its coefficients and their derivatives. Besides the complete results for dimension $d = 4$ summarized here, there is now one complete result known, for arbitrary d, namely the general computation of $H^*(\text{Lie}\ \mathcal{G},\ \Gamma^{loc}(\underline{a}))$, recently carried out by M. Dubois Violette, M. Talon, C.M. Viallet[11)], to which one may apply the various general constructions described in the appendix. For instance, given an n cocycle mod d, $Q^n_p(\omega,\underline{a})^\dagger$ of Lie $\mathcal{G}$ with values in $\Gamma^{loc}(\underline{a})$, we may construct the corresponding Wess Zumino cocycle††

$$\int_{g_t \in T_n(g_o \ldots g_n) \times \mathcal{E}_p} Q^n_p\ (g_t^{-1} d_t g_t,\ {}^{g_t}\underline{a}) \tag{10}$$

where ${}^{g_t}\underline{a}$ is the gauge transform of a :

$${}^{g}\underline{a} = g^{-1}\underline{a}g + g^{-1}dg \tag{11}$$

---

†from now on ω denotes the generator of $H^*(\text{Lie}\ \mathcal{G})$, cf. Appendix, eq.(A9).
††cf. Appendix eq.(A13).

and $\mathcal{C}_p$ is a cycle in the base manifold, of dimension p, the degree of Q_p^n as a differential form.

In particular, for n=1, p=d, the dimension of the base manifold we get the Wess Zumino action

$$\Gamma_{WZ}(g,\underline{a}) = \int_{T_1(e,g)\times M_d} \alpha(g_t^{-1}d_t g_t, {}^{g_t}\underline{a}) \tag{12}$$

which fulfills the Ward identity

$$\delta\,\Gamma_{WZ}(g,\underline{a}) = \int_{M_d} \alpha(\omega,a) \tag{13}$$

where δ is defined by

$$\begin{aligned} \delta\underline{a} &= -\,d\omega - [\underline{a},\omega] \\ \delta g &= \omega g \end{aligned} \tag{14}$$

(cf. Eqs (A27),(A10)).

Note that a non trivial cocycle with values in $\Gamma^{loc}(\underline{a})$ becomes trivial - up to non uniformity - in $\Gamma^{loc}(g,\underline{a})$ where a new variable $g \in \mathfrak{G}$, behaving non linearly under $\mathfrak{G}$, has been introduced. This is one classical procedure to "kill cohomology". The other known procedure, the so called Green Schwartz procedure requires the introduction of a 2-form with suitable transformation properties and will be mentioned in the next section.

The non uniformity of Γ_{WZ} 12) is clearly parametrized by $\pi_1(\mathfrak{G})$, the first homotopy group of the gauge group and, upon proper normalization, can be gotten rid of by exponentiation in most cases of interest[13] (e.g. compact space-time, see MSZ for further details).

We shall now describe in some details the results of M. Dubois Violette, M. Talon, C.M. Viallet[11] (referred to as DTV), namely exhibit the construction of canonical representatives $Q_p^n(\omega,\underline{a})$ for $H^*(\mathrm{Lie}\,\mathfrak{G}, \Gamma^{loc}(\underline{a}))$.

Some technical preparation is in order: Let

$$\underline{F} = d\underline{a} + \frac{1}{2}[\underline{a},\underline{a}] \tag{15}$$

be the curvature of $\underline{a}$.

Let J_n be a symmetric polynomial on Lie G of degree n, invariant under the adjoint action of G.

One has the following transgression formulae inherited from Chern, Weil, Cartan:

$$\begin{aligned} & J_n(F(\underline{a})\ldots F(\underline{a})) - J_n(F(\underline{a}_o)\ldots F(\underline{a}_o)) \\ &= n\,(d+\delta)\int_0^1 J_n(d_t A_t,\, F(\underline{A}_t)\ldots F(\underline{A}_t)) \\ &= (d+\delta)\,Q_n \end{aligned} \tag{16}$$

where $\underline{a}_o$ is a fixed background connection ($\neq 0$ if connections live on a non trivial bundle, cf. MSZ), and

$$\underline{A}_t = t\,\underline{a}_o + (1-t)\,(\underline{a}+\omega) \tag{17}$$

is a family interpolating between $\underline{a}_o$ and $\underline{a}+\omega$. Q_n is an element of total degree $2n+1$ bigraded by the form degree p and the degree in ω,g ($p+g = 2n-1$):

$$Q_n = \sum_{g=0}^{g=2n-1} Q^g_{2n-1-g} \tag{18}$$

Expanding Eq.(16) in powers of ω we get the hierarchy of identities

$$\begin{aligned} J_n(F(\underline{a})) - J_n(F(\underline{a}_o)) &= d\,Q^o_{2n-1} \\ 0 &= \delta\,Q^o_{2n-1} + d\,Q^1_{2n-2} \\ \vdots \quad & \\ &= \delta\,Q^g_{2n-1-g} + d\,Q^{g+1}_{2n-1-g-1} \\ \vdots \quad & \\ 0 &= \delta\,Q^{2n-1}_o \end{aligned} \tag{19}$$

If J_n is irreducible, namely is not the product of several invariant polynomials, Q_o^{2n+1} is an irreducible antisymmetric invariant polynomial in ω. Conversely, one can show that these invariant antisymmetric polynomials generate $H^*(\text{Lie } G)$ when J_n spans a system of generators of invariant symmetric polynomials. The Q_n's appear as building blocks for $H^*(\text{Lie } \mathfrak{G}, \Gamma^{loc}(\underline{a}))$:

Th (D.T.V.): A system of representative cocycles for $H^*(\text{Lie } \mathfrak{G}, \Gamma^{loc}(\underline{a}))$ is given by expanding in powers of ω all expressions of the type:

$$\prod_{\substack{n_i \\ n_1 \leq n_2 \leq \ldots \leq n_k}} \sum_{g_i=0}^{g_i=\min(2n_i+1)} Q^{g_i}_{2n_i-1-g_i} \prod_{m_i \geq n_1} (J_{m_i}(F(\underline{a})) - J_{m_i}(F(\underline{a}_o))), \tag{20}$$

integrating them over cycles of the base manifold, considering as distinct those for which the expressions

$$\prod_{\substack{n_i \\ n_1 \leq n_2 \leq \ldots \leq n_k}} \sum_{g_i=0}^{g_i=\min 2n_i+2} Q^{g_i}_{2n_i-1-g_i} \prod_{m_i \geq n_1} (J_{m_i}(F(\underline{a})) - J_{m_i}(F(\underline{a}_o))) \tag{21}$$

are independent, with the notation

$$Q^{2n_i+2}_{-1} = J_{n_i+1}(F(\underline{a})) - J_{n_i+1}(F(\underline{a}_o)) . \tag{22}$$

The system of generators of $H^*(\text{Lie } \mathfrak{G}, \Gamma^{loc}(\underline{a}))$ has first been written down by J. Thierry-Mieg[14]. That these expressions are δ cocycles mod d is a consequence of eq.(19). This was partly conjectured by J. Dixon, as reported in ref. 10). On the other hand, the independence statement proved by D.T.V. requires heavy use of homological algebra which we cannot report on in any detail here.

For practical purpose this analysis shows that anomalies are all of the form Q^1_{2n-2} (eq.19), and in one to one correspondence with non necessarily irreducible invariants J_n, for even dimensional space time, but it also provides amusing examples in odd dimensional space time when G contains at least two U(1) factors[11),14)].

III. MORE GENERAL DIFFERENTIAL ALGEBRAS[15)]

It must be clear from the preceding section what anomalies are in other similar situations which are of essentially two types:

- the anomaly is still associated with the cohomology of a Lie algebra;
- the anomaly is associated with a differential algebra which is not necessarily - known to be or to contain - the cohomology algebra of a Lie algebra.

Of course the first type is a particular case of the second. It covers the following situations of interest:

- gravitational and mixed gauge anomalies[4),8),16)]; this is both useful and educative because the problem is often set up in such a way that it is not apparent whether one is dealing with the cohomology of a Lie algebra, a situation which still prevails in a number of cases touching supergravity theories;
- supersymmetric gauge anomalies[17)]: the interesting feature here is that the non polynomial structure of the anomaly is rather obscure from the point of view of differential geometry. The reduction of the anomaly to the Wess Zumino gauge, on the other hand, poses a problem of the second type[18),32)];
- anomalies in σ models (see in particular P. Ginsparg's lectures[7),19)]). In this case there remains to compute the relevant cohomology with values local functionals of the σ model fields.

In the second class one finds for instance:

- the full BRST cohomology relevant to the perturbative renormalization of gauge theories; $H^1(\text{BRST}, \Gamma^{loc})$ was computed by BRS under restrictions provided by power counting. H. Kluberg Stern, J.B. Zuber, B.W. Lee, S.D. Joglekar, John Dixon[20] performed a number of additional steps but this problem may have to be taken up again in order to investigate the structure of the Ward identity characterizing a fully quantized anomalous gauge theory[2),21)];
- supergravity theories: in terms of various sets of component fields restricted by gauge choices, these theories are characterized by differential algebras[22] which are not obvious extensions of Lie algebra cohomology algebras. In most cases there remains to check that, extending the gravitational case, local supersymmetry does not modify the classification of anomalies; e.g. the Green Schwartz[23] cancellation mechanism can be performed in a way compatible with supersymmetry[24].

This list is certainly not exhaustive; a number of other differential algebras are in sight e.g. the algebra associated with differential forms[25], the algebras associated with constrained dynamical systems[26].

In the following, we shall gather some results pertaining to a few examples of recent interest we have listed above.

III.1 Gravitational and mixed anomalies[4),8b),16)]

An external gravitational field on space time M may be represented by a vielbein form

$$e^a = e^a_\mu \, dx^\mu \qquad \begin{array}{l} a = 1 \ldots d = \dim M \\ \mu = 1 \ldots d = \dim M \end{array} \tag{23}$$

and a spin connection

$$\theta^a_b = \theta^a_{b\mu} \, dx^\mu \tag{24}$$

which is a Lie(SOd) valued one form. At the classical level, the action describing the coupling of matter with external gravity is invariant under SO(d) gauge transformations and diffeomorphisms of M. At the quantum level the question is whether this is true for the vacuum functional $\Gamma(e,\theta)$. The action of SO(d) gauge transformations is clear but that of the diffeomorphisms of M is not uniquely defined: a sensible definition requires the introduction of a fixed background spin connection $\overset{o}{\theta}$.

The structure equations of the corresponding differential algebra are[16]:

$$\begin{aligned}
\delta e &= -\Omega e - \overset{o}{\mathcal{L}}_{\xi} e \\
\delta\theta &= -D(\theta)\Omega - i_{\xi}R(\theta) + D(\theta)i_{\xi}(\theta-\overset{o}{\theta}) \\
\delta\Omega &= -\frac{1}{2}[\Omega,\Omega] - \overset{o}{\mathcal{L}}_{\xi}\Omega - \frac{1}{2}i_{\xi}i_{\xi}R(\overset{o}{\theta}) \\
\delta\xi &= -\frac{1}{2}[\xi,\xi]
\end{aligned} \qquad (25)$$

where Ω and ξ are the cohomology generators corresponding to SO(d) gauge transformations and diffeomorphisms of M, $D(\theta) = d.+[\theta,.]$, $R(\theta) = d\theta + \frac{1}{2}[\theta,\theta]$, i_{ξ} denotes saturation of a differential form by the vector field ξ and

$$\overset{o}{\mathcal{L}}_{\xi} = i_{\xi}D(\overset{o}{\theta}) - D(\overset{o}{\theta})i_{\xi} \qquad (26)$$

is the covariant Lie derivative operator acting on forms. These are the structure equations for the cohomology algebra of a Lie algebra $\varepsilon(\overset{o}{\theta})$ with values in $\Gamma^{loc}(e,\theta)$. The commutation relations defining $\varepsilon(\overset{o}{\theta})$ are:

$$\begin{aligned}
[W(\Omega), W(\Omega')] &= W([\Omega,\Omega']) \\
[W(\xi), W(\Omega)] &= W(\overset{o}{\mathcal{L}}_{\xi}\Omega) = W(i_{\xi}\overset{o}{D}\Omega) \\
[W(\xi), W(\xi')] &= W([\xi,\xi']) - W(i_{\xi}i_{\xi'}R(\overset{o}{\theta}))
\end{aligned} \qquad (27)$$

If M is parallelizable one may of course choose $\overset{o}{\theta} = 0$.
Given two background connections $\overset{o}{\theta}, \overset{1}{\theta}$, one has an isomorphism

$$\overset{o}{\varepsilon}(\theta) \simeq \overset{1}{\varepsilon}(\theta) \simeq \varepsilon \tag{28}$$

given by

$$\overset{o}{W}(\Omega) = \overset{1}{W}(\Omega) \overset{def}{=} W(\Omega)$$

$$\overset{o}{W}(\xi) = \overset{1}{W}(\xi) + W(i_\xi(\overset{o}{\theta}-\overset{1}{\theta})) \; . \tag{29}$$

A change of generators in the differential algebra (Eq.25):

$$\tilde{\Omega} = \Omega - i_\xi(\theta-\overset{o}{\theta}) \tag{30}$$

allows to cast the structure equations into the form

$$\begin{aligned} \delta e &= -\tilde{\Omega}e - (i_\xi D(\theta) - D(\theta)i_\xi)\, e \\ \delta\theta &= -D(\theta)\tilde{\Omega} - i_\xi R(\theta) \\ \delta\tilde{\Omega} &= -\frac{1}{2}[\tilde{\Omega},\tilde{\Omega}] + \frac{1}{2} i_\xi i_\xi R(\theta) \\ \delta\xi &= -\frac{1}{2}[\xi,\xi] \end{aligned} \tag{31}$$

from which $\overset{o}{\theta}$ has dropped out but which is no more obviously the set of structure equations of a Lie algebra; as one often says, it is related to a "field dependent" Lie algebra in which $W(\xi)$ is interpreted as generating "parallel transport". This may be physically appealing but is mathematically obscure. The cohomology $H^*(\varepsilon, \Gamma^{loc}(\underline{e},\underline{\theta}))$ has not yet been computed, but it is a relatively easy exercise[16] to construct some representatives of $H^*(\varepsilon, \Gamma^{loc}(\underline{\theta}))$: it turns out that

$$\begin{aligned} \mathcal{R}(\theta+\tilde{\Omega}) &= (d+\delta)(\theta+\tilde{\Omega}) + \frac{1}{2}[\theta+\tilde{\Omega}, \theta+\tilde{\Omega}] \\ &= e^{-i\xi} R(\theta) \end{aligned} \tag{32}$$

It follows that

$$J_n(\mathcal{R}^n) - J_n(R(\overset{o}{\theta})^n) = (d+\delta)\, Q_{2n-1} \tag{33}$$

where

$$Q_{2n-1} = n \int_0^1 J_n(\theta-\overset{o}{\theta}+\tilde{\Omega}, \mathcal{R}_t^{n-1}) \tag{34}$$

with

$$\mathcal{R}_t = \mathcal{R}(\overset{o}{\theta} + t\, e^{-i\xi}(\theta-\overset{o}{\theta}+\Omega)) \tag{35}$$

So,

$$\begin{aligned} &e^{-i_\xi} J_n(R(\theta)^n) - J_n(R(\overset{o}{\theta})^n) \\ &= (d+\delta)\, n \int_0^1 J_n\, (e^{-i_\xi}(\theta-\overset{o}{\theta}+\Omega),\, \mathcal{R}(\overset{o}{\theta} + t\, e^{-i_\xi}(\theta-\overset{o}{\theta}+\Omega))^{n-1} \end{aligned} \tag{36}$$

Now, in dimension $\leq 2n-2$

$$(i_\xi)^k J_n(R(\theta)^n) - J_n(R(\overset{o}{\theta})^n) = 0 \tag{37}$$

$\forall$ $k \geq 1$. The terms Q^g_{2n-1-g}, $g \geq 1$, of the expansion of Q_{2n-1} thus yields representatives for $H^*(\varepsilon, \Gamma^{loc}(\theta))$ upon integration over cycles of the correct dimension. One may thus conjecture[16] that $H^*(\varepsilon, \Gamma^{loc}(\underline{\theta}))$ $\simeq H^*(\text{Lie}\, \mathcal{G}, \Gamma^{loc}(\underline{\theta}))$ where Lie$\mathcal{G} \subset \varepsilon$ is the relevant gauge group (corresponding to $SO(d) \times G$ in the mixed gravitational gauge case).

In the language of ref. 8a) the derivation i_ξ is a homotopy which intertwines $d+\delta$. and $\delta+\delta_{\tilde{\Omega}}$, where $\delta_{\tilde{\Omega}} = \delta|_{\xi=0}$.

The first cohomology has been more extensively studied, since it is of immediate physical interest. In particular other particular representatives have been found by W. Bardeen and B. Zumino[27] for $H^1(\varepsilon, \Gamma^{loc}(e,\theta))$ from which the Ω component has been eliminated via a Wess Zumino counterterm.

We now turn to:

III.2 Supersymmetric gauge anomalies in flat d=4 dimensional superspace[17)28)]

The previous examples have lead to results rather compactly expressed in terms of differential forms, most of which can be obtained by topological arguments, via the index theorems,[39] through the isomorphisms $H^{*}(\text{Lie } \mathfrak{g}, ..) \simeq H^{*}_{DR,inv.}(\mathfrak{G}, ..)$ as explained in the appendix. Once the role of locality gets incorporated into the index type analysis, the latter offers of course more accurate results in each specific case since it provides an identification of the relevant invariant polynomials.

The supersymmetric case to be described now is one in which the geometry is still in a complete state of obscurity and in which the algebraic set up yields an answer through an equally obscure route.

Whereas the connection form $\underline{a}$ naturally appears in the classical case, the corresponding gauge superfield V appears by solving constraints to be imposed upon the super-connection forms $\varphi, \bar{\varphi}$ which respectively transform under the chiral and antichiral gauge groups. The search for a manifestly supersymmetric solution of the consistency conditions is ascertained to be a difficult task in view of the theorem that, if the structure group is semi-simple, the anomaly cannot be polynomial in the components of $\varphi, \bar{\varphi}$[29]. It is known to be parametrized in terms of invariant polynomials of degree 3, J_3, by a theorem of O. Piguet and K. Sibold[30] (the P.S. theorem).

On the other hand, remarkably enough, it has been shown by L. Bonora, P. Pasti, M. Tonin (B.P.T.)[17] that a representative, polynomial in the components of $\varphi, \bar{\varphi}$, can be found for the first cohomology of the graded Lie algebra generated by gauge transformations, space time translations and rigid supersymmetry transformations: the algebra used to find gravitational anomalies applies, with ξ restricted to vector fields generating space time translations and rigid supersymmetry transformations. In the present case however the vanishing of $e^{-i_\xi} J_n(R(\theta)^n)$ (cf. eq.(36)), in the useful dimensions, does not apply in rigid superspace. However, remarkably enough, the use of the constraints on $\varphi, \bar{\varphi}$, allow to throw the l.h.s. of Eq.(33) into the r.h.s. under the $d+\delta$ symbol! There follow polynomial expressions involving an explicitly non vanishing supersymmetry anomaly. By a theorem of O. Piguet, M. Schweda, K. Sibold[31], tha latter is trivial. However it turns out to be the coboundary of a non polynomial Wess Zumino type local functional and its elimination provides the manifestly supersymmetric, non polynomial form of the anomaly precedingly announced, which can also be directly obtained by superspace analysis.[17] We refer the reader to the original articles, since the final formulae are neither appetizing nor methodologically illuminating.

Clearly, more work has to be done in order to reach some decent understanding of this structure, which should reasonably emerge from a clear description of the geometry.

Another interesting exercise is to find the anomaly directly in the Wess Zumino gauge. This has been done by H. Itoyama, V.P. Nair, H.C. Ren[18] who find both a gauge and a supersymmetry component, with vanishing translation component. The interest of the exercise lies in the fact that the corresponding differential algebra (cf. ref. 32, eqs 6.2, 6.3) is the prototype of one for which it is not known whether it is an extension of the cohomology algebra of some - in this case graded - Lie algebra.

III.3 σ models[7),19)]

Anomalies in σ models are treated in detail in P. Ginsparg lectures[7]. They have been investigated by topological methods first by G. Moore and P. Nelson, then, together with A. Manohar. This work has now been somewhat deepened and generalized by L. Alvarez Gaumé and P. Ginsparg.

The Lie algebra defining the anomalies is generated by the isometries of the target space and the gauge transformations of the corresponding bundle of frames. It is a relatively simple matter to find representative 1-cocycles by substituting into the usual formulae the expressions of the gauge field in terms of the σ field. But there is at the moment no complete result concerning the non triviality of these expressions (there is no proof either that these are the only possible anomalies). At the moment, anomalies associated with topological obstructions are known to be non trivial in the local sense. In the particular case where the target space is a homogeneous space G/H sufficient triviality conditions on the known candidates have been found by L. Alvarez Gaumé and P. Ginsparg, namely that the H representation of the chiral fermion fields originates from the reduction to H of a representation of G.

More complete results would be welcome in particular because of the role played by σ models in supergravity theories[19)].

III.4 B.R.S.T. cohomology[9),33)]

At the classical level, the action for a fully quantized renormalizable gauge theory reads

$$\Gamma^{cl.} = \int (\mathcal{L}_{inv}(a,\phi) + b.\mathcal{G}^{(a,\phi)} + \bar{\omega}\, s\, \mathcal{G} + Q(b)) \tag{38}$$

where a is the gauge field, ϕ the matter field, b the Stueckelberg field, ω the Faddeev-Popov $(\Phi\pi)$ cohomology generator, $\bar{\omega}$ the $\Phi\pi$ multiplier, $\mathcal{G}$ a gauge function. On has

$$s\Gamma^{cl.} = 0 \tag{39}$$

with

$$sa = -(d\omega + [a,\omega]) \qquad s\omega = -\frac{1}{2}[\omega,\omega]$$

$$s\phi = t(\omega)\phi$$

$$s\bar{\omega} = -b \qquad sb = 0 \tag{40}$$

where $t(.)$ is the representation of Lie $\mathcal{G}$ on the matter field.

In order to allow the invariance property (eq.39) to go through the construction of the perturbative series describing quantum corrections, one introduces sources A, Φ, Ω linearly coupled to $sa, s\phi, s\omega$

$$\Gamma^{source} = \int A \; sa + \Phi \; s\phi + \Omega \; s\omega \tag{41}$$

So that

$$\Gamma^{o} = \Gamma^{d} + \Gamma^{source} \tag{42}$$

fulfills

$$\int \left(\frac{\delta\Gamma}{\delta a}\frac{\delta\Gamma}{\delta A} + \frac{\delta\Gamma}{\delta\phi}\frac{\delta\Gamma}{\delta\Phi} + \frac{\delta\Gamma}{\delta\omega}\frac{\delta\Gamma}{\delta\Omega} - \frac{\delta\Gamma}{\delta\bar{\omega}} \; b \right) = 0 \tag{43}$$

This is the Legendre transform of the linear Ward identity which Z^c, the connected Green's functional fulfills.
Assuming that

$$\Gamma = \Gamma^{o} + \Sigma \, \hbar^{n} \, \Gamma^{(n)} \tag{44}$$

has been proved to fulfill eq.(43) up to and including order $n-1$, the question is whether eq.(43) can be fulfilled at order n:

$$\int \left(\frac{\delta\Gamma^{o}}{\delta a}\frac{\delta\Gamma^{n}}{\delta A} - \frac{\delta\Gamma^{o}}{\delta A}\frac{\delta\Gamma^{n}}{\delta a} + \frac{\delta\Gamma^{o}}{\delta\phi}\frac{\delta\Gamma^{n}}{\delta\Phi} - \frac{\delta\Gamma^{o}}{\delta\Phi}\frac{\delta\Gamma^{n}}{\delta\phi} + \frac{\delta\Gamma^{o}}{\delta\omega}\frac{\delta\Gamma^{n}}{\delta\Omega} - \frac{\delta\Gamma^{o}}{\delta\Omega}\frac{\delta\Gamma^{n}}{\delta\omega} - \frac{\delta\Gamma^{n}}{\delta\bar{\omega}} \; b \right)$$

$$+ \sum_{p=1}^{n-1} \frac{\delta\Gamma^{p}}{\delta a}\frac{\delta\Gamma^{n-p}}{\delta A} + \frac{\delta\Gamma^{p}}{\delta\phi}\frac{\delta\Gamma^{n-p}}{\delta\Phi} = 0 \; . \tag{45}$$

This is an equation for the nth term in the local effective action

$$\Gamma_{eff} = \Gamma^{o}_{eff} + \Sigma \, \hbar^{n}\Gamma^{(n)}_{eff} \overset{e.g}{=} T_4 \, \Gamma \tag{46}$$

obtained, for instance (in the BPHZ framework), by truncating the p-space

Taylor expansions of the coefficients of Γ in such a way that only dimension 4 monomials survive, from which renormalized Feynman graphs are constructed. One may also appeal to a regularization procedure and carry out renormalization by the separation of divergent parts. Equation (45) may be rewritten as

$$\mathcal{S}^o \Gamma^n + R^n = 0 \tag{47}$$

where

$$R^n = \int \sum_{p=1}^{n-1} \frac{\delta\Gamma^p}{\delta a}\frac{\delta\Gamma^{n-p}}{\delta A} + \frac{\delta\Gamma^p}{\delta\phi}\frac{\delta\Gamma^{n-p}}{\delta\Phi} + \frac{\delta\Gamma^p}{\delta\omega}\frac{\delta\Gamma^{n-p}}{\delta\Omega} \tag{48}$$

is fully determined by lower order terms assumed to fulfill the Ward identity and

$$\mathcal{S}^o = \int \frac{\delta\Gamma^o}{\delta a}\frac{\delta}{\delta A} - \frac{\delta\Gamma^o}{\delta A}\frac{\delta}{\delta a} + \frac{\delta\Gamma^o}{\delta\phi}\frac{\delta}{\delta\Phi} - \frac{\delta\Gamma^o}{\delta\Phi}\frac{\delta}{\delta\phi} + \frac{\delta\Gamma^o}{\delta\omega}\frac{\delta}{\delta\Omega} - \frac{\delta\Gamma^o}{\delta\Omega}\frac{\delta}{\delta\omega} - b\frac{\delta}{\delta\overline{\omega}} \tag{49}$$

which fulfills identically

$$(\mathcal{S}^o)^2 = 0 \tag{50}$$

by virtue of

$$\mathcal{S}(\Gamma^o) \equiv \int \frac{\delta\Gamma^o}{\delta a}\frac{\delta\Gamma^o}{\delta A} + \frac{\delta\Gamma^o}{\delta\phi}\frac{\delta\Gamma^o}{\delta\Phi} + \frac{\delta\Gamma^o}{\delta\omega}\frac{\delta\Gamma^o}{\delta\Omega} - b\frac{\delta\Gamma^o}{\delta\overline{\omega}} = 0 \tag{51}$$

Now, it follows from the action principle that

$$\mathcal{S}^o \Gamma^n + R^n = \Delta^n \tag{52}$$

where Δ^n is a local expression.

We will see in a short while that the recursion hypothesis implies

$$\mathcal{S}^o R^n = 0 \tag{53}$$

It follows that

$$\mathcal{S}^o \Delta^n = 0 \tag{54}$$

Thus, if the first local cohomology of $\mathcal{S}^o$ is trivial[41] one has

$$\Delta^n = \mathcal{S}^o \Delta \Gamma^n_{loc} \qquad (55)$$

so that, replacing Γ^n by

$$\Gamma^n - \Delta \Gamma^n_{loc} = \tilde{\Gamma}^n \qquad (56)$$

which amounts to altering $T_4 \Gamma^n = \Gamma^n_{eff}$ by the local counterterm $\Delta \Gamma^n_{loc}$, one fulfills the Ward identity at order n :

$$\mathcal{S}^o \tilde{\Gamma}^n + R^n = 0 . \qquad (57)$$

The first local cohomology of $\mathcal{S}^o$ was computed by B.R.S.[9),33] and shown to boil down to the Adler Bardeen obstruction. Extension of this to the definition of invariant local operators has been performed by S.D. Joglekar and B.W. Lee, H. Kluberg Stern and J.B. Zuber and a general analysis undertaken by J. Dixon[20] and pursued by J. Thierry Mieg[14], L. Baulieu and others, but at the moment there is no complete result on the local cohomology of $\mathcal{S}_o$.

We end up this section proving eq.(54) from the recursion hypothesis[33]:

$$\mathcal{S}(\Gamma^o) = 0$$

$$\mathcal{S}^o \Gamma^1 = 0$$

$$\mathcal{S}^o \Gamma^2 + \frac{1}{2} \mathcal{S}^1 \Gamma^1 = 0$$

$$\vdots$$

$$\mathcal{S}^o \Gamma^{n-1} + \frac{1}{2} [\mathcal{S}^1 \Gamma^{n-2} + \mathcal{S}^2 \Gamma^{n-3} + \dots + \mathcal{S}^{n-2} \Gamma^n] = 0 \qquad (58)$$

with the definitions

$$\mathcal{S}^{(p)} = \int \frac{\delta \Gamma^p}{\delta \psi} \frac{\delta}{\delta \Psi} - \frac{\delta \Gamma^p}{\delta \Psi} \frac{\delta}{\delta \psi}$$

$$= \sigma(\Gamma^p) \qquad p \geq 1 \qquad (59)$$

where for any functional F,

$$\sigma(F) = \int \frac{\delta F}{\delta\psi}\frac{\delta}{\delta\Psi} - \frac{\delta F}{\delta\Psi}\frac{\delta}{\delta\psi} \qquad (60)$$

and

$$\psi = (a,\phi,\omega)$$
$$\Psi = (A,\Phi,\Omega) \qquad (61)$$

One has here a graded symplectic structure[42]:

$$B = \int \delta\psi\wedge\delta\Psi \qquad (62)$$

$\sigma(F)$ is a vector field defined by

$$B(\sigma(F)) = \delta F \qquad (63)$$

For two functionals F, G, one has:

$$\sigma(F)G = (-)^{dgF dgG}\,\sigma(G)F \stackrel{Def}{=} [F,G]_{\text{graded Poisson bracket}} \qquad (64)$$

with

$$\sigma([F,G]_{graded}) = [\sigma(F),\sigma(G)]_{graded} \qquad (65)$$

Thus the graded Poisson bracket fulfills the graded Jacobi identity. With these notations

$$R^n = \frac{1}{2}\sum_{p-1}^{n-1}[\Gamma^p,\Gamma^{n-p}] = \frac{1}{2}\sum_{\substack{(pq),p+q=n\\ p\neq 0,n\ p\leq q}}[\Gamma^p,\Gamma^q] \qquad (66)$$

Now $\mathcal{S}^o$ acts as a graded derivation on the graded Poisson algebra:

$$\mathcal{S}^o[F,G] = \mathcal{S}^o\,\sigma(F)G = [\mathcal{S}^o,\sigma(F)]_{graded}G + (-)^{degF}\,\sigma(F)\,\mathcal{S}^oG \qquad (67)$$

$$= \sigma(\mathcal{S}^oF)G + (-)^{degF}\,\sigma(F)\,\mathcal{S}^oG$$

$$= [\mathcal{S}^oF,G] + (-)^{degF}\,[F,\mathcal{S}^oG], \qquad (67)$$

(The main step of this derivation is to check that $[\delta^o,\sigma(F)] = \sigma(\delta^o F)$).

So,

$$\delta^o R^n = \frac{1}{2}\Sigma\,[\delta^o\Gamma^p,\Gamma^q]_+ + [\Gamma^p,\delta^o\Gamma^q]$$

$$= -\frac{1}{4}\Sigma\,[\delta^{p_1}\Gamma^{p_2},\Gamma^q] - \frac{1}{4}\Sigma\,[\Gamma^p,\delta^{q_1}\Gamma^{q_2}]$$

$$= -\frac{1}{4}\Sigma\,[[\Gamma^{p_1},\Gamma^{p_2}],\Gamma^q] - \frac{1}{4}\Sigma\,[\Gamma^p,[\Gamma^{q_1},\Gamma^{q_2}]]$$

$$= -\frac{1}{2}\Sigma\,[[\Gamma^p,\Gamma^q],\Gamma^r] \qquad (68)$$

where the summations range respectively over:

p,q : $p+q = n$ $\quad p\neq 0,n$;

p_1,p_2,q : $p_1+p_2+q = n$ $\quad q\neq 0,\ p_1\neq 0,\ p_2\neq 0$

p,q_1,q_2 : $p+q_1+q_2 = n$ $\quad p\neq 0,\ q_1\neq 0,\ q_2\neq 0$

p,q,r : $p+q+r = n$ $\quad p\neq 0,\ q\neq 0,\ r\neq 0$.

The final expression vanishes since each triple p,q,r satisfying the correct conditions occurs in three terms which cancel by the Jacobi identity[34]. This concludes the proof which, in the present form, I dedicate to V. Glaser.

As a last remark, let us point out that eq.(43) is also fulfilled by Γ_{eff} and provides in particular a deformation of the classical invariance defined by eq.(40).

III.5 Supergravity theories

We shall limit ourselves to N=1 supergravity, possibly coupled to Yang Mills and chiral matter. In a superspace formalism, the geometrical set up is reasonably clear, namely one has invariance under superdiffeomorphisms, Lorentz supergauge transformations, gauge transformations. The variables are superdifferential forms subject to invariant constraints[35].

However, solving constraints and imposing Wess Zumino - like gauge conditions leads to various systems of component fields and auxiliary fields. The corresponding differential algebras[22] expressing the invariance of the action have been written down explicitly by K. Stelle and P.C. West in the case of the minimal, auxiliary field system, by L. Baulieu and M. Bellon in the case of the Sohnius-West system. The structure equations look like a fairly opaque amplification of those describing the Wess Zumino gauge[32].

This renders the solution of the anomaly problem rather difficult to treat thoroughly. In the case of the Sohnius West system the absence of U(1) anomalies has however been carried out by L. Baulieu and M. Bellon[22].

The question of the cancellation of anomalies in the zero slope limit of string theories has led M.B. Green and J.H. Schwarz[23] to the discovery of a new mechanism which allows to trivialze anomalies in Γ^{loc}(gauge fields) by going to Γ^{loc}(gauge fields, antisymmetric tensor fields[*]) similar to the Wess Zumino trick. However, in order to complete this program, it is necessary to show it can be carried out without violating local supersymmetry[24]. Clearly this requires working within the framework of a well defined differential algebra[40]. There are several indications that supersymmetry, be it local, is cohomologically trivial as it is in the pure gauge case[30], 2-dimensional supergravity[36], (see also ref. 22)). This matter deserves to be pursued.

---

[*] present in supergravity theories.

IV. CONCLUSION AND OUTLOOK

In these notes we have attempted to insist on the structural definition of perturbative anomalies as they show up in a number of geometrical contexts arising from and usually extending symmetries to be implemented at the classical level. The mathematical framework has considerably cleared up, but many of the differential algebras which arise deserve further study. Some of these anomalies are connected with the topological anomalies so far investigated (see P. Ginsparg's lectures) through the canonical constructions summarized in the following appendix. Whereas the local structure of perturbative theories is essential here, its connection with the topological properties of various field spaces seem at the moment rather miraculous. The miracle is presumably related to the structure of the local formulae which express the various index theorems. Both the local cohomologies envisaged here and the topological cohomologies which are the objects considered by the index theorems are awaiting some common cohomological denominator[39].

ACKNOWLEDGEMENTS

The author wishes to express collectively his gratitude to all those who communicated their results prior to publication. He apologizes to those who have been only implicitly referred to, in the bibliography. Setting up a complete list of relevant references would have gone beyond the reasonably modest goal of these notes. On the other hand, the author wishes to thank all his collaborators and many colleagues for discussions and correspondence, among which: O. Alvarez, M. Asorey, M.F. Atiyah, M. Bellon, L.D. Faddeev, P.K. Mitter, D. Quillen, A.G. Reiman, K. Sibold, M. Talon, C.M. Viallet, J. Wess, E. Witten. Special thanks are due to L. Alvarez Gaumé, L. Baulieu, L. Bonora, M. Dubois Violette, G. Girardi, R. Grimm, D. Kastler, O. Piguet, I.M. Singer, M. Tonin, B. Zumino.

APPENDIX

COHOMOLOGY OF LIE ALGEBRAS, LIE GROUPS[37)]

Let G be a Lie group, Lie G its Lie algebra, V a representation space for G, and, consequently, Lie G.

C^n(Lie G, V), the vector space of n cochains of Lie G with values in V is the vector space of multilinear alternate n forms on Lie G, with values in V. Evaluated at $X_o, \ldots, X_{n-1} \in$ Lie G such an n-cochain $\vec{f}_n$ is an element of V : $\vec{f}_n(X_o, \ldots, X_{n-1})$.

On $C^*(\text{Lie } G, V) = \bigoplus_{n=o}^{\infty} C^n(\text{Lie } G, V)$, one defines the coboundary operator δ: $C^n(\text{Lie } G, V) \overset{\delta}{\to} C^{n+1}(\text{Lie } G, V)$, such that $\delta^2 = 0$, by

$$(\delta\vec{f}_n)(X_o \ldots X_n) = \sum_{i=o}^{n} (-)^i \, t(X_i) \, \vec{f}\,(X_o \ldots \hat{X}_i \ldots X_n)$$

$$+ \sum_{o \le i < j \le n} (-)^{i+j} \, \vec{f}\,([X_i, X_j], X_o \ldots \hat{X}_i \ldots \hat{X}_j \ldots X_n) \qquad \text{(A1)}$$

where $t(X_i)$ denotes the representation of Lie G in V.
As is usual one then defines

$$Z^n(\text{Lie } G, V) = \{ f_n \,/\, \delta f_n = 0 \}, \qquad \text{(A2)}$$

$$B^n(\text{Lie } G, V) = \{ f_n \,/\, f_n = \delta f_{n-1}, \quad f_{n-1} \in C^{n-1} \} \qquad \text{(A3)}$$

and, since obviously $B^n(\text{Lie } G, V) \subset Z^n(\text{Lie } G, V)$,

$$H^n(\text{Lie } G, V) = Z^n(\text{Lie } G, V) \,/\, B^n(\text{Lie } G, V) \,. \qquad \text{(A4)}$$

The elements of Z^n are called cocycles, those of B^n are called coboundaries, and those of H^n are called cohomology classes. Sometimes

δ is appended as a subscript or after V, within parentheses e.g. $Z^n_\delta(\text{Lie } G, V)$, or $Z^n(\text{Lie } G, V; \delta)$.

The easiest way to check† that $\delta^2 = 0$ is to observe that the antisymmetry of $\vec{f}$ in its n arguments allows to substitute them through the Maurer Cartan form $\omega = g^{-1}dg$ thus obtaining a left invariant differential form on G with values in V : $f_n(\omega,\ldots\omega)$. Conversely, $f_n(X_o\ldots X_n)$ is recovered by evaluating $f_n(\omega,\ldots\omega)$ at the set of left invariant vector fields $X^*_o\ldots X^*_n$ and substituting X^*_i through the corresponding element X_i of Lie G. One then finds

$$(\delta f_n)(\omega,\ldots\omega) = [d+t(\omega)]f_n(\omega,\ldots\omega) \tag{A5}$$

and

$$[d+t(\omega)]^2 = 0 \tag{A6}$$

follows from the structural equation

$$d\omega + \frac{1}{2}[\omega,\omega] = 0 \tag{A7}$$

The substitution $X_i \to \omega$ transforms $C^n(\text{Lie } G, V)$ into $\Omega^n_{\text{inv.}}(G,V)$, the space of invariant differential forms on G, with values in V.

$C^*(\text{Lie } G, V)$ is then isomorphic to $\Omega^*_{\text{inv.}}(G,V) = \oplus\, \Omega^n_{\text{inv.}}(G,V)$.

If V, besides being a vector space has a graded commutative algebra structure, $\Omega^*_{\text{inv.}}(G,V)$ becomes also a graded commutative algebra $\Omega^*_{\text{inv.}}(G) \underset{\text{graded}}{\otimes} V$, and so does $C^*(\text{Lie } G, V) \simeq C^*(\text{Lie } G) \otimes V$. $C^*(\text{Lie } G)$ is nothing else than the exterior algebra of $\widetilde{\text{Lie } G}$, the dual of Lie G (as a vector space). As an algebra, it is generated in dimension 1 - the product being the exterior product. In terms of a basis e_α of Lie G and a dual basis e^β of $\widetilde{\text{Lie } G}$ $(e^\beta(e_\alpha) = \delta^\beta_\alpha)$, the structure equations read[15)]

---

† see however the remark after equation (A12).

$$\delta e = -\frac{1}{2}[e,e] \tag{A8}$$

where e is the "Lie algebra valued generator of $\widetilde{\text{Lie } G}$"

$$e = \sum e^{\alpha} e_{\alpha} . \tag{A9}$$

e is named by physicists the geometrical Faddeev Popov ghost of Lie G. It is a purely algebraic concept. Note that the covariant differential $d+t(\omega)$ can be transformed into the ordinary differential d by the transformation

$$f \to \tilde{f} = g.f \tag{A10}$$

where g.f. denotes the action of g in V, so that

$$d\tilde{f} = \widetilde{(d+t(\omega))f} \tag{A11}$$

Finally, the complete structure equations of $C^*(\text{Lie } G, V)$ are given by eq.(A8) together with

$$\delta u = \theta(e)u \quad \text{resp.} \quad \tilde{\delta}\tilde{f} = 0 \tag{A12}$$

where u resp. $\tilde{u}$ denotes a set of generators of V and θ the corresponding representation of Lie G.

Thus every $f \in C^*(\text{Lie } G, V)$ can be represented as a polynomial in e and u, and δf is computed from the equations (A9), (A12).

The differential form version of cohomology is denoted

$H^*_{\text{inv.}}(G,V) \simeq H^*_{DR,\text{inv.}}(G) \otimes V$

where DR is an abbreviation for de Rham.

An isomorphic representation[37a)38)] is given by integrating n-forms over n-simplexes in G: Let T_n be a simplex in G, with vertices $g_o \ldots g_{n-1}$, and let

$$\tilde{f}_n(T_n) = \int_{T_n} f_n(\omega,\ldots\omega) \tag{A13}$$

Then, by Stokes theorem

$$(\delta\tilde{f}_n)(T_{n+1}) = \tilde{f}_n(\partial T_{n+1})$$

$$= \sum_{i=o}^{n} (-)^i f_n(\partial T^i_{n+1}) \tag{A14}$$

where ∂T^i_{n+1} is the oriented ith n-face of T_{n+1}. In terms of the corresponding vertices

$$(\tilde{\delta}\tilde{f}_n)(T_{n+1}(g_o\ldots g_{n+1})) = \sum_{i=o}^{n} (-)^i \tilde{f}_n(T_n(g_o\ldots\hat{g}_i\ldots g_n)) \tag{A15}$$

Note that under a smooth deformation of T_n with fixed vertices, a cocycle $\tilde{f}_n(T_n)$ changes by a coboundary

$$\tilde{f}_n(T_n) - \tilde{f}_n(T'_n) = \int_{(\tau)\times T_n(\tau)} d_\tau\, f_n(\omega_\tau\ldots\omega_\tau) \tag{A16}$$

where $\omega_\tau = g_\tau^{-1} d\, g_\tau$ where g_τ interpolates between T_n and T'_n, $0\leq\tau\leq 1$.

Using the cocycle condition in the form

$$(d_t+d_\tau)\tilde{f}_n(g_\tau^{-1}d_\tau g_\tau + g_t^{-1}d_t g_t,\ldots) = 0 \tag{A17}$$

yields

$$\tilde{f}_n(T_n) - \tilde{f}_n(T'_n) = \sum_{i=o}^{n} (-)^i \tilde{\phi}_{n-1}(T_n(g_o,\ldots\hat{g}_i\ldots g_n)) \tag{A18}$$

with

$$\tilde{\phi}_{n-1}(T_n(g_o,\ldots\hat{g}_i\ldots g_n)) = -\int_{T_n(\tau)(g_o\ldots\hat{g}_i\ldots g_{n-1})\times(\tau)} n\,\tilde{f}_n(g_\tau^{-1}d_\tau g_\tau, g_t^{-1}d_t g_t\ldots g_t^{-1}d_t g_t) \tag{A19}$$

The corresponding version of cohomology is denoted $H^*_{top}(G,V)$, and the corresponding cocycles are usually referred to as Wess-Zumino cocycles and sometimes named (not everywhere defined, possibly non uniform) discrete group cocycles. This is because "discrete" group cohomology is defined via V valued cochains $C^n_{discr.}(G,V)$ which are functions of n-1 group elements, with the coboundary operation

$$(\delta f_n)(g_o \ldots g_{n+1}) = \sum_{i=o}^{n+1} (-)^i f_n(g_o \ldots \hat{g}_i \ldots g_{n+1}) . \qquad \text{(A20)}$$

These so-called homogeneous cochains[+] are furthermore supposed to be invariant in the sense that

$$f_n(g_o \ldots g_n) = \gamma \; f_n(\gamma^{-1} g_o, \ldots \gamma^{-1} g_n) . \qquad \text{(A21)}$$

Invariance implies the possible non triviality of cohomology defined by eq.(A20) (which is otherwise trivial).

If it so happens that every n+1-tuple of points in G are the vertices of some simplex $T_n(g_o \ldots g_n)$, unique up to homotopy, then one is allowed to write

$$\tilde{f}_n(T_n(g_o \ldots g_n)) = \tilde{f}_{n\ discr.}(g_o \ldots g_n) \qquad \text{(A22)}$$

and one may identify the Wess Zumino elements of $H^n_{top,inv.}(G,V)$ with those of $H^n_{discr.}(G,V)$, since the Wess Zumino cocycles do fulfill the invariance property. It is however safer to consider Wess Zumino cocycles as elements of $H^n_{top}(G,V)$.

Although no general analysis of the non uniformity of the Wess Zumino cocycles is known to the author, it can be reduced via the exponential map[13]:

[+]We shall not go here into the description of discrete group cohomology via inhomogeneous cochains, which, also more widespread in the current literature, is less suitable for the present purpose.

$$a \in \tilde{V}, \quad f_n \in Z^n_{top,inv.}(G,V)$$

$$\rightarrow \quad \exp_a f_n \;=\; \exp i\, a(f_n) \tag{A23}$$

whenever, upon proper normalization the change in homotopy class of T_n results into the addition of an integer. Of course, then, cohomology definitions have to be written in multiplicative, rather than additive, notation.

It is finally worthwhile describing the procedure which allows to convert a discrete group cocycle into a de Rham cocycle[37a)]:

Let $f_n(g_o \ldots g_n)$ be a smooth discrete group cocycle. Then construct

$$\omega_n(f_n) \;=\; \Delta^* d_{g_1} \ldots d_{g_n}\; f_n(g_o \ldots g_n) \tag{A24}$$

where Δ^* is the pull back of the diagonal application $G \rightarrow G \times \ldots G$: $g \rightarrow (g, \ldots g)$.

It is a simple matter to show that

$$\omega_{n+1}(\delta f_n) \;=\; d\, \omega_n(f) \tag{A25}$$

In particular, applying this construction to a Wess Zumino cocycle

$$f_n(g_o \ldots g_n) \;=\; \int_{T_n(g_o \ldots g_n)} \tilde{f}_n(\omega, \ldots \omega) \tag{A26}$$

yields

$$\Delta^*(d_{g_1} \ldots d_{g_n} \int_{T_n(g_o \ldots g_n)} \tilde{f}(\omega, \ldots \omega) \;=\; \tilde{f}(\omega, \ldots \omega) \text{ mod. exact form.} \tag{A27}$$

REFERENCES AND FOOTNOTES

1) A thorough review of and bibliography on the early period can for instance be found in:
R. Jackiw: "Topological investigations of quantized gauge theories" in: Relativity, Groups, Topology II, Les Houches lectures 1983, B.S. de Witt, R. Stora eds., North Holland Publ. Amsterdam 1984.

2) L.D. Faddeev, private communication.

3) R. Jackiw, R. Rajaraman, P.L. 154B, 305, 1985.

4) A good example of the complexity involved is given by:
L. Alvarez Gaumé, E. Witten, N.P.B. 234, 269-330, 1984.

5) See, for instance O. Alvarez, I.M. Singer, B.Zumino, C.M.P. 94, 409, 1984.
M.F. Atiyah, I.M. Singer, P.N.A.S. 81, 2597, 1984.

6) Typically, the family's index treatment is related to the topology of the field configuration space whereas the algebraic treatment based on locality is not.

7) P. Ginsparg, Lectures in this volume.

8) a) J. Mañes, R. Stora, B. Zumino, "Algebraic study of chiral anomalies", submitted to C.M.P.;
b) L. Alvarez Gaumé, P. Ginsparg, Ann. Phys. 161, 423, 1985.

9) C. Becchi, A. Rouet, R. Stora: "Renormalizable theories with symmetry breaking" in: Field theory quantization and statistical mechanics, E. Tirapegui ed., Reidel 1981.

10) R. Stora: "Continuum gauge theories" in: New developments in quantum field theory and statistical mechanics, 1976 Cargèse Lectures, M. Levy, P. Mitter eds, Plenum New York 1977.

11) M. Dubois Violette, M. Talon, C.M. Viallet, "BRS cohomologies", to appear in C.M.P.

12) E. Witten, "Global aspects of current algebra", N.P.B. 223, 422, 1983.

13) G. Falqui, C. Reina, "BRS cohomology and topological anomalies", Annales IHP to be published.

14) J. Thierry-Mieg, P.L. 147B, 430, 1984.

15) I am indebted to P. Van Nieuwenhuizen for pointing out and discussing the article by D. Sullivan, IHES Pub. n°47, p.269, 1977, whose reading led in particular to the algebraic interpretation of the Faddeev Popov ghost which can be found in the appendix.

16) L. Baulieu, J. Thierry-Mieg, P.L. 145B, 53, 1984;
F. Langouche, T. Schücker, R. Stora, P.L. 145B, 342, 1984;
L. Bonora, P. Pasti, M. Tonin, N.P.B. 1985 to appear.

17) L. Bonora, P. Pasti, M. Tonin, P.L. 155B, 341, 1985 and to appear in NPB;
G. Girardi, R. Grimm, R. Stora, P.L. 156B, 203, 1985);
J. Wess, B. Zumino, in preparation.

18) H. Itoyama, V.P. Nair, H.C. Ren, IAS Princeton preprint 1985.

19) G. Moore, P. Nelson, P.R.L. 53, 1519, 1984;
G. Moore, P. Nelson, HUTP-84/A076;
A. Manohar, G. Moore, P. Nelson, HUTP-84/A088;
L. Alvarez Gaumé, P. Ginsparg, HUTP 85/A015;
P. di Vecchia, S. Ferrara, L. Girardello, P.L. 151B, 199, 1985;
E. Cohen, C. Gomez, N.P.B. 254, 235, 1985;
J. Bagger, D. Nemechansky, S. Yankielowicz in Anomalies, Geometry, Topology, Argonne-Chicago 1985.

20) H. Kluberg Stern, J.B. Zuber, P.R.D. 12, 467, 1975; P.R.D. 12, 3159, 1975;
S.D. Joglekar, B.W. Lee, Ann. Phys. 97, 160, 1976;
J. Dixon, 1976-1978, unpublished.

21) L.D. Faddeev, S.L. Shatashvili, TMϕ 60, 206, 1984.

22) K.S. Stelle, P.C. West, N.P.B. 140, 285, 1978;
L. Baulieu, M. Bellon, to be published.

23) M.B. Green, J.H. Schwarz, P.L. 149B, 117, 1984.

24) S. Cecotti, S. Ferrara, L. Girardello, CERN.TH.4253/85.

25) J. Thierry-Mieg, unpublished, 1979;

L. Baulieu, J. Thierry-Mieg, N.P.B. 228, 259, 1983;

Although both BRS and $\overline{\text{BRS}}$ are used in these papers, BRS alone is sufficient if one does not want to restrict oneself to linear gauge functions.

26) E.S. Fradkin, G.A. Vilkovisky, P.L. 55B, 224, 1975;

I.A. Batalin, G.A. Vilkovisky, P.L. 69B, 309, 1977.

27) W.A. Bardeen, B. Zumino, N.P.B. 244, 421, 1984.

28) L. Bonora, P. Pasti, M. Tonin, P.L.B. to appear.

29) S. Ferrara, L. Girardello, O. Piguet, R. Stora, P.L. 157B, 179, 1985.

30) O. Piguet, K. Sibold, N.P.B. 247, 484, 1984.

31) O. Piguet, M. Schweda, K. Sibold, N.P.B. 174, 183, 1980.

32) B. Zumino in Symposium on Anomalies, Geometry and Topology, Argonne-Chicago, 1985.

33) a) G. Bandelloni, C. Becchi, A. Blasi, R. Collina, Ann. IHP XXVIII, 225-254 and 255-285, 1978.

b) O. Piguet, A. Rouet, Physics Reports 76, 1-77, 1981.

34) cf. Eq. 6.16 of Ref. 33b).

35) I am indebted to R. Grimm for stimulating discussions on this subject.

36) Y. Tanii, TIT/HEP 85;

P. Howe, P. West, P.L. 156B, 335, 1985.

37) a) A. Guichardet, "Cohomologie des groupes topologiques et des algèbres de Lie", CEDIC-NATHAN Paris 1980.

b) S. Mac Lane, "Homology", Springer 1975.

c) H. Cartan, S. Eilenberg, "Homological algebra", Princeton University Press, 1956.

d) N. Jacobson, "Lie Algebras", Interscience, 1962.

38) A.G. Reiman, M.A. Semenov-Tjan-Shansky, L.D. Faddeev, in Journal of functional analysis and its applications, 1984-1985;
L.D. Faddeev, P.L. 145B, 81, 1984;
B. Zumino, N.P.B. 253, 477, 1985;
and references found in 8a).

39) I.M. Singer, in Colloque Elie Cartan, Lyon, June 25-29, 1984, to appear in Astérisque, and references therein.

40) G. Girardi and R. Grimm have just informed me they have constructed a consistent superspace geometry [namely, a correct system of constraints] which allows to incorporate (in a manifestly supersymmetric way) Chern Simons terms into the curvature of the super two form appearing in the Sohnius West multiplet.

41) When an anomaly is present, one can prove a modified Ward identity of the form

$$\int \frac{\delta\Gamma}{\delta a}\frac{\delta\Gamma}{\delta A} + \frac{\delta\Gamma}{\delta\phi}\frac{\delta\Gamma}{\delta\Phi} + \frac{\delta\Gamma}{\delta\omega}\frac{\delta\Gamma}{\delta\Omega} - \frac{\delta\Gamma}{\delta\bar{\omega}}b - \frac{\partial}{\partial\alpha}\Gamma = 0$$

where α is a constant odd scalar with dimension -1, $\phi\pi$ charge -1. (C. Becchi, private communication).

42) C. Becchi, unpublished.

SEMINARS

ONE-LOOP THERMODYNAMIC POTENTIALS IN GAUGE AND NONGAUGE THEORIES AND THEIR HIGH-TEMPERATURE EXPANSIONS[1)]

Alfred Actor

Departamento de Física Teórica[2)]
Universidad de Salamanca
Salamanca, Spain

and

Department of Physics[3)]
The Pennsylvania State University
Fogelsville, PA 18051 USA

1) Expanded version of lecture notes delivered at the XVI GIFT Seminar on Theoretical Physics, Jaca, June 3-8, 1985

2) Address for October 1984 - December 1985

3) Permanent address

I.- INTRODUCTION

The purpose of these notes is to explain how to calculate, quickly and efficiently, the one-loop thermodynamic potential associated with a finite-temperature quantum field theory. A nice feature will be that two prototypical one-loop thermodynamic potentials Ω_B and Ω_F — one for Bose statistics and one for Fermi statistics — suffice to give the one-loop thermodynamic potential (hereafter called Ω_{one}) for an arbitrary theory with any number of fields. This is true both for gauge and nongauge theories, independently of whether or not the theory has Higgs fields and spontaneous symmetry breakdown. In general, Ω_{one} will be a sum of Ω_B's and Ω_F's with appropriately chosen parameters. Consequently, it is rather easy to do equilibrium statistical mechanics in the language of temperature field theory, at least at the one-loop level. If one wishes to calculate to two or more loops, additional and usually considerable exertions are necessary. The present notes contain little information about anything beyond one-loop.

Before saying more, we may as well write down the prototype thermodynamic potentials from which all Ω_{one}'s can be constructed. They are [1]

$$\beta\Omega_B(\beta, M, \mu, a) \equiv \mathrm{Re}\ln \det_+(-D^2+M^2) \qquad (1.1a)$$

$$= \mathrm{Re}\,\frac{V}{(2\pi)^s}\int d\overset{s}{k}\left\{\ln\left(1-e^{-\beta(\omega+iA_0)}\right) + A_0 \to -A_0\right\} \qquad (1.1b)$$

$$= -2VT^s a_s \sum_{m=1}^{\infty} \frac{1}{m^{s+1}} f_s(m\beta M) \times \cosh m\beta\mu \cos m\beta a\,, \qquad (1.1c)$$

$$\beta\Omega_F(\beta,M,\mu,a) \equiv -\frac{d}{2}\,\mathrm{Re}\ln\det{}_-(-D^2+M^2) \qquad (1.2a)$$

$$= -\frac{d}{2}\,\mathrm{Re}\frac{V}{(2\pi)^s}\int d^s k\left\{\ln\left(1+e^{-\beta(\omega+iA_0)}\right) + A_0 \to -A_0\right\} \qquad (1.2b)$$

$$= -dVT^s a_s \sum_{m=1}^{\infty}(-)^{m+1}\frac{1}{m^{s+1}} f_s(m\beta M) \times \cosh m\beta\mu \cos m\beta a. \qquad (1.2c)$$

Here

$$D_\nu = \partial_\nu - i\delta_{\nu 0}A_0\ , \quad A_0 = a - i\mu \qquad (1.3)$$

and space-time has s+1 dimensions so that the dimension d of the Dirac representation is a variable. The functional determinants $\det_\pm$ are defined on spaces of functions which are periodic/antiperiodic in Euclidean time (Sec.II). These determinants have to be regularized (Sec. III) to obtain the formulas (1.1b) and (1.2b). The momentum integrals in the latter have been performed in Eqs. (1.1c), (1.2c) by expanding the logarithms and the result (Sec. III) is an infinite Bessel function series in which

$$a_s f_s(m\beta M) = 2\left(\frac{m\beta M}{2\pi}\right)^{\frac{s+1}{2}} K_{\frac{s+1}{2}}(m\beta M). \qquad (1.4)$$

K_ν is the modified Bessel function of the second kind and

$$a_s = 2\pi^{s/2}/\Gamma(\tfrac{s}{2})$$

is the area of the unit sphere in s dimensions.

Eqs. (1.1b), (1.2b) look like the thermodynamic potentials for the relativistic Bose and Fermi gases; indeed this is essentially what $\Omega_{B,F}$ are. Hence our claim — that Ω_{one} can be expressed as a sum of Ω_B's and Ω_F's — really means that Ω_{one} is a sum of ideal gas contributions, all interactions between particles having been switched off by the neglect of two-loop and higher diagrams. When phrased this way, the claim is obviously true. It might also seem trivial, but this is not the case for several reasons. First, bosonic and fermionic chemical potentials μ are included in Eqs. (1.1), (1.2). Thus we shall solve the problem of expressing Ω_{one} as a function of arbitrary chemical potentials. This is not a remarkable achievement, but it is a very useful piece of technical equipment. Second, when one is dealing with internal symmetries, Yang-mills theories, etc., it is not entirely obvious how to obtain the individual "ideal gas" contributions. This problem has to be solved.

A third important point, peculiar to gauge theories, is a bit more subtle. At finite temperature T , the electric and magnetic sectors of any gauge theory become kinematically distinct through the existence of a preferred frame (the rest frame of the thermal background). These sectors are well-defined, in contrast to the situation at T = 0 , because the boosts which mix electric and magnetic fields do not preserve the preferred frame. Since it is valid to work in this frame, boosts are effectively suppressed, rendering the electric and magnetic sectors distinct. Let the effective actions for these sectors be called $S_{el}[A_0]$ and $S_{mag}[A_i]$ where the relevant dynamical variables have been indicated. Generally $S_{mag}[A_i]$ is a gauge theory (not a theory with conventional action however). We have little to say about the magnetic sector and its action in these notes. Our concern will be the electric sector, which is described by its scalar effective action $S_{el}[A_0]$. This effective theory is scalar because the suppression of boosts detaches A_0 from the remainder A_i of the gauge potential and restores it to its pre-relativistic status as scalar field. All electrical properties of the gauge theory (e.g. confinement in QCD) are determined by S_{el}. Being a scalar theory, the electric sector is comparatively easy to investigate. Some important recent advances in temperature gauge theories have only been possible through the recognition of this fact. One key observation is that a temperature gauge theory depends nontrivially even on constant eigenvalues of A_0 : $A_0 = \text{diag}(a_b)$. This dependence can be isolated by integrating S_{el} over fluctuations around these constant eigenvalues to obtain an electric thermodynamic potential Ω_{el} depending on the a_b as well as the usual thermodynamic parameters. (From now on these numbers a_b will be referred to as vacuum parameters.) Coming finally to the point of this paragraph: The one-loop contribution to Ω_{el} is a sum of Ω_B's and Ω_F's just as for any other thermodynamic potential. Eqs. (1.1), (1.2) give $\Omega_{B,F}$ as a function of (Abelian) vacuum parameter a. This suffices to give Ω_{one} as a function of Abelian or nonAbelian vacuum parameters in all situations, including ones where chemical potentials play a role. The inclusion of chemical potentials in the vacuum parameter analysis is reported here (and in Ref. [1]) for the first time. This analysis is sufficiently powerful to provide us with partial understanding of high-temperature gauge theories: their plasma behaviour in the electrical sector, and how this is affected by chemical potentials of any type.

To summarize: we explain how to obtain Ω_{one} for the following catagories of field theories:

(a) Nongauge theories with arbitrary masses and chemical potentials, but with no Higgs fields and no spontaneous symmetry breakdown (hereafter SSB). The simplest examples are the relativistic Bose and Fermi gases, for which $\Omega_{one} = \Omega_{B,F}$ is exact because higher loops do not exist. In general, the one-loop thermodynamic potential Ω_{one} neglects particle interactions, while Bose or Fermi statistics are taken into account for the quantum particle moving in the thermal background, in equilibrium with this background. Ω_{one} is simply a sum of contributions from ideal Bose and Fermi gases. The interactions between the gas particles first appear at the two-loop level.

(b) Nongauge theories with Higgs fields and SSB. This case is more interesting, although mathematically things work exactly as in case (a). Ω_{one} is a sum of ideal gas terms $\Omega_{B,F}$. But now the Higgs-generated particle masses are functions of the Higgs vacuum expectation value $\langle\phi\rangle$, and hence the ideal gas contributions to Ω_{one} depend on $\langle\phi\rangle$ and are precisely the one-loop corrections to the Higgs classical potential $U_H(\langle\phi\rangle)$. As is well-known, those one-loop corrections change the shape of U_H in such a way that SSB disappears above some critical temperature T_c. This is one of the most important properties of the Higgs mechanism. As always, chemical potentials are incorporated in the analysis.

(c) Gauge theories without SSB, including arbitrary matter field masses and chemical potentials. Here the theory depends on vacuum parameters a_b (i.e. constant diagonal terms in A_o). Ω_{one} is a sum of ideal gas terms where the latter depend on the a_b, as well as other thermodynamic parameters. From the dependence of Ω_{one} on the a_b we immediately obtain a crucial result: the electrical sector of a temperature gauge theory is massive, with a quantum-mechanical mass $m_{el}(T)$ which increases with T. This mass is an inverse screening length: the electrical sector of any gauge theory is plasmalike (at least for sufficiently high temperature). This is true in particular for QCD, and is the reason why QCD must deconfine at high T. The effects of chemical potentials on the plasma behaviour of gauge theories have received limited study previously, and certain results appear in these notes for the first time.

(d) Gauge theories with Higgs fields and SSB. The mathematies is similar to case (c); however there are important differences. For one thing, only the unbroken sector of the gauge theory has vacuum parameters associated with it. Moreover, there are Higgs vacuum expectation values which change with temperature and chemical potential due to corections to the Higgs potential.Our one-loop calculation takes all of these things into account, as always with the neglect of particle interactions.

In all of the preceeding,the words "calculation of Ω_{one}" can be understood to mean the derivation of a formula for Ω_{one} like the formulas conventionally presented for the thermodynamic potentials of the relativistic Bose and Fermi gases. As in Eqs. (1.1), (1.2), these formulas are either momentum integrals in which the Matsubara sum has been evaluated explicitly, or (less commonly) with these integrals presented as an infinite Bessel function series. Neither form is particularly good at exposing the dependence of Ω_{one} on the thermodynamic parameters T , M_i , μ_α ,... Generally, low temperature is easy to analyse: the leading contributions to $\Omega_{B,F}$ have exponential corrections. High-T is much more of a problem. For small β, a power series expansion in the dimensionless variables βM_i , $\beta\mu_\alpha$, βa_b ,... ($\beta = 1/_T$) is appropiate. Until 1981 the exact high-T series for the relativistic Bose and Fermi gases were unknown . Then, the Mellin transformation method was used [2] to resum the Bessel function series (1.1c) with a = 0 to a high-T expansion. Subsequently, other authors applied the same method to the Fermi gas. More recently, the author has discovered a different and simpler method for calculating high-T series. In Sec. IV this will be used to obtain the power series expansions of Eqs. (1.1c) and (1.2c). Hence the high-T expansions of Ω_{one}

for all theories in catagories (a) - (d) above can be said to be known.

The Table of Contents exhibits the organization of these notes in a self-explanatory fashion. We wish to emphasize that the notes were written not only to explain one-loop thermodynamics, but also to provide an introduction to finite-temperature field theory suitable for physicists with little knowledge of this subject. This is not a comprehensive introduction, but rather one which reaches many of the essential points quickly. We hope that both purposes will be served. Finally, the references given (mainly to original or classic papers) are far from complete.

II.- FUNDAMENTALS OF TEMPERATURE FIELD THEORY

It is common knowledge that in finite temperature field theory, boson fields are periodic in Euclidean time while fermion fields are antiperiodic;

$$\phi_{\pm}(x_0+\beta,\vec{x}) = \pm\,\phi_{\pm}(x_0,\vec{x}) \;, \quad \beta = 1/T \;. \qquad (2.1)$$

(Here we begin using a label ± to distinguish between the boson and fermion cases.) The periodicity condition (2.1) is properly regarded as the mechanism by which one introduces temperature T into a field theory. A geometrical view of this mechanism would be that Euclidean space-time E^{s+1} has been deformed into the cylinder $E^s \times S^1$ by the compactification $0 \le x_0 \le \beta$ of Euclidean time. One cannot understand finite-temperature field theory without understanding the origin of the periodicity condition (2.1), and we first explain this point. Then we mention some of the consequences of defining field theory on a cylinder. Finally, the incorporation of chemical potentials into a field theory is discussed.

A. Connection Between Statistical Mechanics and Field Theory

The basic thing to understand is where temperature field theory comes from. One begins with a statistical system in equilibrium at temperature T. This system consists of specified types of particles, and we imagine that the forces between these particles are known. Given the characteristics of the various particles involved, one can introduce a set of fields $\{\phi_{\pm}\}$ whose elementary excitations are precisely these particles. In this way a set of fields can be associated with the arbitrary statistical system. To elevate this set of fields to a field theory one needs the interactions. These, of course, will be determined by the forces acting between the particles of the statistical system. In this reasonably obvious and straightforward fashion, a field theory can be associated with any statistical system.

This association has to be put on a mathematical basis. In 1953 Feynman [3] explained how one would do the corresponding thing in quantum mechanics, and two years later Feynman's idea was generalized to field theory by Matsubara [4]. The idea is to write the partition function for the equilibrium statistical system as a path integral. Naturally the integration variables for the path integral are fields: in fact precisely the fields of the temperature field theory associated with the statistical system. Working through the details of this in the best way to learn why temperature field theories are Euclidean, why they have periodic boundary conditions, etc... For the time being we follow the treatment given by Bernard [5] in one of the classic papers on this subject.

Suppose the system only contains one type of scalar particle. Then there will be one scalar field ϕ in the field theory. The Hilbert space of states can be spanned by a Schwinger basis set $|\phi(x_0, \vec{x})\rangle$. Note that we are in Euclidean space-time because, as will be seen immediately, that is the appropriate place to be working. Given the states $|\phi(x_0, \vec{x})\rangle$ one can define the Feynman transition amplitude

$$\langle\phi_\beta|e^{-\beta\hat{H}}|\phi_0\rangle = \int[d\pi]\int[d\phi]\, e^{-\int_0^\beta dx_0\int d^s\vec{x}\,[i\pi\partial_0\phi-\mathcal{H}]} \,. \qquad (2.2)$$

$$\phi(\beta,\vec{x})=\phi_\beta$$
$$\phi(0,\vec{x})=\phi_0$$

Here ϕ_β, ϕ_0 refer to classical fields evaluated at Euclidean time $x_0 = \beta, 0$ respectively. Note that the integral over canonical momentum π is unconstrained, whereas the integral $[d\phi]$ is constrained by the boundary condition that $\phi = \phi_\beta, \phi_0$ at $x_0 = \beta, 0$. For conventional theories (i.e. with Hamiltonian quadratic in π) the momentum integral is done in the usual way by completing the square. Thus $\pi' = \pi - i\partial_0\phi$ decouples the ϕ and π' integrations, and the latter is simply discarded leaving

$$\langle \phi_\beta | e^{-\beta\hat{H}} | \phi_0 \rangle = \int_{\substack{\phi(\beta,\vec{x}) = \phi_\beta \\ \phi(0,\vec{x}) = \phi_0}} [d\phi]\, e^{-S[\phi]} \tag{2.3}$$

where $S[\phi]$ is the Euclidean action. Now we have an expression for arbitrary matrix elements of the operator $\exp(-\beta\hat{H})$. The obvious thing to do is to interpret $\beta = 1/kT$ as the inverse temperature (units with $k = 1$ will be assumed) and to define the partition function by taking the trace of this operator:

$$\begin{aligned} Z = e^{-\beta\Omega} &= \mathrm{Tr}\, e^{-\beta\hat{H}} \\ &= \sum_{\phi_0} \langle \phi_0 | e^{-\beta\hat{H}} | \phi_0 \rangle \\ &= \int_{\phi_\beta = \phi_0} [d\phi]\, e^{-S[\phi]} . \end{aligned} \tag{2.4}$$

In this way we obtain a path-integral formula for the partition function which can be readily generalized to an arbitrary boson theory. We would like to emphasize two general features of this formula: (1) In real time (Minkowski space-time) one would replace exp $(-\beta \hat{H})$ by exp $(i \Delta t \hat{H})$ for some real time difference Δt, and Eq. (2.4) would not be a partition function. To obtain a formula for Z we were forced to go to Euclidean space-time in the functional integral. In general: All fields used as integration variables in formulas like (2.4) are intrinsically Euclidean fields.

(2) The periodic boundary condition $\phi_\beta = \phi_0$ on the boson field ϕ arises automatically from the definition of Z as a (functional) trace. This is the origin of the periodicity condition in finite-temperature field theory.

Consider next a system consisting of one type of fermion. Then the field theory contains a single spinor field ψ. In this case the boundary condition on the path integral is antiperiodic. We can easily show that this change in boundary condition is related to the Grassmann nature of spinor variables in path integrals. To this end, consider the expectation value

$$
\begin{aligned}
& Z \langle \phi_\pm(\beta,\vec{y})\, \phi_\pm(z_0,\vec{z}) \rangle \\
& \equiv \int\limits_{\phi_{\pm\beta}=\eta\phi_{\pm 0}} [d\phi_\pm]\, \phi_\pm(\beta,\vec{y})\, \phi_\pm(z_0,\vec{z})\, e^{-S} \\
& = \eta \int\limits_{\phi_{\pm\beta}=\eta\phi_{\pm 0}} [d\phi_\pm]\, \phi_\pm(0,\vec{y})\, \phi_\pm(z_0,\vec{z})\, e^{-S} \\
& = \mathrm{Tr}\, e^{-\beta\hat{H}}\, \hat{\phi}_\pm(\beta,\vec{y})\, \hat{\phi}_\pm(z_0,\vec{z})
\end{aligned}
$$

$$= Tr\, \hat{\phi}_\pm(0,\vec{y})\, e^{-\beta\hat{H}}\, \hat{\phi}_\pm(z_0,\vec{z})$$

$$= Tr\, e^{-\beta\hat{H}}\, \hat{\phi}_\pm(z_0,\vec{z})\, \hat{\phi}_\pm(0,\vec{y})$$

$$= \int [d\phi_\pm]\, \phi_\pm(z_0,\vec{z})\, \phi_\pm(0,\vec{y})\, e^{-S} .$$

$$\phi_{\pm\beta} = \eta\phi_{\pm 0} \qquad (2.5)$$

Comparing the path integrals in the second and sixth equations, one concludes that the c-number integration variables satisfy

$$\phi_\pm(z_0,\vec{z})\, \phi_\pm(0,\vec{y}) = \eta\, \phi_\pm(0,\vec{y})\, \phi_\pm(z_0,\vec{z}) \quad (2.6)$$

where $\eta = \pm$ for ordinary (Grassmann) variables. Thus $\eta = \pm$ in the boundary condition for boson (fermion) fields, as stated in Eq. (2.1).

Putting the boson and Fermion cases together one can write down the general formula linking field theory with equilibrium statistical mechanics:

$$e^{-\beta\Omega} = \int_+ [d\phi_+] \int_- [d\phi_-]\, e^{-S[\phi_+,\phi_-]} . \quad (2.7)$$

(From now on, the ± label on the integral will mean periodic/antiperiodic boundary conditions.) This equation, like any other, can be read in both directions. Given an interesting field theory such as QCD, one can define and investigate its thermodynamics.

Conversely, given an interesting statistical system, one can associate with it a field theory, and use the powerful methods of quantum field theory to investigate the thermodynamics properties of the system. It is the latter viewpoint we have been stressing here. The definition of a temperature field theory is based on Eq. (2.7). It is this formula which gives physical meaning to a Euclidean field theory with periodic (antiperiodic) boundary conditions on its boson (fermion) fields. Any such a theory provides the integration variables or quantum fields for a path integral representation of the partition function of some equilibrium statistical system. This field theory is intrinsically Euclidean (and not from a desire to have convergent path integrals) More precisely, it is an imaginary time theory, but not quite Euclidean because space-time is cylindrical rather than flat. Nevertheless, we shall continue to refer to these theories as being basically Euclidean (as distinct from Minkowskian.) In the zero-temperature limit, the space-time cylinder $E^s \times S^1$ becomes Euclidean space-time E^{s+1} , and finite - T field theories become Euclidean ones.

B.- Feynman Rules for $T > 0$

Perturbative field theory at finite temperature is handled much as at zero temperature, but with one very important change. Quantum fluctuations, or particles travelling around closed loops, cannot have arbitrary energy. The periodicity condition (2.1) (or the compactification of Euclidean time) only permits a discrete set of energies. To find these energies a Fourier decomposition in time is appropriate;

$$\phi_\pm(x_0,\vec{x}) = \sum_n e^{i\omega_n^\pm x_0}\,\phi_\pm(n,\vec{x}) \,, \quad (2.8)$$

$$\omega_n^+ = 2n\pi/\beta \,, \quad \omega_n^- = (2n+1)\pi/\beta \,, \quad (2.9)$$

where n is any integer. Thus finite temperature only allows the discrete energies ω_n^+ for bosons, ω_n^- for fermions. In Feynman diagrams all closed loops are represented by integrals over spatial loop momenta exactly as at T = 0 , but sums over loop energy. The latter are often referred to as Matsubara sums [4] . It is this one change — the substitution of Matsubara sums for loop energy integrals — that gives finite - temperature field theory its distinctive features. Other aspects of Feynman diagrams are handled essentially as at T = 0 (see e.q. Refs. [6,7,11]).

The Fourier expansion (2.8) is important for other reasons. One can regard this equation as a way of getting rid of the compactified time variable x_0 , at the cost of replacing $\phi_\pm(x_0, \vec{x})$ by an infinite set of static fields $\phi_\pm(n, \vec{x})$. There are computational advantages as well as insight to be gained from this viewpoint which we wish to describe. First let us define static and nonstatic modes:

$$\phi_+(n,\vec{x}) \begin{array}{l} = \text{static mode} \quad n=0 \\ = \text{nonstatic mode} \quad n\neq 0, \end{array}$$

$$\phi_-(n,\vec{x}) = \text{nonstatic mode} \quad \text{All } n . \qquad (2.10)$$

Here the terminology static and nonstatic refers to Eq. (2.8).The boson static mode has zero energy $\omega_0^+ = 0$, and only this term in Eq. (2.8) is independent of x_0 . All other bosonic modes, and all fermionic modes, have nonzero energy $\omega_n^\pm \neq 0$ and contribute a time-dependent term to $\phi_\pm(x_0, \vec{x})$.

The distinction between static and nonstatic modes becomes increasingly important as the temperature goes up. There is a simple reason why. Nonstatic modes aquire classical masses of O(T) while static modes do not. This is seen by eliminating time derivatives to obtain an explicit action in the static and non-static modes $\phi_{\pm}(n, \vec{x})$. Because of their 0(T) classical masses, all nonstatic modes decouple in the limit $T \to \infty$ (see e.g. Ref. [8]) and in general only the static modes of a temperature field theory survive this limit. Fermion fields, which have no static modes, never survive the $T \to \infty$ limit. Note that this limit is a dimensional reduction problem of particularly simple type. The $T > 0$ space-time cylinder $E^s \times S^1$ becomes flat E^s when S^1 shrinks to a point ($T = \infty$). The $T > 0$ field theory defined on $E^s \times S^1$ becomes a Euclidean field theory defined on E^s. This latter $T = \infty$ theory is static from the viewpoint of the original theory in one higher dimension, and its fields are precisely the static boson modes of the higher-dimensional theory. No fermion field survives this particular dimensional reduction. Continuing the sequence to lower dimension, one might wish to compactify a chosen new time direction in E^s, obtaining a boson theory on $E^{s-1} \times S^1$ with a new temperature T'. The limit $T' \to \infty$ could be taken, with only static modes surviving this limit, and so on.

The properties of static and nonstatic modes just described can be put to practical use. For sufficiently high temperature, the nonstatic modes are very massive and can be treated as perturbations about the "static" theory consisting of all static modes, the latter being light. One can define an effective theory S_{static} for the static modes by integrating over all the nonstatic ones. The effects of the latter can be included to one loop, two loops, etc., while the exact dependence on the static modes is retained.

An interesting application of this idea to gauge theories with SSB has been made by Ginsparg [9] . For weak coupling, the critical temperature T_c at which the full gauge symmetry is restored will be large, and nonstatic modes can be treated perturbatively in the critical region. Renormalization group methods for dealing with critical behaviour are known; in particular, there is the ϵ-expansion. However, such methods require a three-dimensional formulation of the four-dimensional theory, i.e. in terms of the static and nonstatic modes. Only when the latter can be dealt with perturbatively does one have a managable situation. As mentioned, this will be the case for weak coupling, and the analysis indicates a weakly first-order symmetry restoration.

C.- Incorporating Chemical Potentials

All the preceeding was for zero chemical potential, i.e. for systems living in zero background charge density. But often it is neccessary to consider systems situated in environments where there are non zero charge distributions. The effects of any number of background charge densities on the system can be determined with the help of a set of chemical potentials μ_i introduced through the formula

$$e^{-\beta\Omega(\mu_i)} = \mathrm{Tr}\, e^{-\beta(\hat{H}-\mu_i\hat{N}_i)}$$
$$= \int_{\pm}[d\phi_\pm]\, e^{-S[\phi_+,\phi_-;\mu_i]} \quad . \tag{2.11}$$

Here $\hat{N}_i$ are conserved charges, and we imagine that external charge densities ρ_i (of type $\hat{N}_i$) are maintained from outside the system. The chemical potential μ_i will depend on ρ_i and is determined implicity by

$$\rho_i V = -\frac{\partial\Omega}{\partial\mu_i} \quad . \tag{2.12}$$

By definition, the action $S[\phi_+, \phi_-; \mu_i]$ of the temperature field theory, modified to include chemical potentials, is determined by evaluating the trace (2.11) as a path integral. One actually has to start with the Feynman formula (2.2) because the charge operators $\hat{N}_i$ may depend on the canonical momenta π. These calculations are discussed in Refs. [10,11] and [2,12] for fermion and boson chemical potentials respectively. We are not going to review these calculations here, but simply give the result which is extremely neat, and not difficult to obtain. The effect of a given chemical potential μ_i on the action of a temperature field theory is simply to change its kinetic terms in the following way:

$$\bar{\phi}(-\partial_0^2)\phi \rightarrow \bar{\phi}(-D_0^2)\phi$$
$$\bar{\Psi}\gamma_0\partial_0\Psi \rightarrow \bar{\Psi}\gamma_0 D_0\Psi \qquad (2.13)$$

where $D_0 = \partial_0 - \mu_i$ and ϕ, Ψ are boson, fermion fields coupled to the charge $\hat{N}_i$. (Note that Eq. (2.13) has to be slightly amended for Higgs fields, see Sec. V.) There are no other changes in the action. By making the replacement (2.13) in the action $S[\phi_+, \phi_-]$ defined by Eq. (2.7) for every chemical potential in the problem, one obtains the modified action $S[\phi_+, \phi_-; \mu_i]$. Then this modified action can be used for the calculation of the thermodynamic potential $\Omega(\mu_i)$ for the system as a function of chemical potential.

III.- PROTOTYPE THERMODYNAMIC POTENTIALS

In this section we calculate the ideal bose and Fermi gas thermodynamic potentials $\Omega_{B,F}$ in Eqs. (1.1) and (1.2).The calculation proceeds in two stages. First, the functional determinants have to be regularized. A simple way to do this is provided. Second, the momentum integrals in Eqs. (1.1b) and (1.2b) are evaluated in the form of the infinite Bessel function series (1.1c) and (1.2c). The low-T behaviour of $\Omega_{B,F}$ is given. Some elementary examples are considered which illustrate the construction of Ω_{one} from the ideal gas potentials $\Omega_{B,F}$. The calculation of Ω_{one} takes into account that thermal equilibrium exists, with each type of particle propagating in a thermal background of identical particles of the same type. Quantum statistics is incorporated in Ω_{one}, but not direct particle-particle interactions. These first appear at the two-loop level.

A.-Bose Potential Ω_B

Let us begin with the boson case. The fermion case will turn out to be a simple modification of this one. From the definition (1.1a) one has in momentum space

$$\beta\Omega_B = \mathrm{Re}\frac{V}{(2\pi)^s}\int d^s k \sum_n \ln\left[(\omega_n^+ - A_0)^2 + \omega^2\right], \qquad \omega^2 = \vec{k}^2 + M^2 \tag{3.1}$$

where $\omega_n^+ = 2\pi n/\beta$ are the discrete energies accessible to the quantum bose particle traveling around the closed loop. Eventually ω is going to be the Fock space energy of a single particle. The calculation of Ω_B involves the evaluation of the Matsubara sum over n and the evaluation of the momentum integral.

The Matsubara sum is easy to evaluate in closed form. Consider the related problem

$$\sum_n \left[(n-a)^2+b^2\right]^{-1}$$

$$= \frac{1}{2i}\oint_C dz\, \frac{\cot \pi z}{(z-a)^2+b^2}$$

$$= \frac{1}{2i}\oint_C dz'\, \frac{\cot\pi(z'+a)}{(z'+ib)(z'-ib)} \qquad z'=z-a$$

$$= -\frac{\pi}{2ib}\left[\cot\pi(a+ib)-\cot\pi(a-ib)\right]$$

$$= \frac{\pi}{2b}\,\frac{\sinh 2\pi b}{\sin^2\pi a+\sinh^2\pi b}$$

where the Cauchy theorem has been used to rewrite the sum as an integral (C encloses the real axis counterclockwise). Then the contour is expanded to infinity and the Cauchy theorem used again to obtain the final expression. If we integrate the above identity with respect to b we find

$$\sum_n \ln\left[1+\frac{b^2}{(n-a)^2}\right] = \ln\left[1+\frac{\sinh^2\pi b}{\sin^2\pi a}\right]$$

up to a constant which we disregard. Now let a = $\beta A_0/2\pi$ and b = $\beta\omega/2\pi$; then

$$\sum_n \ln\left[1+\frac{\omega^2}{(\omega_n^+-A_0)^2}\right]=\ln\left[1+\frac{\sinh^2\beta\omega/2}{\sin^2\beta A_0/2}\right]$$

which can be rearranged to read

$$\sum_n \ln\left[(\omega_n^+-A_0)^2+\omega^2\right]-C(A_0)$$
$$=\ln 4\left[\sin^2\frac{\beta A_0}{2}+\sinh^2\frac{\beta\omega}{2}\right] \qquad (3.2)$$

where

$$C(A_0)=\sum_n \ln(\omega_n^+-A_0)^2-\ln 4\sin^2\frac{\beta A_0}{2}\,.$$

The identity (3.2) evaluates the Matsubara sum in Eq. (3.1) for us. Moreover, we see that the constant $C(A_0)$ can be ignored because it does not depend on the particle energy ω . Thus we have found

$$\beta\Omega_B=\mathrm{Re}\frac{V}{(2\pi)^s}\int d\overset{s}{k}\ln 4\left[\sin^2\frac{\beta A_0}{2}+\sinh^2\frac{\beta\omega}{2}\right]$$
$$=\mathrm{Re}\frac{V}{(2\pi)^s}\int d\overset{s}{k}\left\{\beta\omega+\ln\left(1-e^{-\beta(\omega+iA_0)}\right)\right.$$
$$\left.+\,A_0\to -A_0\right\} \qquad (3.3)$$

which is Eq. (1.1b) except for the infinite vacuum energy term which we discard. Eq. (3.3) is the formula usually given in textbooks (setting $A_0 = -i\mu$) for the thermodynamic potential of the relativistic bose gas with chemical potential μ . Writing $A_0 = a - i\mu$ we see how the formula is to be modified to include nonzero vacuum parameter a.

Even though Eq. (3.3) looks familiar, it is not a very explicit function of T,M, μ and a. To make some progress in this direction the momentum integration has to be performed. This can be done by expanding the logarithms to find [13]

$$
\begin{aligned}
\beta\Omega_B &= -\mathrm{Re}\frac{V}{(2\pi)^s}\int_0^\infty a_s k^{s-1}dk \sum_{m=1}^\infty \frac{1}{m} e^{-m\beta\omega}\cos m\beta A_0 \\
&= -\frac{2Va_s}{(2\pi)^s}\sum_{m=1}^\infty \frac{1}{m}\cos m\beta a \cosh m\beta\mu \\
&\qquad \times \int_0^\infty dk\, k^{s-1} e^{-m\beta\omega} \qquad (3.4)\\
&= -2VT^s a_s \sum_{m=1}^\infty \frac{1}{m^{s+1}} f_s(m\beta M) \\
&\qquad \times \cosh m\beta\mu \cos m\beta a
\end{aligned}
$$

where

$$
\begin{aligned}
f_s(m\beta M) &\equiv \left(\frac{m\beta M}{2\pi}\right)^s \int_0^\infty dx\, x^{s-1} e^{-m\beta M\sqrt{1+x^2}} \qquad (3.5)\\
&= \Gamma\left(\frac{s}{2}\right)\frac{1}{\pi^{s/2}}\left(\frac{m\beta M}{2\pi}\right)^{\frac{s+1}{2}} K_{\frac{s+1}{2}}(m\beta M)
\end{aligned}
$$

and K_ν is the modified Bessel function of the second kind [14,15]. Furthermore

$$a_s = 2\pi^{s/2} / \Gamma\left(\frac{s}{2}\right) \qquad (3.6)$$

is the area of the unit sphere in s dimensions and we wrote $d^s k = a_s\, k^{s-1} dk$ in the first equality in Eq. (3.4). Thus we have obtained the Bessel function series representation (1.1c) of Ω_B.

Note that for M = 0

$$f_s(0) = (s-1)! / (2\pi)^s \qquad (3.7)$$

and setting also $\mu = 0$ leads to [16]

$$\Omega_B(\beta,0,0,a) = -2VT^{s+1} a_s\, (s-1)!\, (2\pi)^{-s} \times \sum_{m=1}^{\infty} \frac{1}{m^{s+1}} \cos m\beta a . \qquad (3.8)$$

(We explain in a moment why μ cannot be nonzero in this formula for all values of a.) Using the ζ-function method in Sec. IV below, one can easily show that this expression is a polynomial of degree s + 1 (or an infinite series) in β a for s + 1 = even (or odd) integer.

Eq. (3.8) is the high-T limit of $\Omega_B(\beta, M, \mu, a)$, with high-T defined by $\beta M \ll 1$. In this limit Ω_B is also independent of the bosonic chemical potential μ, the reason being that μ is constrained by $M^2 > \mu^2$. The mathematical origin of this constraint can be found in Eq. (1.1b), where the logarithms under the momentum integral are

$$\int d^s k \ln\left[1 - e^{-\beta(\omega \pm (\mu + ia))}\right] \qquad (3.9)$$

with $\omega^2 = \vec{k}^2 + M^2$. Clearly, $\Omega_B(\beta, M, \mu, a)$ has branch points in the μ-plane at

$$\mu = \pm M - ia \; .$$

For $a \neq 0$ these branch points do not affect real μ , and there is no constraint limiting real μ . However, for $a = 0$ the branch points move to the real axis $\mu = \pm M$, leading to the constraint $M^2 > \mu^2$. Physically, this constraint means that number densities are nonnegative [2a] .

The constraint $M^2 > \mu^2$ is connected with Bose-Einstein condensation [2a,12] . Recall that μ is determined implicitly from the charge density ρ by

$$\rho V = -\frac{\partial}{\partial \mu}\Omega_B \; . \qquad (3.10)$$

There exists a critical temperature T_c , above which this equation has a solution μ satisfying the constraint. Below T_c there is no such solution. The critical temperature T_c at which Bose-Einstein condensation occurs is determined implicitly by $M^2 = \mu^2$ in the equation above.

B.- Fermi Potential Ω_F

The Fermi thermodynamic potential Ω_F defined by Eq. (1.2a) has the momentum space form

$$\beta\Omega_F = -\frac{d}{2} \mathrm{Re} \frac{V}{(2\pi)^s} \int d\vec{k}^s \sum_n \ln\left[(\omega_n^- - A_0)^2 + \omega^2\right] \qquad (3.11)$$

where $\omega_n^- = (2n + 1)\pi/\beta$ are the allowed energies for the quantum fermi particle running around the closed loop. Otherwise, Eq. (3.11) is very much like the definition (3.1) of Ω_B. In fact, we can change Ω_F into Ω_B simply by shifting A_0 by

π/β. This is so because $\omega_n^- = \omega_n^+ + \pi/\beta$, and therefore

$$\omega_n^- - A_0 = \omega_n^+ - (A_0 - \pi/\beta) \quad (3.12)$$

which amounts to a shift a $\rightarrow$ a $- \pi/\beta$ in the vacuum parameter. Even more explicitly,

$$\Omega_F(\beta, M, \mu, a) = -\frac{d}{2}\,\Omega_B\left(\beta, M, \mu, a + \frac{\pi}{\beta}\right). \quad (3.13)$$

Changing a into a $+ \pi/\beta$ has the effect of converting Eqs. (1.1b,c) into Eqs. (1.2b,c) respectively. This completes the derivation of these formulas.

Setting M = 0 we obtain the high-T limit of Ω_F [1],

$$\begin{aligned}\Omega_F(\beta, 0, \mu, a) &= -dVT^{s+1} a_s \frac{(s-1)!}{(2\pi)^s} \sum_{m=1}^{\infty} (-)^{m+1} \frac{1}{m^{s+1}} \\ &\times \cosh m\beta\mu \cos m\beta a .\end{aligned} \quad (3.14)$$

The ζ-function method in Sec.IV enables this to be expressed as a polynomial of degree s + 1 (or an infinite series) in $\beta\mu$ and βa for s + 1 an even (or odd) integer. In the fermion case there is no constraint M $> |\mu|$ and we are free to set M = 0 while keeping μ arbitrary.

C.- Low-T Behaviour

The low-T behaviour of $\Omega_{B,F}$ is readily obtained from Eq. (3.5) and properties [14,15] of the modified Bessel function. For large βM

$$a_s f_s(m\beta M) \approx \left(\frac{m\beta M}{2\pi}\right)^{s/2} e^{-m\beta M} \qquad (3.15)$$

and

$$\begin{aligned}
&\Omega_B(\beta, M, \mu, a) \\
&\approx -2VT\left(\frac{TM}{2\pi}\right)^{s/2} e^{-\beta M} \cosh\beta\mu \cos\beta a \\
&\approx -VT\left(\frac{TM}{2\pi}\right)^{s/2} e^{\beta\mu_{NR}} \cos\beta a \\
&\approx \frac{2}{d}\,\Omega_F(\beta, M, \mu, a) \qquad (3.16)
\end{aligned}$$

where $\mu_{NR} = \mu - M$ is the nonrelativistic chemical potential. Eq. (3.16) agrees with textbook results [17,18] in four space--time dimensions, except that there is a dependence here on the vacuum parameter. If desired, one can give exact series expressions [2,19] for the low-T thermodynamic potentials.

D.- Some One-Loop Calculations

1. Ideal Bose Gas:

$\Omega_B(\beta, M, \mu, 0)$ is the exact thermodynamic potential for the relativistic Bose gas with chemical potential. To derive this fact from the basic formula (2.11) connecting field theory with statistics, let S be the action for a noninteracting complex scalar field,

$$S=\int_\beta \bar{\phi}(-\partial^2+M^2)\phi \tag{3.17}$$

where the integration is over the space-time cylinder (Appendix A). Replacing ∂_μ by $D_\mu=\partial_\mu - iA_0\delta_{\mu 0}$ we find

$$\begin{aligned} e^{-\beta\Omega_B} &= \int_+ [d\phi]\, e^{-\int_\beta \bar{\phi}(-D^2+M^2)\phi} \\ &= 1/\det_+(-D^2+M^2) \end{aligned} \tag{3.18}$$

as in Eq. (1.1). Diagramatically, Ω_B corresponds to the scalar particle travelling around a closed loop. There are no higher loops because this particle has no self interaction. Hence Ω_B is exact. Quantum effects in the form of Bose statistics have been incorporated into Ω_B. The scalar particle propagates in a thermal background of identical particles, and it is in thermal equilibrium with this background. Mathematically, it is the periodic boundary condition in Eq. (3.18) which enforces Bose statistics. Note that if the field ϕ were real the determinant in Eq. (3.19) would be replaced by its square root, and Ω_B gets multiplied by 1/2.

2.- Ideal Fermi Gas:

$\Omega_F(\beta, M, \mu, 0)$ is the exact thermodynamic potential for the relativistic fermi gas with chemical potential. From the action for a noninteracting spinor theory

$$S=\int_\beta \bar{\Psi}(\partial\!\!\!/+M)\Psi\,, \tag{3.19}$$

replacing ∂_μ by D_μ, we find from Eq. (2.11)

$$e^{-\beta\Omega_F} = \int [d\psi]\, e^{-\int_\beta \bar{\psi}(\not{D}+M)\psi}$$
$$= \det{}_D \det{}_- (\not{D}+M) \qquad (3.20)$$
$$= \left[\det{}_-(-D^2+M^2)\right]^{d/2} \text{ if } A_\mu \text{ is CONSTANT}$$

as in Eq. (1.2). Here $\det_D$ means a determinant over the Dirac indices and d is the dimension of the Dirac matrices in s + 1 dimensions. The fermi particle propagating around the closed loop is in thermal equilibrium with its background of identical particles, and fermi statistics are enforced by the antiperiodic boundary condition.

3. Scalar-Spinor Theory:

Consider an arbitrary field theory with scalar fields ϕ_b and spinor fields ψ_f coupled in any fashion whatever. For the calculation of Ω_{one} the interaction terms play no role and

$$\Omega_{ONE} = \sum_b \Omega_B(\beta, M_b, \mu_b, 0) + \sum_f \Omega_F(\beta, M_F, \mu_F, 0) \qquad (3.21)$$

is simply the thermodynamic potential of a system of noninteracting fields.

4. Matter Fields in the Fundamental Representation of a Gauge Group

Things become a bit more interesting when we encounter gauge theories. These will be discussed in some detail in Sec. VI. Here, for later use, let us consider typical one-loop calculations with matter fields in the presence of a constant background gauge potential A_0. Suppose the theory has one scalar and one spinor field, each in the fundamental representation:

$$S=\int_\beta \sum_{a=1}^{N}\left[|D_\mu\phi_a|^2+M_b^2|\phi_a|^2+\bar{\Psi}_a(\not{D}+M_f)\Psi_a + \text{interaction}\right]. \quad (3.22)$$

It is trivially verified that

$$\Omega_{\text{ONE}}=\sum_{a=1}^{N}\left[\Omega_B(\beta,M_b,\mu_b,a) + \Omega_F(\beta,M_F,\mu_F,a)\right]. \quad (3.23)$$

Here we assume that ϕ and Ψ are coupled to distinct charges (distinct chemical potentials) whereas the vacuum parameter is the same for both. It is worth mentioning that the covariant derivative

$$D_\mu=\partial_\mu-iA_\mu\,,\quad A_\mu=\text{const}$$

would lead to the same result (3.23) because A_i = const gives no contribution to Ω while A_0 = const does contribute. The reason is that the energies $\omega_n^{\pm}$ allowed for quantum fluctuations at finite temperature are discrete while the spatial momentum $\vec{k}$ carried around loops is continuous. Thus in

$$\ln \det_{\pm}(-D^2+M^2)$$
$$= \frac{V}{(2\pi)^s}\int d\vec{k}^s \sum_{n} \ln\left[(\omega_n^{\pm}-A_0)^2+(k_i-A_i)^2 + M^2\right]$$

one can redefine k_i to absorb A_i = const , but A_0 cannot be eliminated in this fashion.

5. Scalar Field in the Adjoint Representation

Finally, consider a scalar field in the adjoint representation of the gauge group: $\phi = \phi^+ = (\phi_{ab})$, a and b taking values 1 to N , with the constraint tr ϕ = 0 on the diagonal elements ϕ_{aa} . As in Appendix D of Ref. [20] we find

$$e^{-\beta\Omega_{\rm ONE}}$$
$$= \int_+ [d\phi_{ab}]\,\delta[\mathrm{tr}\,\phi]\, e^{-\int_\beta \mathrm{tr}\,\phi(-D^2+M^2)\phi}$$
$$= \int_+ [d\phi_{ab}][d\lambda]\, e^{-i\int_\beta \lambda \sum_a \phi_{aa}} \qquad (3.24)$$
$$\times\, e^{-\int_\beta \sum_{a,b} \phi_{ab}(-D^2_{ab}+M^2)\phi_{ab}} ,$$

where D_μ for adjoint ϕ (given in Appendix A) leads to

$$D_{\mu ab} = \partial_\mu - i(A_{0a}-A_{0b})\delta_{\mu 0} ,$$
$$(D_\mu\phi)_{ab} = D_{\mu ab}\,\phi_{ab} . \qquad (3.25)$$

Only the integrations over diagonal elements ϕ_{aa} are affected by the constraint $\mathrm{Tr}\,\phi = 0$. Consider one of these integrations:

$$\int_+ [d\phi_{11}][d\lambda]\, e^{-\int_\beta [\phi_{11}(-\Box+M^2)\phi_{11} - i\lambda\phi_{11}]}$$

$$= \int_+ [d\chi_{11}][d\lambda]\, e^{-\int_\beta [\chi_{11}(-\Box+M^2)\chi_{11} - \frac{1}{4}\lambda(-\Box+M^2)^{-1}\lambda]}$$

where a change of variables

$$\phi_{11} = \chi_{11} - \frac{i}{2}(-\Box+M^2)^{-1}\lambda$$

has been made. All of the diagonal elements together give

$$\int_+ [d\phi_{11}\ldots d\phi_{NN}][d\lambda] \times e^{-\int_\beta \sum_a [\phi_{aa}(-\Box+M^2)\phi_{aa} + i\lambda\phi_{aa}]}$$

$$= [\det{}_+(-\Box+M^2)]^{-N/2} \times \int_+ [d\lambda]\, e^{\frac{N}{4}\int_\beta \lambda(-\Box+M^2)^{-1}\lambda}$$

$$= [\det{}_+(-\Box+M^2)]^{-\frac{1}{2}(N-1)} .$$

The remaining integrations are trivial gaussians, and the final answer is

$$e^{-\beta\Omega_{one}} = \left\{ \prod_{a,b} \left[\det{}_+ \left(-D_{ab}^2 + M^2 \right) \right]^{-1/2} \right\} \times \left[\det{}_+ \left(-\Box + M^2 \right) \right]^{1/2}$$

$$\equiv \left[\det{}_+ \left(-D^2 + M^2 \right) \right]_{adj}^{-1/2}$$

or

$$\beta\Omega_{ONE} = \frac{1}{2} \sum_{a,b} \ln \det{}_+ \left(-D_{ab}^2 + M^2 \right) - \frac{1}{2} \ln \det{}_+ \left(-\Box + M^2 \right) . \qquad (3.26)$$

The factors of 1/2 here are due to the independent variables in the functional being real, rather than complex as in Eq.(3.18).

IV.- High-Temperature Expansions

The thermodynamic potential associated with a temperature field theory is a function of the temperature, of all chemical potentials μ_α , all physical masses M_k , all couplings g_i , and (in the case of a gauge theory) all vacuum parameters a_n :

$$\Omega = \Omega(\beta, M_k, \mu_\alpha, a_n, g_i) .$$

The one loop approximation to this disregards all couplings (Higgs generated masses are an exception),

$$\Omega_{one} = \Omega_{one}(\beta, M_k, \mu_\alpha, a_n) .$$

Sooner or later, in any physical problem, one will want to know Ω as an explicit function of the independent dimensionless parameters βM_k , $\beta\mu_\alpha$, βa_n ,... If these parameters are small (as at high temperature) the most useful form would be a power series. Such a series will be called a high-T expansion. In the past, even for Ω_{one} , exact high-T expansions have proven quite difficult to obtain. (The low-temperature region, in contrast, is easy to analyse; see Sec.III.) Here we are going to calculate the exact high-T series for $\Omega_{B,F}$ in four dimensions. (The high-T expansions of $\Omega_{B,F}$ in arbitrary dimension for $a \neq 0$ will be given elsewhere, and for $a = 0$ they can be found in Ref. [19] .) From these results the high-T series for arbitrary Ω_{one} in four-dimensions are obtained. Other authors have calculated the high-T series for Ω_B with $a = 0$ [2] and Ω_F with $\mu = a = 0$ [21] using a different method which is described in Appendix C. These references should also be consulted.

Beyond the one-loop level, all diagrams depend on couplings. In general, the contribution Ω_{two} , $\Omega_{three}, \cdots$ of a given loop order will have its own high-T expansion. The series $\Omega_{one} + \Omega_{two}$ will be some modification of the series Ω_{one}, and two types of modification can be anticipated: (1) corrections to the individual terms in Ω_{one}; and (2) new terms (i.e. new functions of the variables βM_k, $\beta\mu_\alpha$, βa_n). The series $\Omega_{one} + \Omega_{two} + \Omega_{three}$ will involve additional modifications, and so on. General machinery for constructing Ω_{two} comparable to what we are able to do for Ω_{one} does not exist at present. However, one could imagine assembing the two-loop machinery needed to deal with all conventional gauge theories, for example. In other words, a basic set of two-loop diagrams could be calculated from which Ω_{two} could be constructed. From published two-loop calculations (see e.q. Refs. [12,21,22] it seems clear that the high-T expansion for Ω_{two} is straightforward to obtain, and in fact depends substantially on the results obtained at one loop.

A.- Zeta-Function Regularization

The infinite series (1.1c), (1.2c) have a very characteristic form: the summand is a product of functions, each having argument mx_i where m is the summation index and x_i is a dimensionless variable ($x_i = \beta M$, $\beta\mu$ or βa). Two methods are known for resumming such series to obtain power series in the variables x_i. One of these involves a mellin transformation. This was the approach used by Haber and Weldon [2] to obtain for the first time (in 1981 !) the high-T expansion of $\Omega_B(\beta, M, \mu, 0)$. Independently, the same method was applied to $\Omega_F(\beta, M, 0, 0)$ [21] to obtain its high-T expansion. The Mellin transform method is briefly described in Appendix C. In this section we outline another, simpler, calculation [1,19] which quickly leads to the high-T expansions of $\Omega_{B,F}(\beta, M, \mu, a)$. As is also true of the Mellin transformation approach, this second method has a very wide range of applicability.

Consider a general series of the type described above,

$$F_\pm(x_1,\ldots x_A) = \sum_{m=1}^{\infty} (\pm)^{m+1} \frac{1}{m^s} f_1(mx_1) \times \ldots f_A(mx_A), \quad (4.1)$$

where the x_i are dimensionless variables. Alternating (nonalternating) sign under the sum will be relevant for boson (fermion) calculations. If the arbitrary functions f_i are all set equal to one then

$$F_\pm \rightarrow \zeta^\pm(s) \equiv \sum_{m=1}^{\infty} (\pm)^{m+1} \frac{1}{m^s} \quad (4.2)$$

and $\zeta^+(s) = \zeta(s)$ will be recognized as the Riemann ζ-function (see e.q. Refs. [14,15] and Appendix B while)

$$\zeta^-(s) = \eta(s) = (1-2^{1-s})\,\zeta(s) \quad (4.3)$$

is the alternating sign ζ-function. For the sum (4.2) to be well-defined it is necessary that Re s > 1 . However, when this sum is replaced by a contour integral, a more general definition of $\zeta^\pm(s)$ is obtained which is regular everywhere in the s-plane except that $\zeta^+(s)$ has a simple pole at s = 1 with residue one. In particular, $\zeta^\pm(s)$ is well-behaved along the negative s-axis where the sum in Eq. (4.2) is badly divergent. From now on, whenever we encounter such a sum, we shall identify it with the well-behaved ζ-function $\zeta^\pm(s)$. This is essentially the entire regularization procedure for the high-T expansions which follow. One apparent ambiguity which can arise is when $\zeta^+(s)$ has to be evaluated at s = 1 in some particular term. Such terms have to be dealt with individually. Thus far, the author has not encountered a problem in which this particular ambiguity could not be resolved. However, implicit in the ζ-function regularization procedure is the assumption the necessary analytic continuation can be performed in Eq. (4.1). Weldon [40] has studied this question and established criteria for when the method functions perfectly, and when it requires correction terms. More will be said about this below.

The only input we need for calculating the desired power-series expansion of $F_{\pm}(x_1, \ldots x_A)$ are the individual power series expansions of the functions $f_i(mx_i)$,

$$f_i(mx_i) = \sum_{b_i} (mx_i)^{b_i} c_i(b_i) . \qquad (4.4)$$

Inserting these power series in Eq. (4.1) and commuting the sum over m to the far right we find immediately

$$\begin{aligned} F_{\pm}(x_1, \ldots x_A) = \sum_{\{b_i\}} & x_1^{b_1} \ldots x_A^{b_A} \qquad (4.5) \\ & \times c_1(b_1) \ldots c_A(b_A) \\ & \times \zeta^{\pm}(s - b_1 - \ldots - b_A) . \end{aligned}$$

Here we have $F_{\pm}$ expressed as a power series in the variables $x_1 \ldots x_A$. The coefficients in this power series are finite (excepting always $\zeta^{+}(1)$). This would not be true if $\zeta^{\pm}(s - x_1 - \ldots - x_A)$ were interpreted as a sum, of course, for then nearly every coefficient in Eq. (4.5) would be badly divergent.

While ζ-function regularization of the power series (4.5) verges on simplicity itself, the reader might worry about the range of applicability of this method. Not all functions have power series as simple as Eq. (4.4). There exist functions which cannot be expanded as power series in their arguments, the logarithm being a prominant example. In particular, coefficients containing $\ln m$ may occur in the series expansion (4.4). (This is true of the modified Bessel functions $K_\nu(m\beta M)$ encountered in the calculation of $\Omega_{B,F}$.) Such coefficients can be dealt with, as we show in Appendix B. Any power of $\ln m$ can be accomodated within the minicalculus of ζ-function regularization. As for as other nonpower functions of m are concerned, there are presently no results.

Let us now proceed to our physical example, the calculation of the high-T series for $\Omega_{B,F}(\beta, M, \mu, a)$. The cases of even and odd space-time dimension have to be done separately, as discussed at length in Refs. [2,19] . Rather than go into all this detail here, it seems appropriate to restrict the calculation to four dimensions. Thus s = 3 for the remainder of this section.

B.- High-T Series for Ω_B

In Ref. [19] we established that the method described above works very well for fermionic calculations with alternating sign in Eq. (4.1). However, for bosonic calculations (non-alternating sign) care must be exercised. When the high-T series $\Omega_B(\beta, M, \mu, 0)$ is calculated by the ζ-function method, the branch points at $\mu = \pm M$ and associated cuts in this function are not obtained. A single term $\Delta\Omega_B$ in the series, which represents the branch points and cuts, is not reproduced by the ζ-function method, while the Mellin transform approach [2] does give this important term. There is no other discrepancy between the two end results. In the following we use the ζ-function method to obtain the nonsingular high-T expansion of $\Omega_B(\beta, M, \mu, a)$. The missing branch point term will be supplied at the end.

From Eq. (1.1c) we have

$$\Omega_B(\beta, M, \mu, a) = -\frac{4V}{\pi^2} T^4 \sum_{m=1}^{\infty} (-)^m \frac{1}{m^4} \left(\frac{m\beta M}{2}\right)^2 K_2(m\beta M) \times \cosh m\beta\mu \cos m(\beta a - \pi) \tag{4.6}$$

where $\cos m\beta a = (-)^m \cos m(\beta a - \pi)$ has been used. The input needed for the ζ-function method are the power series expansions of cosh and cos , and that of K_2 [14,15] which is

$$\left(\frac{m\beta M}{2}\right)^2 K_2(m\beta M) = \left\{\frac{1}{2} - \frac{1}{8}(m\beta M)^2\right\} + \left\{\sum_{b=0}^{\infty}\left(\frac{m\beta M}{2}\right)^{4+2b} \frac{-\ln\frac{\beta M}{4\pi} + C(2,b)}{b!\,(b+2)!}\right\} - \left\{\sum_{b=0}^{\infty}\left(\frac{m\beta M}{2}\right)^{4+2b} \frac{\ln 2\pi m}{b!\,(b+2)!}\right\} \tag{4.7}$$

with the constants

$$C(2,b) = -\gamma + \frac{1}{2}\sum_{n=1}^{b}\frac{1}{n} + \frac{1}{2}\sum_{n=1}^{b+2}\frac{1}{n} .$$

Inserting the appropriate power series in Eq. (4.6) we find immediately

$$\Omega_B = -\frac{4V}{\pi^2} T^4 \left\{ P(M,\mu,a) + Q(M) + R(M,\mu,a) \right\} \tag{4.8}$$

where P , Q and R correspond to the three curly brackets in the Bessel function formula (4.7):

$$
\begin{aligned}
&P(M,\mu,a)\\
&\equiv \sum_m \frac{1}{m^4}\left[\frac{1}{2}-\frac{1}{8}(m\beta M)^2\right]\sum_{c=0}^{\infty}\frac{(m\beta\mu)^{2c}}{(2c)!}\\
&\quad\times(-)^m\sum_{d=0}^{\infty}(-)^d\frac{1}{(2d)!}m^{2d}(\beta a-\pi)^{2d}\\
&=\sum_{c=0}^{\infty}\frac{(\beta\mu)^{2c}}{(2c)!}\sum_{d=0}^{\infty}(-)^d\frac{(\beta a-\pi)^{2d}}{(2d)!}\\
&\quad\times\left[-\frac{1}{2}\eta(4-2c-2d)+\frac{1}{8}(\beta M)^2\eta(2-2c-2d)\right]\\
&=-\frac{1}{2}\Big\{\eta(4)+\frac{1}{2}\eta(2)\left[(\beta\mu)^2-(\beta a-\pi)^2\right]\\
&\qquad+\frac{1}{4!}\eta(0)\left[(\beta\mu)^4+(\beta a-\pi)^4-6(\beta\mu)^2(\beta a-\pi)^2\right]\Big\}\\
&\quad+\frac{1}{8}(\beta M)^2\Big\{\eta(2)+\frac{1}{2}\eta(0)\left[(\beta\mu)^2-(\beta a-\pi)^2\right]\Big\}\\
&=-\frac{1}{48}\Big\{\frac{7\pi^4}{30}+\pi^2\left[(\beta\mu)^2-(\beta a-\pi)^2\right]\\
&\qquad+\frac{1}{2}\left[(\beta\mu)^4+(\beta a-\pi)^4-6(\beta\mu)^2(\beta a-\pi)^2\right]\Big\}\\
&\quad+\frac{1}{32}(\beta M)^2\Big\{\frac{\pi^3}{3}+\left[(\beta\mu)^2-(\beta a-\pi)^2\right]\Big\},
\end{aligned}
\tag{4.9}
$$

$$
\begin{aligned}
Q(M) &\equiv \sum_m \frac{1}{m^4} \sum_{b=0}^{\infty} \left(\frac{m\beta M}{2}\right)^{4+2b} \\
&\quad \times \frac{-\ln \frac{\beta M}{4\pi} + C(2,b)}{b!\,(b+2)!} \\
&\quad \times \cosh m\beta\mu \,(-)^m \cos m(\beta a - \pi) \\
&= -\sum_{b=0}^{\infty} \left(\frac{\beta M}{2}\right)^{4+2b} \frac{-\ln \frac{\beta M}{4\pi} + C(2,b)}{b!\,(b+2)!} \\
&\quad \times \sum_{c=0}^{\infty} \frac{(\beta\mu)^{2c}}{(2c)!} \sum_{d=0}^{\infty} (-)^d \frac{(\beta a - \pi)^{2d}}{(2d)!} \\
&\quad \times \eta(2b+2c+2d) \qquad (4.10) \\
&= -\frac{1}{2}\left(\frac{\beta M}{2}\right)^4 \left[-\ln \frac{\beta M}{4\pi} + C(2,0)\right] \eta(0) \\
&= \frac{1}{4}\left(\frac{\beta M}{2}\right)^4 \left[\ln \frac{\beta M}{4\pi} + \gamma - 3/4\right],
\end{aligned}
$$

$$
\begin{aligned}
R(M,\mu,a) & \qquad (4.11) \\
&\equiv -\sum_m \frac{1}{m^4} \sum_{b=0}^{\infty} \left(\frac{m\beta M}{2}\right)^{4+2b} \frac{\ln 2\pi m}{b!\,(b+2)!} \\
&\quad \times \cosh m\beta\mu \,(-)^m \cos m(\beta a - \pi) \\
&= -\sum_{b,c,d} \left(\frac{\beta M}{2}\right)^{4+2b} \frac{1}{b!\,(b+2)!} \frac{(\beta\mu)^{2c}}{(2c)!} (-)^d \\
&\quad \times \frac{(\beta a - \pi)^{2d}}{(2d)!} \sum_{m=1}^{\infty} (-)^m \frac{\ln 2\pi m}{m^{-2b-2c-2d}}
\end{aligned}
$$

where (see Appendix B)

$$\sum_{m=1}^{\infty} (-)^m m^{2N} \ln 2\pi m \tag{4.12}$$
$$= -\ln 2 \qquad N=0$$
$$= (-)^N \left(1-2^{1+2N}\right)\frac{1}{2}\frac{(2N)!}{(2\pi)^{2N}}\zeta(1+2N)$$
$$N=1,2,3\ldots$$

In Eqs. (4.9), (4.10) the only numerical values of ζ - functions needed are $\eta(0) = 1/2$, $\eta(2) = \pi^2/12$ and $\eta(4) = 7\pi^4/720$.

Comments:

(1) In Eqs. (4.9) - (4.11) the function $\cos m\beta a$ was expanded about the point $\beta a = \pi$,

$$\cos m\beta a = (-)^m \sum_{d=0}^{\infty} (-)^d \frac{m^{2d}}{(2d)!}(\beta a - \pi)^{2d},$$

rather than about the point a = 0 . The reason is the angular nature of the parameter βa . In Ref. [19] we learned that to correctly obtain the Bernoulli polynomials, e.g.

$$B_4(x) = x^2(x-1)^2 - 1/30$$
$$= \frac{3}{\pi^4}\sum_{m=1}^{\infty}\frac{1}{m^4}\cos 2\pi m x$$

(where $0 \leq x \leq 1$) by the ζ-function method, it is necessary to expand about $x = 1/2$. An expansion about the endpoint $x = 0$ misses the odd term $-2x^3$ in $B_4(x)$. The present calculation is just an extension of the Bernoulli polynomial calculation, and the same precaution must be observed. Note that a factor $(-)^m$ is thereby generated under the sum over m, which effectively converts the boson sum into an alternating sign fermion sum and leads to the η-function rather than the ζ-function. Note also that replacing βa by $\beta a + \pi$ changes Ω_B in Eq. (4.8) into Ω_F in accordance with Eq. (3.13).

(2) The limit $a \longrightarrow 0$ in Eq. (4.9) is straightforward. One obtains the polynomial given in Ref. [2,19]. A formula useful for verifying this is

$$\zeta(s) = \sum_{d=0}^{\infty} (-)^{d+1} \frac{\pi^{2d}}{(2d)!} \eta(s-2d) ,$$

which converts the appropriate sum over η-functions into an ζ-function. If instead we set $M = \mu = 0$ the result is

$$P(0,0,a) = \frac{1}{12}\left[\frac{\pi^2}{15} + \frac{\pi}{2}(\beta a)^3 - \frac{\pi^2}{2}(\beta a)^2 - \frac{1}{8}(\beta a)^4\right]$$

which is the Bernoulli polynomial result in Ref. [16] (see originally Ref. [20]), while Q , R vanish in this limit.

(3) The limit $a \longrightarrow 0$ in Eq. (4.11) requires caution. One finds

$$R(M,\mu,0) = -\sum_{b,c}\left(\frac{\beta M}{2}\right)^{4+2b} \frac{(\beta\mu)^{2c}}{b!(b+2)!(2c)!}$$
$$\times \sum_{m=1}^{\infty}\left\{\sum_{d=0}^{\infty}(-)^d \frac{(m\pi)^{2d}}{(2d)!}\right\}(-)^m \frac{\ln 2\pi m}{m^{-2b-2c}}$$

$$= -\sum_{b,c} \left(\frac{\beta M}{2}\right)^{4+2b} \frac{(\beta\mu)^{2c}}{b!\,(b+2)!\,(2c)!} \times \sum_{m=1}^{\infty} \frac{\ln 2\pi m}{m^{-2b-2c}}$$

where the sum over d leads to $\cos m\pi = (-)^m$ so that the alternating sign $(-)^m$ is cancelled. Then one can use Eq. (B10) to evaluate the sum over m. Consider $\mu = 0$. Whereas Eq. (4.11) has a term coming from b = c = d = 0 (N = 0 in Eq. (4.12)),

$$R(M,0,a\neq 0) = \frac{1}{2}\ln 2\left(\frac{\beta M}{2}\right)^4 + \ldots$$

this term does not occur in R(M,0,0) above (see Eq. (B6)), indicating that it is inappropriate to commute the sums over m and d in Eq. (4.11) in this limit.

(4) As already mentioned, Eqs. (4.8) - (4.11) do not contain any term representing the branch points at $M \pm i A_0 = 0$ in Eq. (1.1b). The Mellin transform method (Appendix C) does yield an appropriate branch point term [2,19]. The explanation of this descrepancy is to be found in the analytic continuation implicit in ζ-function regularization. Weldon [40] has analysed this question and shown that the missing branch point term originates in Eq. (4.9) when the continuation is performed explicitly. For a = 0 it is [2,19]

$$\Delta\Omega_B(\beta, M, \mu, 0) = -\frac{VT}{6\pi}\left(M^2 - \mu^2\right)^{3/2}.$$

The methods employed to calculate this term admit continuation in μ to the complex value μ + ia , and taking the real part we have

$$\Delta\Omega_B(\beta,M,\mu,a) = -\frac{VT}{6\pi}\mathrm{Re}\left[M^2-(\mu+ia)^2\right]^{3/2} \quad (4.13)$$

where $0 \leq \beta a < \pi$. For $\beta a = \pi$ the boson potential Ω_B becomes the fermion potential Ω_F with a = 0 (see Eq. (3.13)) which does not have a branch point. Hence the term (4.13) goes discontinuously to zero.

C.- High-T series for Ω_F

From Eq. (1.2c) and Eqs. (4.6), (4.8) we easily find

$$\Omega_F(\beta,M,\mu,a) \quad (4.14)$$

$$= -\frac{8V}{\pi^2}T^4\sum_{m=1}^{\infty}(-)^{m+1}\frac{1}{m^4}\left(\frac{m\beta M}{2}\right)^2 \times K_2(m\beta M)\cosh m\beta\mu\cos m\beta a$$

$$= \frac{8V}{\pi^2}T^4\left[P(M,\mu,a+\pi/\beta)+Q(M) + R(M,\mu,a+\pi/\beta)\right]$$

where P , Q , R are the functions defined in Eqs. (4.9)-(4.11). Eqs. (4.14) and (4.8) are related by the symmetry (3.13) as they should be.

Remarks:

(1) In the calculation of Eq. (4.14) we expand $\cos m\beta a$ about $a = 0$, rather than $\beta a = \pi$ as in the boson case;

$$\cos m\beta a = \sum_{d=0}^{\infty} (-)^d \frac{m^{2d}}{(2d)!} (\beta a)^{2d} .$$

This is dictated by another lesson concerning Bernoulli polynomials in Ref. [19] ;

$$B_4\left(x \pm \frac{1}{2}\right) = \frac{\pi^4}{3} \sum_{m=1}^{\infty} (-)^{m+1} \frac{1}{m^4} \cos 2\pi m x ,$$

where the sign is chosen to keep the argument of the polynomial between 0 and 1 , and $-1/2 \leq x \leq 1/2$. The middle point appropriate for the expansion is $x = 0$ in this case. As a result, the limit $a \to 0$ is smooth; there is no ambiguity in, for example, the infinite series $R(M, \mu, a+\pi/\beta)$ as $a \to 0$.

(2) The series (4.14) has been calculated by the Mellin transform method for $\mu = a = 0$ in Ref. [21] . For this case the results of the two methods agree perfectly, except that one nonleading term in Eq. (4.14), namely

$$\frac{V}{4\pi^2} M^4 \ln 2$$

which comes from $b = c = d = 0$ in Eq. (4.11) ($N = 0$ in (4.12)), is not produced by the Mellin method. While this term may, therefore, be suspect, it is an unimportant term in every respect.

(3) More important are the branch point singularities in Ω_F for $a \neq 0$. In order to satisfy the symmetry (3.13) we must include the extra term

$$\Delta\Omega_F(\beta, M, \mu, a) = -\frac{d}{2}\Delta\Omega_B(\beta, M, \mu, a+\pi/\beta) \tag{4.15}$$

in the high-T series (4.14), where $\Delta\Omega_B$ is the function (4.13). However, for $a = 0$ this term is replaced by zero.

V.- Thermodynamic Potential in Higgs Theories

One very important application of finite-temperature field theory is to models in which the spontaneous breakdown of a symmetry is realized by the Higgs mechanism. There are reasons (see below) why one would expect that a spontaneously broken symmetry must eventually be restored as the temperature grows large, no matter how the symmetry breakdown mechanism works. To be convincing, the Higgs mechanism should exhibit this behaviour [23] . It does [24-28] , as can be shown by straightforward one-loop calculations of the thermodynamic potential.

In this section we give two examples to show how one-loop calculations are done for theories with SSB. Both examples —a pure Higgs model and the Abelian Higgs model— have been studied previously. Our treatment goes beyond the standard results for these models in certain respects: We work in arbitrary dimension. A chemical potential μ for the Higgs field is introduced which permits the condensate to be modified independently of temperature. Our one-loop corrections include the effects of Higgs chemical potential. Generally the temperature at which symmetry restoration occurs is increased by Higgs chemical potential [29,12] , i.e. symmetry breakdown is enhanced. Thus Higgs chemical potential works against temperature. This is to be expected. An increase in μ represents an increase in the condensate responsible for spontaneous symmetry breakdown, and the symmetry restoration temperature goes up.

Chemical potentials for nonHiggs fields can of course be introduced as well. Their effects depend on the fields involved. In some cases SSB is enhanced, in other cases weakened. For brevity we do not go into these matters here although it would be interesting to do so. Many details are given by Linde [28] .

Let us repeat the standard argument why spontaneous symmetry breakdown (SSB) should only be a low-T phenomenen , with symmetry restoration above some critical temperature T_c . In the real world, superconductors, and other physical systems possessing an order-disorder transition, always exhibit this behaviour. The low-T phase is the ordered phase. At high T the system will choose the disordered phase to minimize its free energy F(T,V) = E - TS . An increase in T always favors larger entropy, and no matter how the internal energy E changes with temperature, the system must eventually be driven out of the ordered phase as T increases indefinately.

A.- Pure Higgs theory

For orientation, consider a pure Higgs model with real scalar field ϕ ,

$$S = \int_\beta \left[\frac{1}{2}(\partial_\mu\phi)^2 + U_H(\phi^2)\right]$$

where U_H is the Higgs potential

$$U_H(\phi^2) = -\frac{\alpha^2}{2}\phi^2 + \frac{\lambda}{4}\phi^4 \, . \qquad (5.1)$$

As usual, the quantum field σ will be introduced through $\phi = \langle\phi\rangle + \sigma$ where $\langle\phi\rangle$ is the nonvanishing vacuum expectation value of ϕ . We also wish to include a bosonic chemical potential μ . A prescription for doing this is given by Eq. (2.13). Normally this prescription is unambigu ous. However, in dealing with Higgs field a bit of care is necessary. The correct prescription is [12]

$$\phi(-\partial^2)\phi \rightarrow \sigma(-D^2)\sigma - \mu^2\left[\langle\phi\rangle^2 + 2\langle\phi\rangle\sigma\right] ,$$

leading to the physical action

$$S = \int_\beta \Big\{ \frac{1}{2}\sigma(-D^2)\sigma + \frac{1}{2}M^2\sigma^2 + \langle\phi\rangle\sigma\left[-\alpha^2-\mu^2+\lambda\langle\phi\rangle^2\right] + \lambda\langle\phi\rangle\sigma^3 + \frac{\lambda}{4}\sigma^4 - \frac{1}{2}\mu^2\langle\phi\rangle^2 + U_H(\langle\phi\rangle^2)\Big\} \qquad (5.2)$$

where the Higgs-generated mass of σ is

$$M^2 = 3\lambda\langle\phi\rangle^2 - \alpha^2 . \qquad (5.3)$$

Comments:

(1) The term linear in σ is eliminated by choosing

$$\langle\phi\rangle^2 = \frac{1}{\lambda}(\alpha^2+\mu^2) \qquad (5.4)$$

which fixes the value of $\langle\phi\rangle$ at the tree level. The bosonic chemical potential μ affects $\langle\phi\rangle$ even classically, and its effect is to strengthen SSB [2,12]. Even if $\alpha^2 < 0$ (which eliminates SSB at zero chemical potential), when μ^2 exceeds $|\alpha^2|$ the vacuum expectation value $\langle\phi\rangle$ will become nonzero. Thus bosonic chemical potentials apparently can induce SSB.

(2) Another way to see this to observe that the classical potential in Eq. (5.2) is not U_H but rather

$$\tilde{U}_H(\phi^2) \equiv -\frac{\mu^2}{2}\phi^2 + U_H(\phi^2)$$
$$= -\frac{1}{2}(\alpha^2+\mu^2)\phi^2 + \frac{\lambda}{4}\phi^4. \qquad (5.5)$$

Effectively, the symmetry breaking parameter α^2 gets replaced by $\alpha^2+\mu^2$ when the chemical potential is introduced.From this point on, it will be the modified potential (5.5) rather than U_H that appears in our formulas.

(3) Quantum corrections will modify the classical value of $\langle\phi\rangle$. To investigate this, one can use the physical action (5.2) in Eq.(2.11) to define a thermodynamic potential $\Omega(\beta, M, \mu, \langle\phi\rangle)$. Immediately below we calculate the one-loop contribution to Ω, obtaining $\langle\phi\rangle$ to this order as a function of T and μ. The result is that temperature and chemical potential pull in opposite directions: Increasing T weakens SSB , while increassing μ strengthens SSB as we have just seen. (This is an interesting physical situation, with some of the features of Bose-Einstein condensation. See Refs. [2,12] for a discussion.) Above some critical temperature (with μ fixed) the symmetry will be restored. Thus the Higgs mechanism behaves with temperature as we expect it to.

Proceeding now to the one-loop calculation, we have

$$S_{ONE} = \int_\beta \left[\frac{1}{2}\sigma(-D^2+M^2)\sigma + \tilde{U}_H(\langle\phi\rangle^2)\right]$$

where the covariant derivative $D_\mu = \partial_\mu - i\,\delta_{\mu 0}A_0$ has been introduced, and the classical Higgs potential term must be retained because it depends on $\langle\phi\rangle$. From Eq. (2.11) we find immediately

$$e^{-\beta\Omega_{ONE}} = \left[\det{}_+(-D^2+M^2)\right]^{-1/2} \times e^{-\beta V \tilde{U}_H(\langle\phi\rangle^2)} \qquad (5.6)$$

or

$$\Omega_{ONE} = \frac{1}{2}\Omega_B(\beta, M, \mu, 0) + V\tilde{U}_H(\langle\phi\rangle^2). \qquad (5.7)$$

Assuming that Ω_B is the dominant quantum correction, the vacuum expectation value $\langle\phi\rangle$ will be a solution of

$$0 = \frac{\partial\Omega_{ONE}}{\partial\langle\phi\rangle} = \langle\phi\rangle\left[3\lambda\frac{\partial\Omega_B}{\partial M^2} + V(\lambda\langle\phi\rangle^2 - \alpha^2 - \mu^2)\right]. \qquad (5.8)$$

$\langle\phi\rangle = 0$ is always a solution. However, another solution is obtained by requiring that the square bracket vanish:

$$\langle\phi\rangle^2 = \frac{\alpha^2+\mu^2}{\lambda} - \frac{3}{V}\frac{\partial\Omega_B}{\partial M^2}. \qquad (5.9)$$

Here Ω_B is an increasing function of M^2, and the second term is positive and increasing with T. Therefore, the one--loop corrections <u>reduce</u> $\langle\phi\rangle^2$ below its classical value (5.4). The stage is set for symmetry restoration with increasing temperature. Before we discuss this in more detail, let us point out one problem with the calculation in progress.

We are computing the quantum corrections to a classical potential $U_H(\langle\phi\rangle^2)$ which is <u>nonconvex</u> in its central part: $\lambda \langle\phi\rangle^2 \le \alpha^2 + \mu^2$. In general, quantum corrections to a nonconvex classical potential are not well-defined, and certainly that is the case here. The one-loop term in Eq. (5.7) is $\Omega_B(\beta, M, \mu, 0)$, with M given by Eq. (5.3). M becomes imaginary for $\langle\phi\rangle^2 < \alpha^2/3\lambda^2$, and hence for $\langle\phi\rangle$ in this range the one-loop calculation fails. In other words, we cannot follow $\langle\phi\rangle$ in Eq. (5.9) all the way to $\langle\phi\rangle = 0$, and symmetry restoration cannot be observed — not at the one-loop level anyway.

The preceeding comment would be disasterous if there were no way to circumvent it. Fortunately, at high temperature, the one-loop calculation does provide self-consistent evidence for symmetry restoration [24-28] . We need the leading terms in the high-T expansion of Ω_B (Sec.IV):

$$\Omega_B(\beta, M, \mu, 0) = V\{F_1(\beta,\mu) + M^2 F_2(\beta,\mu) + \dots\} \quad (5.10)$$

where terms of higher order in M are discarded. Then from Eq. (5.9) one finds the high-T result

$$\langle\phi\rangle^2 \approx \frac{\alpha^2+\mu^2}{\lambda} - 3F_2(\beta,\mu) \, . \quad (5.11)$$

Here there is no mention of M , and this equation can be used to estimate the temperature T_c at which symmetry restoration occurs. T_c is the temperature at which the right-hand side vanishes. To be certain that T_c is real one needs the functions $F_2(\beta, \mu)$, but these are known for arbitrary space-time dimension [2,19] .

Example: four dimensions

From Eq. (4.8) one finds

$$F_2(\beta,\mu) = \frac{T^2}{12}\left[1 - \frac{3}{2\pi^2}(\beta\mu)^2\right] \qquad (5.12)$$

and hence

$$\langle\phi\rangle^2 \approx \frac{\alpha^2+\mu^2}{\lambda} - \frac{T^2}{4}\left[1-\frac{3}{2\pi^2}(\beta\mu)^2\right], \qquad (5.13)$$

$$T_c^2 \approx \frac{4}{\lambda}(\alpha^2+\mu^2) + \frac{3\mu^2}{2\pi^2}\,. \qquad (5.14)$$

These are just the classic results, modified to take bosonic chemical potential into account. The assumption of high temperature is justified in the small coupling domain $\alpha^2 \gg \lambda$, when the one-loop approximation can be expected to be valid. Thus everything seems to be consistent. Eq. (5.14) shows explicitly that a critical temperature exists, above which the symmetry is restored. It shows that a Higgs chemical potential increases T_c. First at the classical level (the first term) and again at the quantum level (second term). Of course, one may question the validity of including the chemical potential term in the one-loop function (5.12), since terms of similar order in M have been neglected. However, it can be argued that the constraint $M^2 > \mu^2$ has ceased to be valid in the present situation. Strictly speaking, the mass $M^2 \approx -\alpha^2$ is not even real near $\langle\phi\rangle \approx 0$. Indeed, if we believe in the relevance of the branch points at $\mu = \pm M$ in the one-loop thermodynamic potential Ω_B in Eq. (5.7), then these would presumably squeeze μ to the value zero as the critical temperature is approached from below, because M vanishes. This would mean that $\mu = 0$ even in the classical term in Eq. (5.14) — i.e. that

Higgs chemical potential plays no role in symmetry restoration.

B.- Abelian Higgs Model

We proceed to a more complicated Higgs theory which has also been studied in the classic papers on symmetry restoration [24-28]. Again we work with a Higgs chemical potential.

The Abelian Higgs model has action

$$S = \int_\beta \left[\frac{1}{4e^2} F_{\mu\nu}^2 + \frac{1}{2} |\overline{D}_\mu \phi|^2 + U_H(\phi^2) \right] \tag{5.15}$$

where $\overline{D}_\mu = \partial_\mu - iA_\mu$ and ϕ is complex. To expand about the minimum $\phi = \langle\phi\rangle$ of the Higgs potential one can write

$$\phi = (\langle\phi\rangle + \sigma) e^{i\chi} \tag{5.16}$$

where σ and χ are real. The phase χ turns out to be a massless Goldstone field, destined to be absorbed into the gauge field (which thereby is provided with a third degree of freedom and becomes massive). While one can demonstrate this explicitly, let us simply quote the standard Higgs-Kibble result, that seting $\phi = \langle\phi\rangle + \sigma$ leads to the correct unitary gauge action

$$\begin{aligned} S = \int_\beta \Big\{ & \frac{1}{e^2} \left[\frac{1}{4} F_{\mu\nu}^2 + \frac{1}{2} m_A^2 A_\mu^2 \right] \\ & + \frac{1}{2} \left[|\overline{D}_\mu \sigma|^2 + M_\sigma^2 \sigma^2 \right] + U_H(\langle\phi\rangle^2) \\ & + \langle\phi\rangle \left[\lambda \langle\phi\rangle^2 - \alpha^2 + A_\mu^2 \right] \sigma \\ & + \lambda \langle\phi\rangle \sigma^3 + \frac{\lambda}{4} \sigma^4 \Big\} , \end{aligned} \tag{5.17}$$

where

$$m_A^2 = e^2 \langle\phi\rangle^2 , \quad M^2 = 3\lambda\langle\phi\rangle^2 - \alpha^2 . \quad (5.18)$$

In this gauge the four physical degrees of freedom A_μ and σ are manifest, and there are no ghosts. Parenthetically, we remark that the Abelian Higgs model has no vacuum parameter (i.e. a constant term in the potential A_0) because local gauge invariance is completely broken. (In gauge theories with SSB , only the unbroken sector of the gauge theory can be given vacuum parameters. See Sec. VI.)

Quantum corrections to the Higgs potential U_H in gauge theories with SSB are known to be gauge dependent [24] . If we use the quadratic part of Eq. (5.17) for the one-loop calculation we will get a certain answer. If we go to a another gauge we may get a different answer. (This slightly disturbing feature is a concomitant of SSB ; in an unbroken gauge theory there is no such ambiguity.) We therefore had better look for some criterion to guide us. Fortunately, there is a very good criterion available. The action (5.17) does not represent a renormalizable theory, and we therefore have no reason to trust the perturbation theory based on it. However, the famous R-gauges exist in which renormalizability can be established for gauge theories with SSB . Perturbation theory is believed to be reliable in these gauges; presumably also the loop expansion. Therefore, we proceed to these gauges to perform the one-loop calculation.

Let us write $\phi = \phi_1 + i\phi_2$ which changes the action (5.17) into

$$S = \int_\beta \Big\{ \frac{1}{4e^2} F_{\mu\nu}^2 + \frac{1}{2}(\partial_\mu \phi_a)^2 + \frac{1}{2}\phi^2 A_\mu^2 + \epsilon_{ab} A_\mu (\partial_\mu \phi_a)\phi_b + U_H(\phi^2) \Big\} \quad (5.19)$$

where indices a,b take values 1 or 2 and $\phi^2 = \phi_a^2$. Next let

$$\phi_a = \langle\phi_a\rangle + \sigma_a \qquad (5.20)$$

and incorporate bosonic chemical potential as in the preceeding example to obtain the one-loop action (compare with Eq. (5.2))

$$S_{ONE} = \int_\beta \Big\{ \frac{1}{4e^2} F_{\mu\nu}^2 + \frac{1}{2}\langle\phi\rangle^2 A_\mu^2 + \widetilde{U}_H(\langle\phi\rangle^2) + \frac{1}{2}\sigma_a(-D^2)\sigma_a + \epsilon_{ab} A_\mu \partial_\mu \sigma_a \langle\phi_b\rangle + \frac{1}{2}\sigma_a\sigma_b\Big[(\lambda\langle\phi\rangle^2 - \alpha^2)\delta_{ab} + 2\lambda\langle\phi_a\rangle\langle\phi_b\rangle\Big]\Big\} \qquad (5.21)$$

where now $D_\nu = \partial_\nu - \delta_{\nu 0}\mu$ and the linear terms

$$\sigma_a\langle\phi_a\rangle\left[-\alpha^2 - \mu^2 + \lambda\langle\phi\rangle^2\right] \qquad (5.22)$$

have been discarded. Some other comments are:

(1) The cross term

$$\epsilon_{ab} A_\mu (D_\mu \phi_a)\phi_b = \epsilon_{ab} A_\mu (\partial_\mu \sigma_a)\langle\phi_b\rangle$$

is independent of the bosonic chemical potential μ.

(2) The tree level vacuum expectation value

$$\langle\phi\rangle^2 = \frac{1}{\lambda}(\alpha^2 + \mu^2)$$

is shifted up by nonvanishing μ just as in the preceeding example. This will always be the effect of Higgs-field chemical potentials.

(3) U_H has been replaced by $\tilde{U}_H$ in Eq. (5.5), i.e. $\alpha^2 \to \alpha^2 + \mu^2$.

(4) α^2 does <u>not</u> get replaced by $\alpha^2 + \mu^2$ in the $\sigma_a \sigma_b$ term in Eq. (5.21).

It remains to fix the gauge. This can be done by introducing the factor (see Ref. [29])

$$1 = \Delta[A_\mu, \phi_a] \int [dU]\, \delta\left[F(A_\mu^U, \phi_a^U - f(x)\right] \tag{5.23}$$

in the integrand of the partition function Z . Here $F(A_\mu, \phi_a)$ is the gauge-fixing function (see Sec. VI) and f(x) is an arbitrary function of x . Both Δ and Z are independent of this function f(x). One can integrate Z multiplied by any functional h [f] of f over f and the result is still Z . In particular,

$$\begin{aligned} Z &= \int [df]\, Z\, e^{-\int_\beta \frac{1}{2} f^2} \\ &= \int [df]\, e^{-\int_\beta \frac{1}{2} f^2} \int_+ [dA_\mu d\phi_a] \\ &\quad \times \Delta \int [dU]\, \delta[F^U - f]\, e^{-S} \\ &= \int_+ [dA_\mu d\phi_a]\, \Delta \int [dU]\, e^{-S - \int_\beta \frac{1}{2}(F^U)^2} , \end{aligned} \tag{5.24}$$

where $F^U = F(A_\mu^U, \phi_a^U)$. Now, using the gauge-invariance of the integrand (see Sec. VI) we can rewrite this as

$$Z = \int_+ [dA_\mu d\phi_a] \Delta[A_\mu, \phi_a] e^{-\bar{S}} \quad (5.25)$$

where

$$\bar{S} = S + \int_\beta \frac{1}{2} F^2(A_\mu, \phi_a) . \quad (5.26)$$

Here we have assumed that the gauge chosen is a periodic one; i.e. the periodic boundary conditions are not "twisted" by gauge fixing (Sec. VI). This is true of the R-gauges.

The cross term $A_\mu \partial_\mu \sigma_a$ in Eq. (5.21) is neatly removed by the R-gauge choice [30]

$$F = \sqrt{\xi} \left[\partial_\mu A_\mu + \frac{1}{\xi} e^2 \epsilon_{ab} \sigma_a \langle \phi_b \rangle \right] . \quad (5.27)$$

Then the one-loop part of $\bar{S}$ is

$$\bar{S}_{ONE} = \int_\beta \left\{ \frac{1}{2e^2} A_\mu M_{\mu\nu} A_\nu + \frac{1}{2} \sigma_a M_{ab} \sigma_b + \tilde{U}_H(\langle \phi \rangle^2) \right. \quad (5.28)$$

where

$$M_{\mu\nu} = \delta_{\mu\nu} \left[-\Box + e^2 \langle \phi \rangle^2 \right] + (1 - \xi) \partial_\mu \partial_\nu , \quad (5.29)$$

$$M_{ab} = \delta_{ab}\left[-D^2+\left(\lambda+\frac{e^2}{\xi}\right)\langle\phi\rangle^2-\alpha^2\right] + \left(2\lambda-\frac{e^2}{\xi}\right)\langle\phi_a\rangle\langle\phi_b\rangle . \quad (5.30)$$

Moreover, the Faddeev-Popov factor Δ for this gauge choice is

$$\Delta = \det_+\left[-\Box+\frac{e^2}{\xi}\left(\langle\phi\rangle^2+\langle\phi_a\rangle\,\sigma_a\right)\right] \quad (5.31)$$

where we have recognized that F defines a periodic gauge. To one-loop the Faddeev-Popov factor is simply

$$\Delta = \det_+\left(-\Box+m_A^2/\xi\right) \quad (5.32)$$

and the one-loop thermodynamic potential is given by

$$e^{-\beta\Omega_{ONE}} = \det_+(-\Box+m_A^2)\left[\det_+(-\Box+m_A^2)\right]^{-\frac{S+1}{2}} \times\left[\det_+ M_{ab}\right]^{-1/2} e^{-\beta V\tilde{U}_H(\langle\phi\rangle^2)} \quad (5.33)$$

where $\xi = 1$ has been chosen in order to diagonalize $M_{\mu\nu}$. Noting that

$$\det{}_+ M_{ab} \qquad (5.34)$$
$$= \det{}_+(-D^2+M_1)^2 \det{}_+(-D^2+M_2^2)$$

where

$$M_1^2 = 3\lambda\langle\phi\rangle^2 - \alpha^2$$
$$M_2^2 = (\lambda+e^2)\langle\phi\rangle^2 - \alpha^2 \qquad (5.35)$$

and $D_\mu = \partial_\mu - \delta_{\mu 0}\mu$ we obtain the final answer

$$\Omega_{ONE} = \frac{1}{2}(S-1)\Omega_B(\beta, m_A, 0, 0)$$
$$+ \frac{1}{2}\Omega_B(\beta, M_1, \mu, 0)$$
$$+ \frac{1}{2}\Omega_B(\beta, M_2, \mu, 0)$$
$$+ V\tilde{U}_H(\langle\phi\rangle^2) . \qquad (5.36)$$

The first three terms in Eq. (5.36) comprise the one--loop corrections to the Higgs potential $\tilde{U}_H$. Assuming these are the dominant corrections, the vacuum expectation value will be a solution of

$$0 = \frac{\partial\Omega_{ONE}}{\partial\langle\phi\rangle}$$
$$= \langle\phi\rangle\left\{(S-1)e^2\frac{\partial}{\partial m_A^2}\Omega_B(\beta, m_A, 0, 0)\right.$$

$$+ 3\lambda \frac{\partial}{\partial M_1^2} \Omega_B(\beta, M_1, \mu, 0)$$
$$+ (\lambda + e^2) \frac{\partial}{\partial M_2^2} \Omega_B(\beta, M_2, \mu, 0)$$
$$+ V(-\alpha^2 - \mu^2 + \lambda \langle\phi\rangle^2)\Big\} . \qquad (5.37)$$

The interesting solution is the one for which the curly bracket vanishes. $M_{1,2}$ become imaginary as $\langle\phi\rangle \longrightarrow 0$, and again we only find consistent evidence for symmetry restoration in the high-T limit. All the remarks made concerning Eq. (5.11) can be repeated here. We obtain an estimate of the symmetry-restoration temperature T_c from the high-T expansions of the Ω_B's in Eq. (5.37). This can be done for arbitrary space-time dimension, but let us just consider four dimensions.

From Eqs. (5.10), (5.12) we find

$$\langle\phi\rangle^2 \approx \frac{\alpha^2 + \mu^2}{\lambda} - \frac{T^2}{12}\left(4 + \frac{3e^2}{\lambda}\right) + \frac{\mu^2}{8\pi^2}\left(4 + \frac{e^2}{\lambda}\right) \qquad (5.38)$$

and critical temperature

$$T_c^2 \approx \frac{12\left(\frac{\alpha^2 + \mu^2}{\lambda}\right) + \frac{3\mu^2}{2\pi^2}\left(4 + \frac{e^2}{\lambda}\right)}{4 + 3\frac{e^2}{\lambda}} . \qquad (5.39)$$

To obtain these formulas it was assumed that all of βm_A, βM_1 and βM_2 are small.

$\underline{\mu = 0}$:

For consistency T_c must be large in the limit of weak coupling. But here we see it is necessary to stipulate that e^2/λ cannot be arbitrarily large, for this would make T_c small. Thus the relative size of e^2 and λ enters the analysis. If e^2/λ is allowed to be large the masses cannot be ignored and the analysis becomes much less straightforward. See Refs. [27,28] for some details of the $\mu = 0$ case.

$\mu \neq 0$:

As long as e^2/λ is not very large, the above picture changes as expected when Higgs chemical potential is included. T_c increases both classically and quantum mechanically. However, it is also possible to consider the limit of large e^2/λ. Taking $e^2/\lambda \gg 4$ we have

$$T_c^2 \approx \frac{4}{e^2}(\alpha^2+\mu^2) + \frac{\mu^2}{2\pi^2} \quad . \qquad (5.40)$$

For $\mu = 0$ the first term shows that T_c^2 is small for e^2 large—the problem mentioned above. But if μ is sufficiently large, T_c can remain large and the calculation remains consistent.

C.- Remarks

(1) The main features of symmetry restoration at high-T illustrated by the preceeding examples are to be found in nonAbelian gauge theories with SSB as well. The Weinberg-Salam model provides an important example [12,28] . Above some critical temperature T_c the full SU(2) x U(1) local gauge symmetry will be restored. All Higgs-generated masses vanish for $T > T_c$. In the past it has occasionally been claimed that this means the weak interactions become long-range. This is in fact not true. Even though the $W^{\pm}$ and Z masses are zero for $T > T_c$, there is strong screening of all components of the SU(2) x U(1) electric field. This is because every unbroken gauge theory is an

electric plasma at high-T . (The term "electric" being understood in a non Abelian sense if the gauge theory is nonAbelian.) Every electric charge (Abelian or nonAbelian) is screened at high-T with a screening length which decreases as T increases (see Sec.VI). Therefore, even though the $W^{\pm}$ and Z masses vanish, the gauge fields do not become long-range. Models of Grand Unification provide another example. Higgs fields are used to break an exact gauge symmetry relating strong, electromagnetic and weak forces, yielding Weinberg-Salam plus QCD. In the very early universe, with T so large that symmetry is restored in these models, one has a plasma in which no distinction exists between the strong, electromagnetic and weak forces. The electric fields associated with all of these forces are equally strongly screened in the primordial plasma.

(2) In Sec. VI we describe the behaviour of unbroken gauge theories as a function of temperature. The reader may be curious to know how the picture described there can be applied to a partially-broken gauge theory. The answer is just what one would expect. SSB will give masses to some components — call these W_μ — of the original gauge potential and leave some other components massless — let us call these A_μ . The potentials A_μ take values in the algebra of the unbroken subgroup $H \subset G$ of the original gauge group G . Thus SSB leaves a residual gauge theory with action $S[A_\mu, W_\mu, \ldots]$ in which the W_μ are effectively matter fields along with whatever scalars and spinors were present initially. This gauge theory has exactly the behaviour described in Sec. VI. However, the presence of SSB does introduce certain technical limitations. One-loop calculations of the electric action have to be done in the 't Hooft gauge, for example, much as in the Abelian Higgs model example above.

VI.- THERMODYNAMIC POTENTIAL IN GAUGE THEORIES

The effects of finite temperature on unbroken gauge theories like QCD are important for physics, and in the author's view exhibit elements of mathematical beauty. Two basic facts determine the general picture: (1) Gauge theories are formulated in terms of a vector potential A_μ ; (2) finite T introduces a preferred frame (the rest frame of the thermal background and the walls which maintain this background at the equilibrium temperature). To retain explicit Lorentz covariance in the physics one can introduce a four-velocity for this preferred frame. Nevertheless, it is entirely permissible to work in the preferred frame (as is almost always done in statistical mechanics and thermodynamics). From now on we perform <u>all</u> calculations in this frame. Once the validity of this restriction has been accepted, Lorentz invariance is <u>effectively</u> reduced to rotation invariance: boosts may not be performed. This, in turn, effectively detaches the time component A_0 from the spatial components A_i of any space-time vector A_μ ;

$$\underset{(s+1)\text{-vector}}{A_\mu = (A_0, A_i)} \longrightarrow \underset{\text{scalar}}{A_0} \quad \text{and} \quad \underset{s\text{-vector}}{(A_i)} . \qquad (6.1)$$

The scalar A_0 and the O(s) rotation vector A_i are kinematically unrelated in the preferred frame because of the suppression of boosts. In effect, A_0 is not one component of a vector but rather a scalar field (!) at finite T . Obviously A_i remains an s-vector under rotations. At T = 0 none of this is possible because there is no preferred frame to build upon.

Next, observe that A_0 and A_i are the dynamical variables of the electric and magnetic sectors of the gauge theory respectively. The meaning of Eq. (6.1) is that the electric and magnetic sectors of a gauge theory become kinematically distinct at finite T ; one can meaningfully define these sectors.(This ceases to be possible when T $\longrightarrow$ 0 .) This does not imply that the electric and magnetic sectors are decoupled — certainly they are coupled.

Just to make this point more explicit, let us imagine that we can ignore gauge fixing (which, of course, we cannot do ultimately). If this were the case then one would have the following natural definitions of effective actions S_{el} and S_{mag} for the electric and magnetic sectors respectively:

$$e^{-\beta S_{el}[A_0]} = \int_{+} [dA_i]\, e^{-S[A_\mu]} ,$$

$$e^{-\beta S_{mag}[A_i]} = \int_{+} [dA_0]\, e^{-S[A_\mu]} , \qquad (6.2)$$

where $S[A_\mu]$ is the gauge theory action and dependence on other fields is suppressed. Clearly the actions S_{el} and S_{mag} have very different properties. The most simple way to discuss this is in terms of an Abelian gauge theory. So let us assume for the moment that the gauge group is U(1).

Magnetic sector:

The action $S[A_\mu]$ is invariant under any gauge transformation. Consider a static gauge transformation, leaving A_0 unchanged. The magnetic action $S_{mag}[A_i]$ is invariant under this gauge transformation, and it therefore represents some type of gauge theory. (Not a gauge theory with conventional action, of course, because infinitely many interaction terms have been introduced by integration over A_0 .) In general, one can say that the gauge group G of the original theory is inherited by its magnetic sector.

Electric Sector:

Turning next to the electric sector, we see that the electric action $S_{el}[A_0]$ is clearly not a gauge theory. Rather, it is a scalar theory, with scalar variable A_0. (Note that position-independent gauge transformations leaving A_i invariant might seen to provide S_{el} with a vestige of gauge invariance. This does not survive gauge fixing, however. The electric action has none of the attributes of a gauge theory.)

To summarize: At finite T the existence of a thermal background affects the vectorial structure of a gauge theory in such a way that its electric and magnetic sectors become kinematically distinct (without becoming decoupled). The electric sector is described by an effective scalar theory, while the magnetic sector is some unusual and complicated type of gauge theory. The magnetic sector is hardly involved in the rest of these notes, and we are going to say very little about it. Our attention will be on the electric sector and its action $S_{el}[A_0]$. Much of the important progress in gauge theories in recent years has only been possible through the realization that S_{el} exist and contains accessible information. The earliest papers known to the author which recognize this fact are by Polyakov [31] and by Susskind [32]. In this section we explain how to calculate the electric action for constant A_0 to one loop. This exercise will provide us with certain important results. But before we can do anything we must fix the gauge [5].

A. Gauge Fixing

Let us derive the electric effective action in a careful way by fixing the gauge a la Faddeev-Popov. First we consider a pure Yang-Mills theory. Matter fields can easily be put in later. Begin with the unfixed partition function defined on the manifold of periodic gauge potentials $\tilde{A}_\mu$;

$$Z_{\text{UNFixed}} = \int_{+} [d\tilde{A}_\mu] e^{-S[\tilde{A}_\mu]} . \quad (6.3)$$

Define the Faddeev-Popov factor $\Delta[A_\mu]$ in the standard way [29],

$$1 = \Delta[A_\mu] \int [dU] \delta[F(A_\mu^U)] \quad (6.4)$$

where dU is the Haar measure for the gauge group G and

$$A_\mu^U = U[A_\mu + i\partial_\mu]U^{-1} \quad (6.5)$$

is the gauge transformation rule for A_μ. F in Eq. (6.4) is the gauge - fixing function, and the Faddeev-Popov factor has the meaning expressed by

$$[dU]\Delta[A_\mu] = [dF(A_\mu^U)] . \quad (6.6)$$

In other words, $\Delta[A_\mu]$ is a Jacobian factor or functional determinant. To find out what this factor is, consider a small gauge transformation $U = 1 + iu$,

$$F(A_\mu^U) = F(A_\mu) + M(A_\mu)u + O(u^2) . \quad (6.7)$$

In the gauge fixing problem we only need $M(A_\mu)$ for A_μ which satisfy $F(A_\mu) = 0$. For such A_μ

$$F(A_\mu^U) = M(A_\mu)u + O(u^2) \qquad (6.8)$$

and Eq. (6.4) becomes

$$\begin{aligned} 1 &= \Delta[A_\mu] \int [du]\, \delta[Mu] \qquad (6.9) \\ &= \Delta[A_\mu] (\det M)^{-1} \int [d(Mu)]\, \delta[Mu] \end{aligned}$$

so that

$$\begin{aligned} \Delta[A_\mu] &= \det\nolimits_+ M(A_\mu) \qquad (6.10) \\ &= \int_+ [dc]\, e^{-\int_\beta \bar{c} M(A_\mu) c} \; . \end{aligned}$$

Thus, given the gauge-fixing function F , we can calculate M and Δ . In the second equality in Eq. (6.10) the famous ghost fields have been introduced. These are Grassmannian, but satisfy periodic boundary conditions because they originate in the boson sector.

Now we can fix the gauge. Notice what happens in the boundary conditions when this is done:

$$\begin{aligned} &Z_{\text{unfixed}} \qquad (6.11a) \\ &= \int_+ [d\tilde{A}_\mu]\, e^{-S[\tilde{A}_\mu]}\, \Delta[\tilde{A}_\mu] \int [dU]\, \delta[F(\tilde{A}_\mu^U)] \end{aligned}$$

$$= \int [dU] \int\limits_{(\tilde{A}_\mu^U)_\beta^{U^{-1}} = (\tilde{A}_\mu^U)_0^{U^{-1}}} [d\tilde{A}_\mu^U] \Delta[\tilde{A}_\mu^U] \delta[F(\tilde{A}_\mu^U)] \times e^{-S[\tilde{A}_\mu^U]} \qquad (6.11b)$$

$$= \int [dU] \int\limits_{A_{\mu\beta}^{U_\beta^{-1}} = A_{\mu 0}^{U_0^{-1}}} [dA_\mu] \Delta[A_\mu] \delta[F(A_\mu)] \times e^{-S[A_\mu]} \qquad (6.11c)$$

$$= \int\limits_{x_0 \neq \beta, 0} [dU] \int [dU_\beta dU_0] \int\limits_{A_{\mu\beta} = A_{\mu 0}^{\Omega^{-1}}} [dA_\mu] \Delta[A_\mu] \times \delta[F(A_\mu)] e^{-S[A_\mu]} . \qquad (6.11d)$$

We pause to explain these several steps before going on to the final formula. In Eq. (6.11a) we have inserted 1 under the integral. In (6.11b) gauge invariance has been used to replace $\tilde{A}_\mu$ by $\tilde{A}_\mu^{\ U}$ in S , Δ and $[d\tilde{A}_\mu]$. The boundary condition is not gauge invariant, and we have rewritten it in the only way possible to make it depend on $\tilde{A}_\mu^{\ U}$. The subscripts β , 0 always mean evaluation at times $x_0 = \beta$, 0 respectively. In (6.11c) $\tilde{A}_\mu^{\ U}$ has been renamed A_μ , and in terms of this variable the boundary condition is no longer periodic if $U_\beta \neq U_0$. In (6.11d) we are preparing to throw away the redundant integration over U . Unlike the corresponding T = 0 problem, however, we cannot therow away all of it. The integrand of $\int [dU]$

depends on the "twist" variable

$$\Omega(\vec{x}) \equiv U_0(\vec{x}) U_\beta^{-1}(\vec{x}) \in G \quad (6.12)$$

through the boundary condition.. Therefore, one must retain an integration $\int [d\Omega]$ while discarding the rest of $\int [dU]$. The result is the fixed partition function

$$Z = e^{-\beta\Omega} = \int [d\Omega]\, e^{-\beta S_{el}[\Omega]} \quad (6.13)$$

$$e^{-\beta S_{el}[\Omega]} \equiv \int_{A_{\mu\beta}=A_{\mu 0}^{\Omega^{-1}}} [dA_\mu]\, \Delta[A_\mu]\, \delta[F(A_\mu)]\, e^{-S}. \quad (6.14)$$

Eqs. (6.13), (6.14) are final formulas for the thermo dynamic potential Ω of the gauge theory. Ω has been expressed as a functional integral over the twist variable $\Omega(\vec{x})$ defined by Eq. (6.12). (We regret that the symbols are the same and that both are common usage; from the context it will always be clear which is meant.) Eq. (6.14) defines the electric action for the gauge theory, assuming the gauge chosen is nonperiodic — i.e. $\Omega \neq 1$ for this gauge. Previously we clained that S_{el} is a functional of A_0, and shortly this will be seen to be the case. For the moment, however, let us dwell on certain features of Eq. (6.13), (6.14).

1) Without the boundary condition (i.e. for $T = 0$) there would be no electric action S_{el}.

2) The boundary condition has been twisted (if $\Omega \neq 1$) by the procedure of gauge fixing. In Sec. II we regarded all boson

fields as periodic. This is the only option for nongauge fields. However, a gauge theory is different in that a local gauge transformation has no physical meaning. The twist $A_{\mu\beta} = A_{\mu 0}^{\Omega^{-1}}$ in the boundary condition is therefore an allowable option for a gauge theory.

3) When matter fields are included the only change is to incorporate these fields into the functional integral (6.14) with appropriate boundary conditions

$$\phi_{\pm\beta} = \pm \phi_{\pm 0}^{\Omega^{-1}} \tag{6.15}$$

where ϕ^U is the gauge transformation rule for the field ϕ.

B. Temporal Gauge

It is very useful to choose a particular nonperiodic gauge, the temporal gauge with gauge-fixing function $F(A_\mu) = A_0$. For $U = 1 + iu$ we have

$$F(A_\mu^U) = A_0 + i[u, A_0] + \partial_0 u$$

and consequently $M = \partial_0$ is independent of A_μ. This means ghosts are decoupled from A_μ in this gauge and the factor $\Delta[A_\mu]$ in Eq. (6.14) can simply be dropped, which leaves the temporal gauge formula

$$e^{-\beta S_{el}[\Omega]} = \int_{A_{i\beta} = A_{i0}^{\Omega^{-1}}} [dA_i]\, e^{-S[A_i]} \tag{6.16}$$

where $A_0 = 0$. To calculate the twist Ω in the temporal gauge we need the gauge transformation U which takes an arbitrary periodic potential $\tilde{A}_\mu$ (belonging to the function space integrated over in Eq. (6.3)) to the temporal gauge. A satisfactory U is

$$U^{-1} = P\exp i\int_0^{x_0} dx_0\, \tilde{A}_0 \qquad (6.17)$$

where P means path or Dyson ordering. Then one readily finds the expression

$$\Omega(\vec{x}) = P\exp i\int_0^{\beta} dx_0\, \tilde{A}_0(x_0,\vec{x}) \qquad (6.18)$$

for the temporal gauge twist. We see that $\Omega(\vec{x})$ is a functional of $\tilde{A}_0$, and thus S_{el} in Eq. (6.16) can be regarded either as a functional of the twist, or of $\tilde{A}_0$. This establishes the connection between Eq. (6.14) and the discussion given at the beginning of this section which emphasized the role of $\tilde{A}_0$.

Actually, the electric action $S_{el}[\Omega]$ only depends on symmetric functions of the eigenvalues of Ω . This is because there is some remaining gauge ambiguity in the gauge transformation U in Eq. (6.17). Let this U be replaced by $U' = V(\vec{x})U$ where $V(\vec{x})$ is a static gauge transformation. U' also takes $\tilde{A}_\mu$ to the temporal gauge, but gives a different twist $\Omega' = V\Omega V^{-1}$. Clearly V can be chosen such as to diagonalize Ω , i.e. $\Omega'(\vec{x}) = \mathrm{diag}\,(\omega_a(\vec{x}))$ where ω_a are the eigenvalues of the matrix (6.18). Moreover, the global permutation matrices which permute the ω_a among themselves also belong to the gauge group, and hence the order of the eigenvalues is irrelevant. Thus S_{el} is a symmetric function of the ω_a as claimed.

Note also that the freedom to perform static gauge transformations $V(\vec{x})$ in the temporal gauge enables one to diagonalize static $\tilde{A}_0$,

$$\tilde{A}_0^V = V\tilde{A}_0 V^{-1} = \mathrm{diag}(\tilde{A}_{0b}(\vec{x})) \quad (6.19)$$

Therefore we can regard the electric action S_{el} as a symmetric function of the eigenvalues $A_{0b}(\vec{x})$ of a static potential . This brings us to the vacuum parameters of the gauge theory. These parameters are constant terms a_b in the eigenvalues $A_{0b}(\vec{x})$;

$$\tilde{A}_{0b}(\vec{x}) = a_b + \alpha_b(\vec{x}) . \quad (6.20)$$

Later on we shall calculate the dependence of the theory on these constants, to one loop. This calculation provides crucial physical insight into the electric sector.

C. Periodic Gauges

For completeness, let us explain what happens when the gauge is periodic, i.e. when $U_\beta = U_0$ and the twist $\Omega = 1$ is always unity. Almost all of the familiar gauges (axial, Coulomb, Lorentz,...) are periodic. In these gauges Eqs. (6.13), (6.14) are inappropriate, and the derivation (6.11) leads instead to a different formula for the partition function,

$$Z = \int_{+} [d\tilde{A}_0] e^{-\beta S_{el}[\tilde{A}_0]} , \quad (6.21)$$

$$e^{-\beta S_{el}[\tilde{A}_0]} \equiv \int_{+} [d\tilde{A}_i] \Delta[\tilde{A}_\mu] \delta[F(\tilde{A}_\mu)] e^{-S} . \quad (6.22)$$

D. Electric Thermodynamic Potential

With its periodic boundary condition, Eq. (6.22) looks rather like the definition of a thermodynamic potential. $S_{el}[\tilde{A}_0]$ is far more complicated, of course, because it depends on a field. But we can easily obtain a quantity which has precisely the meaning of a thermodynamic potential by integrating $S_{el}[\tilde{A}_0]$ over fluctuations around a constant value of its argument. Thus

$$Z = e^{-\beta\Omega_{el}(a_b)} \equiv \int_{+} [d\tilde{A}_0]\, e^{-\beta S_{el}[\mathrm{diag}(a_b)+\tilde{A}_0]} . \qquad (6.23)$$

This defines Ω_{el} which we shall refer to as the electric thermodynamic potential. This function is an ordinary thermodynamic potential in every respect, aside from its dependence on the vacuum parameters a_b . It has a loop expansion: the one-loop term in this expansion will be a sum of $\Omega_{B's}$ and $\Omega_{F's}$ just as for a nongauge theory. We now give two examples.

E. Example: QED

Consider QED in s + 1 dimensions with one scalar and one spinor field,

$$S = \int_\beta \Big[\frac{1}{4e^2}F_{\mu\nu}^2 + \frac{1}{2}\bar{\Phi}(-D^2+M_b^2)\phi + \bar{\Psi}(\not{D}+M_f)\Psi\Big] . \qquad (6.24)$$

Write the gauge potential as $A_\mu = (a, \vec{0}) + a_\mu$ to introduce the vacuum parameter a . Observe that the Abelian theory kinetic term does not depend on a. Interactions are necessary to produce a dependence on vacuum parameters. Choose the Lorentz gauge $\partial_\mu a_\mu = 0$ for the quantum gauge field. Then the quadratic action is

$$S_{ONE} = \int_\beta \Big[\frac{1}{2e^2} a_\nu(-\Box) a_\nu + \frac{1}{2}\bar{\phi}(-D^2 + M_b^2)\phi + \bar{\Psi}(\not{D} + M_f)\Psi\Big]. \quad (6.25)$$

Here $D_\nu = \partial_\nu - i\delta_{\nu 0}(a - i\mu)$ in the scalar and spinor terms. To find the Faddeev-Popov factor Δ we note that

$$F(a_\mu^U) = \partial_\mu a_\mu^U = \partial_\mu(a_\mu + \partial_\mu u)$$

and therefore $M = \Box$ and $\Delta = \det_+(-\Box)$. The one-loop thermodynamic potential is

$$\begin{aligned} & e^{-\beta\Omega_{ONE}} \\ &= \int_+ [dA_\mu d\phi d\Psi] \det_+(-\Box)\, e^{-S_{ONE}} \\ &= \det_+(-\Box)\big[\det_+(-\Box)\big]^{-\frac{1}{2}(s+1)} \\ &\quad \times \big[\det_+(-D^2 + M_b^2)\big]^{-1} \det_{D-}(\not{D} + M_f) \end{aligned} \quad (6.26)$$

and therefore

$$\Omega_{ONE} = \frac{1}{2}(s-1)\Omega_B(\beta,0,0,0) + \Omega_B(\beta,M_b,\mu_b,a) + \Omega_F(\beta,M_F,\mu_F,a) . \qquad (6.27)$$

The first term comes from the non interacting photon gas with s - 1 transverse degrees of freedom. From the calculation we see that there is no possibility of introducing a chemical potential for the photon gas. Terms two and three do involve chemical potentials as well as the vacuum parameter, however. These terms come from the free (to one loop) bose and fermi gases, whose particles propagate in a thermal background characterized by μ and a .

F. Example QCD_N

As another physically relevant example, let us consider QCD with gauge group SU(N) and one flavor,

$$S = \int_\beta \left[\frac{1}{2g^2} \operatorname{tr} F_{\mu\nu}^2 + \bar{\Psi}(\not{\partial}+M)\Psi\right] . \qquad (6.28)$$

Our task is to construct Ω_{one}. For the fermion sector this is trivial (Sec. III D.4). However, we have to work a bit harder in the boson sector, because it is necessary to calculate the gluon fluctuations about the background potential

$$C_\mu = \delta_{\mu 0} \operatorname{diag}(A_{0a}) , \quad a=1,2,\ldots N . \qquad (6.29)$$

Writing $A_\mu = C_\mu + a_\mu$, the standard proceedure is to choose the background gauge condition [33]

$$F(a_\mu) \equiv \bar{D}_\mu a_\mu \equiv \partial_\mu a_\mu + i[a_\mu, C_\mu] = 0 . \tag{6.30}$$

The field strength tensor becomes

$$F_{\mu\nu} = \bar{D}_\mu a_\nu - \bar{D}_\nu a_\mu - i[a_\mu, a_\nu] \tag{6.31}$$

and the one-loop action splits into boson and fermion parts

$$S_{ONE} = S_B + S_F . \tag{6.32}$$

In the boson part

$$S_B = \int_\beta \frac{1}{g^2} \mathrm{tr}\, a_\nu[-\bar{D}_\mu \bar{D}_\mu a_\nu + \bar{D}_\mu \bar{D}_\nu a_\mu] \tag{6.33}$$

the second term vanishes by Eq. (6.33) leaving a gaussian integral of the adjoint type in Eq. (3.24) to be performed. This problem was solved in Sec. III D.5. However, there is still the Faddeev-Popov factor to worry about; i.e. we need the determinant of the operator M defined by Eq. (6.7). Let us recall that in the definition of the background gauge [33],

$$C_\mu^U \equiv C_\mu \tag{6.34}$$

is required to be independent of gauge transformations. Hence the gauge transformation rule for the quantum field a_μ is

$$a_\mu^U = -c_\mu + U(c_\mu + a_\mu)U^{-1} + iU\partial_\mu U^{-1},$$
$$(6.35)$$

and for an infinitesimal transformation $U = 1 + iu$

$$a_\mu^U = a_\mu + i[u, a_\mu] + \bar{D}_\mu u \qquad (6.36)$$

where

$$\bar{D}_\mu u = \partial_\mu u + i[u, c_\mu]. \qquad (6.37)$$

Therefore, in Eq. (6.7) we have

$$\begin{aligned} F(a_\mu^U) &= \bar{D}_\mu a_\mu^U \\ &= \bar{D}_\mu a_\mu + \bar{D}^2 u + i\bar{D}_\mu[u, a_\mu] \\ &\equiv \bar{D}_\mu a_\mu + Mu \quad , \end{aligned} \qquad (6.38)$$

and

$$Mu = \bar{D}^2 u + i\bar{D}_\mu[u, a_\mu]. \qquad (6.39)$$

At the one-loop level $M = \bar{D}^2$, and the relevant Faddeev-Popov factor is $\Delta = \det_+ \bar{D}^2$, where the determinant is understood to be in the same adjoint sense as the adjoint gaussian integral in Eq. (3.24). (Note that the background gauge is periodic, i. e. the twist is always $\Omega = 1$. To check this just let

$C_\mu \longrightarrow 0$ and the background gauge becomes the Lorentz gauge which is obviously periodic.) Thus, we have

$$
\begin{aligned}
e^{-\beta\Omega_{\text{ONE}}} &= \Delta \int_+ [da_\mu]\, e^{-\int_\beta \mathrm{tr}\, a_\nu(-\bar{D}^2)a_\nu} \\
&\quad \times \int_- [d\Psi]\, e^{-\int_\beta \bar{\Psi}(\not{D}+M)\Psi} \\
&= [\det{}_+(-\bar{D}^2)]_{\text{adj}} [\det{}_+(-\bar{D}^2)]_{\text{adj}}^{-\frac{1}{2}(s+1)} \\
&\qquad \times [\det{}_-(-\bar{D}^2+M^2)]_{\text{FUND}}^{d/2}
\end{aligned}
\tag{6.40}
$$

and finally

$$
\begin{aligned}
\beta\Omega_{\text{ONE}} = \frac{1}{2}(s-1)\Big\{ &\sum_{a,b} \ln \det{}_+(-D_{ab}^2) \\
&- \ln \det{}_+(-\Box) \Big\} \\
&- \frac{d}{2}\sum_a \ln \det{}_-(-D_a^2+M^2) ,
\end{aligned}
\tag{6.41}
$$

$$
\begin{aligned}
D_{\mu ab} &= \partial_\mu - i(A_{0a}-A_{0b})\delta_{\mu 0} , \\
D_{\mu a} &= \partial_\mu - iA_{0a}\delta_{\mu 0} .
\end{aligned}
\tag{6.42}
$$

In terms of the prototype potentials $\Omega_{B,F}$ the answer is

$$\Omega_{\text{ONE}} = \frac{1}{2}(s-1)\Big\{\sum_{b,c}\Omega_B(\beta,0,0,a_b-a_c) - \Omega_B(\beta,0,0,0)\Big\} + \sum_b \Omega_F(\beta,M,\mu,a_b) \ . \qquad (6.43)$$

In the gluon and ghost determinants there is no chemical potential (i.e. we use $A_{0b} = a_b$) because gluon and ghost number are not conserved. In the fermion determinant, a chemical potential μ can be introduced (i.e. $A_{0b} = a_b - i\mu$) because fermion number is a conserved quantity. If several flavors of quarks are present, then one adds additional fermion terms in Eq. (6.43), each with its own chemical potential. Note that the chemical potential is not a dynamical variable, but rather is introduced by hand into A_0. Hence the chemical potential in $A_{0b} = a_b - i\mu$ is not subject to the constraint $\sum_b A_{0b} = 0$ arising from $\text{tr}A_\mu = 0$.

In all determinants the same set of vacuum parameters a_b are used. As has been repeatedly emphasized, these are dynamical parameters and hence are subject to the constraint $\text{tr}A_0 = 0$. Thus

$$\sum_{b=1}^{N} a_b = 0 \ , \qquad (6.44)$$

and in Yang-Mills problems this constraint must be taken into account.

Let us make contact with some published results. Set the quark mass M = 0 and go to s = 3 spatial dimensions. Then from the results on high-T expansions in Sec. IV we have the exact results

$$\Omega_B(\beta,0,0,0) = -V\,\frac{\pi^2 T^4}{45}\,, \qquad (6.45)$$

$$\Omega_F(\beta,0,\mu,0) \qquad (6.46)$$
$$= -V\,\frac{T^4}{6\pi^2}\left[\frac{7\pi^4}{30} + \pi^2(\beta\mu)^2 + \frac{1}{2}(\beta\mu)^4\right]$$

where the vacuum parameters have been set equal to zero. Then

$$\Omega_{ONE}(a_b=0)$$
$$= (N^2-1)\,\Omega_B(\beta,0,0,0) + N\,\Omega_F(\beta,0,\mu,0)$$
$$= -V\left\{\frac{\pi^2 T^4}{45}\left(N^2-1+\frac{7}{4}N\right)\right. \qquad (6.47)$$
$$\left. + \frac{N}{6}\left(T^2\mu^2 + \frac{\mu^4}{2\pi^2}\right)\right\}.$$

This formula can be compared with results in the literature [12,34] and seen to agree, as it must. By consulting references (e.q. [11]) in which two-loop and higher corrections to Ω_{one} are calculated, a feeling for the amount of work involved will quickly be gained. By comparison, the one-loop calculations here are easy.

In may be emphasized that Eq. (6.43), from which the formula (6.47) was obtained, contains lots of additional information.Not only is the quark mass M arbitrary; the vacuum parameters a_b are also present. This is typical of the formulism developed in these notes. It provides maximum generality at the one-loop level in the sense that all masses, chemical potentials and vacuum parameters are unrestricted throughtout the calculation of Ω_{one} until the end, when the choice of parameters relevant for one's particular problem can be made.

G. Some Physical Properties of Temperature Gauge Theories

During the past few years there have been significant advances in the understanding of temperature gauge theories. In the remainder of this section we sketch some main features of these developments, phrased in the language of our one-loop thermodynamic analysis. Since the Gross-Pisarski-Yaffe review [20] , nothing has appeared in print which offers a comprehensive survey of the more recent material. (An article [35] with some pretensions of completeness is in preparation.) The comments to follow are not intended to bridge this gap, but perhaps they will aid the reader in forming a general picture of this subject.

(1) Electric plasma at high-T :

Perhaps the most important single statement which can be made is that all gauge theories become electric plasmas at high T . This will be true regardless of the nature of the theory at low T —— even for a confining theory like QCD . Note that this is a quantum-mechanical statement: The electric sector is characterized by a quantum-mechanical mass $m_{el}(T)$ which grows with T . This mass has an evident interpretarion as an inverse Debye screening length. The electric sector is plamalike, and electric charge screening intensifies as T increases. Physically this makes very good sense because the thermal background of elementary electric charges will become more dense as temperature goes up.

(2) Electric mass:

The mass m_{el} characteristic of the electric sector of a $T > 0$ gauge theory is generated by quantum effects. One way to calculate this mass is to expand the thermodynamic potential Ω in powers of the vacuum parameters. Recalling that A_μ contains the gauge coupling g in our notation, the appropriate formula is

$$\Omega(\beta, M_k, \mu_\alpha, a_n, g_i) \tag{6.48}$$
$$= \frac{V}{g^2}\left[\text{const} + \sum_n \frac{1}{2} m_{el}^2 a_n^2 + O(a_n^4)\right]$$

where $m_{el}(\beta, M_k, \mu_\alpha, g_i)$ will be the same for all a_n. We know how to calculate the one-loop contribution to Ω in terms of the $\Omega_{B,F}$. Therefore, we know how to calculate the one--loop contribution to m_{el}. It is sufficient to compute the electric masses $m_{B,F}$ associated with $\Omega_{B,F}$. From Eqs.(1.1c), (1.2c) and the prescription (6.48) above we find trivially

$$m_B^2 = 2T^{s-1} g^2 a_s \sum_{m=1}^{\infty} \frac{1}{m^{s-1}} f_s(m\beta M) \times \cosh m\beta\mu, \tag{6.49}$$

$$m_F^2 = dT^{s-1} g^2 a_s \sum_{m=1}^{\infty} (-)^{m+1} \frac{1}{m^{s-1}} \times f_s(m\beta M) \cosh m\beta\mu. \tag{6.50}$$

Because the one-loop term in m_{el}^2 will be a sum of appropriate m_B^2 and m_F^2, one can investigate the general, case by examining $m_{B,F}$ individually.

The high-T expansions of $m_{B,F}$ are particularly interesting. From the results in Sec. IV we find in four dimensions the boson expansion

$$m_B^2 = \Delta m_B^2 + \frac{4}{\pi^2} g^2 T^2 \Big\{ \frac{\pi^2}{12} - \frac{(\beta\mu)^2}{8} + \frac{(\beta M)^2}{16} - \sum_{b,c} \left(\frac{\beta M}{2}\right)^{4+2b} \frac{(\beta\mu)^{2c}}{b!(b+2)!(2c)!} \times \sum_{m=1}^{\infty} \frac{\ln 2\pi m}{m^{-2-2b-2c}} \Big\} \quad (6.51)$$

where the sum over m is given by Eq. (B10) and

$$\Delta m_B^2 = g^2 \frac{T}{2\pi} (2\mu^2 - M^2)(M^2-\mu^2)^{-1/2} \quad (6.52)$$

$$= g^2 \frac{T}{2\pi M}\left[\frac{3}{2}\mu^2 - M^2 + \frac{\mu^4}{M^2} + \ldots\right]$$

is the contribution from the branch-point term $\Delta\Omega_B$ (see Eq. (4.13)).

Comments:

(a) m_B grows with T, and $m_B \approx gT/\sqrt{3}$ in the limit $T \to \infty$. The electric mass always diverges in this limit. All electric fields—indeed the entire electric sector—become infinitely massive and decouple (from the magnetic sector) at $T = \infty$. Thus screening becomes infinitely short-range in the electric plasma, and the electric charges of the original gauge theory disappear completely in the dimensional reduction limit $T \to \infty$. The zero temperature limit is just the contrary: $m_B \longrightarrow 0$ exponentially (Sec. III) and screening rapidly dies out with decreasing temperature.

(b) In Eq. (6.49) it is clear that m_B increases with growing chemical potential. This means plasma behaviour and electric screening are enhanced. This is physically sensible because $\mu \neq 0$ represents an externally maintained background charge density which strengthens the thermal background. In QCD the critical temperature at which deconfinement occurs will therefore be lowered by nonvanishing bosonic chemical potential, since deconfinement signals the onset of plasma behaviour. A mathematical point is worth mentioning. The leading contribution from μ in Eq. (6.51) would be negative if it were not for the extra term (6.52), whose contribution from μ is positive and larger.

In four dimensions, the high-T expansion of m_F is

$$m_F^2 = \frac{8}{\pi^2} g^2 T^2 \Big\{ \frac{\pi^2}{24} + \frac{(\beta\mu)^2}{8} - \frac{(\beta M)^2}{16} + \sum_{b,c} \left(\frac{\beta M}{2}\right)^{4+2b} \frac{(\beta\mu)^{2c}}{b!\,(b+2)!\,(2c)!} \times \sum_{m=1}^{\infty} (-)^m \frac{\ln 2\pi m}{m^{-2-2b-2c}} \tag{6.53}$$

where the sum over m is given by Eq. (4.12).

Comments:

(a) m_F behaves much like m_B as a function of T and eventually reaches the same limiting value $m_{B,F} \approx gT/\sqrt{3}$ as $T \to \infty$.

(b) Plasma behaviour and screening are enhanced by nonvanishing fermionic chemical potential, much as by bosonic chemical potential. This is due to m_F increasing with μ .

(c) In Eq. (6.53) we have not included a contribution from the branch-point term (4.15) in Ω_F . This term would contribute

$$\Delta m_F^2 = \frac{1}{\pi} g^2 T \,\mathrm{Re} \frac{M^2 - 2(\mu + i\pi/\beta)^2}{\left[M^2 - (\mu + i\pi/\beta)^2\right]^{1/2}} \tag{6.54}$$

as shown by a straightforward calculation. Setting $M = \mu = 0$ one obtains

$$\Delta m_F^2 = 2g^2T^2 \tag{6.55}$$

which is nonzero. There are other ways to calculate m_F (see e.g. Ref. [20]) and for $M = \mu = 0$ the answer is $m_F = gT/\sqrt{3}$ as given by Eq. (6.53) without the Δm_F contribution. We conclude that Δm_F has to be discarded. This is not as arbitrary as it may seem, because to obtain m_F we are expanding $\Omega_F(\beta, M, \mu, a)$ about $a = 0$. However, $\Delta\Omega_F$ does not behave continuously at $a = 0$, and this term cannot be reliably expanded about this value. Thus it is good to have an argument that $\Delta\Omega_F$ does not contribute to m_F . Still, one would like to understand this point better.

(3) Infinite-Temperature Limit

The limit $T \to \infty$ in temperature field theory is an interesting dimensional reduction problem. As $\beta \to 0$ the space-

-time cylinder $E^s \times S^1$ shrinks to E^s. What survives, in general, is the static part of the original theory — i.e. the part independent of the S^1 coordinate. For a gauge theory there is one extra feature to consider: namely that the basic field is an (s + 1)-vector A_μ. At T = ∞ the component A_0 no longer has any place in the theory. As we have seen, the theory properly arranges itself by providing A_0 with a quantum mass m_{el} which becomes infinite as $T \to \infty$. (Incidently, this is how one can be sure the exact $m_{el} \to \infty$ and not merely the one-loop mass.) A_0 is eliminated as a dynamical variable, and what remains is the (static) magnetic sector with dynamical variables A_i. For example, QCD_{s+1} $(T = \infty)$ is Euclidean $QCD_s(T'=0)$ at a new zero temperature. While the latter theory can be regarded as the static magnetic sector of the former, the connection between those two theories is less tight than one might think. The two theories have totally different electrical sectors: e.g. confinement in one theory has no immediate connection with confinement in the other because quite distinct electrical charges are involved. The decoupling of fermions at T = ∞ is another important point. The absence of fermions in the T = ∞ theory has no direct bearing on the importance of these particles for the original theory (which does have the fermions) at high T.

(4) Deconfinement in QCD :

A very important physical result is that QCD is a plasma at sufficiently high T, and a plasma is not a confining environment. Therefore, QCD must experience a deconfinement transition at some temperature T_c. Perhaps this is a phase transition; possibly something smoother is involved. What matters is that quarks become unconfined above T_c. The proof of this is implicit in the previous statement that QCD at high T is an electric (color) plasma. But let us give a little of the mathematics relevant for static deconfinement. Static confinement means that an isolated static quark in pure QCD (i.e. without dynamical quarks) has infinite energy. The energy F of the static quark is introduced via the "Polyakov line"

$$L \equiv P \exp i \int_0^\beta dx_0 A_0 \qquad (6.56)$$

whose statistical average is

$$\langle L \rangle = e^{-\beta F} . \qquad (6.57)$$

This identification derives from the following picture of L . A static quark-antiquark pair are created somewhere. The antiquark is held fixed while the quark travels in time around the space--time cylinder and returns to the original position, where the pair annihilates. The time-travelling quark is alone except for the brief instant of creation/annihilation: hence Eq. (6.57), with F the energy of an isolated static quark. For $T > T_c$ it is clear that F will be finite and the theory is nonconfining. This follows from the short-range nature of A_0 in Eq. (6.56). One can think of m_{el} as the mass of A_0 , and since m_{el} is large, nearly all fields will contribute $A_0 \approx 0$ in Eq. (6.56), yielding $\langle L \rangle \neq 0$ and F finite. The only way that $\langle L \rangle$ can vanish is for many different values of A_0 to contribute, with strong cancellations occurring in Eq. (6.57). For a different approach and more details see Ref. [36] .

(5) Analogy with spin systems:

A great advantage of S_{el} is its simplicity — in comparison with a gauge theory. This scalar theory may eventually yield its secrets to existing analytic methods. One can regard S_{el} as some kind of "generalized spin theory", perhaps not intrinsically different from more familiar spin theories which have been studied for years in statistical mechanics. S_{el} will have symmetries which may suffice to place it in a definite universality class. If other spin theories in this class have been solved to the extent that their critical behaviour is known, then one could infer from this the critical behaviour of the electrical sector of the gauge theory. (See the Svetitsky-Yaffe paper [37] for a detailed presentation of this idea.) This analogy with spin systems could be regarded as a program designed to probe the connection between gauge theories and far simpler (yet perhaps analogous) statistical systems. Many efforts in this direction are being made through numerical simulation of lattice gauge theories.

(6) Magnetic sector:

Although we have nothing concrete to say about the magnetic sector in this paper, it would seem wrong to ignore it altogether. As we have stressed repeatedly, the properties of the magnetic sector have relatively little to do with the electrical properties of the same gauge theory, particularly at high-T. The magnetic sector may be massless and long-range, or massive (screening) and short-range. Abelian theories fall into the former catagory; non Abelian theories like QCD may belong to the latter. It is clear that only a self-interacting gauge theory could have nonzero magnetic mass m_{mag}. If $m_{mag} > 0$ then somehow the theory is able to behave like a magnetic plasma, capable of screening magnetic charge. However gauge theories do not contain elementary magnetic charges (monopoles); only elementary electric charges. Hence one does not expect to find monopoles in the thermal background. Only through complicated self--interactions could a gauge theory produce monopoles in its own thermal background, and thereby generate magnetic plasma behaviour. This is an extremely difficult problem to analyse. At present it is not known it $m_{mag} > 0$ for QCD_4, for example. The solution of this problem may take some time.

Acknowledgements -

I have profited from discussions with many persons on the subject of these lecture notes. In particular, José Maria Cerveró and César Gómez have been helpful on numerous occasions. The notes themselves originated in lectures given at the Universities of Salamanca, Zaragoza and Dortmund. I am grateful to the theory groups at these institutions for their warm hospitality, and for discussions which led to the clarification of many details in these notes. I thank all organizers of the 1985 GIFT Seminar on Theoretical Physics for the opportunity to deliver an abbreviated version of this material in Jaca. The award of a grant from the Comisión Asesora de Investigación Científica y Técnica enabling the author to spend one year in Salamanca is acknowledged with gratitude. A second grant, from the Comite Conjunto Hispano-NorteAmericano Para Cooperacion Cientifica y Tecnologica, enabling this stay to be prolonged, is acknowleged with equal gratitude. The continuing support of the Penn State FSSF is greatly appreciated.

APPENDIX A - NOTATION

Our matrix notation for Euclidean field theories is summarized in the following formulas:

$$S = \int_\beta \left[\frac{1}{2g^2} \mathrm{tr} F_{\mu\nu}^2 + \mathrm{tr}\left(|D_\mu \phi|^2 + M^2 |\phi|^2\right) + \bar{\Psi}(\not{D} + M')\Psi + \ldots \right],$$

$$F_{\mu\nu} = \partial_\mu A_\nu - \partial_\nu A_\mu - i[A_\mu, A_\nu],$$

$$\left.\begin{array}{l} D_\mu \phi = (\partial_\mu - iA_\mu)\phi \\ D_\mu \Psi = (\partial_\mu - iA_\mu)\Psi \end{array}\right\} \text{fundamental}$$

$$D_\mu \phi = \partial_\mu \phi + i[\phi, A_\mu] \quad \text{adjoint}$$

$$A_\mu = \tfrac{1}{2} T_a A_\mu^a, \quad F_{\mu\nu} = \tfrac{1}{2} T_a F_{\mu\nu}^a,$$

$$\mathrm{tr}\, T_a T_b = 2\delta_{ab}, \quad T_a^\dagger = T_a,$$

$$[T_a, T_b] = 2i f_{abc} T_c,$$

$$\int_\beta \equiv \int_0^\beta dx_0 \int d^s x .$$

APPENDIX B - ZETA FUNCTIONS

The Riemann zeta function $\zeta(s)$ and the related function $\eta(s)$ are defined by

$$\zeta(s) = \sum_{m=1}^{\infty} \frac{1}{m^s} \quad , \qquad \text{(B1)}$$

$$\eta(s) = \sum_{m=1}^{\infty} (-)^{m+1} \frac{1}{m^s} = (1-2^{1-s})\zeta(s) . \qquad \text{(B2)}$$

Both series definitions are valid for Res $>$ 1 . However, when these series are reexpressed as contour integrals, the resulting functions are defined throughout the s-plane [14,15] . The same symbols $\zeta(s)$, $\eta(s)$ are used for these latter functions. $\zeta(s)$ has a pole at $s = 1$,

$$\zeta(s) = \frac{1}{s-1} + \gamma + O(s-1) \qquad \text{(B3)}$$

while $\eta(s)$ is regular everywhere including $s = 1$: $\eta(1) = \ln 2$. Some useful formulas for integral $n > 0$ are

$$\begin{aligned} \zeta(-2n) &= \eta(-2n) = 0 , \\ \zeta(1-2n) &= -\frac{1}{2n} B_{2n} , \\ \eta(1-2n) &= (2^{2n}-1)\frac{1}{2n} B_{2n} , \\ \zeta(2n) &= (-)^{n+1} \frac{(2\pi)^{2n}}{4n(2n-1)!} B_{2n} , \end{aligned} \qquad \text{(B4)}$$

where the B's are Bernoulli constants. From Eq. (B1) we find

$$\zeta'(s) = -\sum_{m=1}^{\infty} \frac{\ln m}{m^s} = -\sum_{m=1}^{\infty} \frac{\ln 2\pi m}{m^s} + \zeta(s)\ln 2\pi \,. \tag{B5}$$

Because $\zeta'(0) = \zeta(0)\ln 2\pi$ (see below) it follows that

$$\sum_{m=1}^{\infty} \ln 2\pi m = 0 \,. \tag{B6}$$

The reflection formula

$$\zeta(s) = \frac{1}{\pi} e^{s\ln 2\pi} \sin\frac{\pi s}{2}\, \Gamma(1-s)\,\zeta(1-s) \tag{B7}$$

can be differentiated to find

$$\zeta'(s) = \frac{1}{\pi} e^{s\ln 2\pi}\, \Gamma(1-s) \sin\frac{\pi s}{2}\, \zeta(1-s) \times \left\{ \ln 2\pi + \frac{\pi}{2}\cot\frac{\pi s}{2} - \Psi(1-s) - \frac{\zeta'(1-s)}{\zeta(1-s)} \right\} \tag{B8}$$

where $\Psi(1-s) = \Gamma'(1-s) / \Gamma(1-s)$.

Given this formula it is straightforward to verify

$$\zeta'(-2n) = (-)^n \frac{1}{2}\frac{(2n)!}{(2\pi)^{2n}}\zeta(1+2n), \qquad n=1,2,\ldots \tag{B9}$$

and also $\zeta'(0) = \zeta(0) \ln 2\pi$. From Eq. (B5) we then find

$$\sum_{m=1}^{\infty} \frac{\ln 2\pi m}{m^{-2n}} \qquad n=1,2,\ldots$$

$$= (-)^{n+1}\frac{1}{2}\frac{(2n)!}{(2\pi)^{2n}}\zeta(1+2n). \tag{B10}$$

Corresponding results for $\eta(s)$ are

$$\eta'(s) = \sum_{m=1}^{\infty}(-)^m \frac{\ln m}{m^s} \tag{B11}$$

$$= \zeta(s)\ln 2 + (1-2^{1-s})\zeta'(s),$$

$$\eta'(0) = \frac{1}{2}\ln\pi, \tag{B12}$$

$$\sum_{m=1}^{\infty}(-)^m \ln 2\pi m = -\ln 2, \tag{B13}$$

$$\sum_{m=1}^{\infty}(-)^m \frac{\ln 2\pi m}{m^{-2n}} = (1-2^{1+2n})(-)^n \tag{B14}$$

$$\times \frac{1}{2}\frac{(2n)!}{(2\pi)^{2n}}\zeta(1+2n).$$

APPENDIX C - MELLIN TRANSFORMATION METHOD

There exists an alternative to the ζ-function method for evaluating the high-T expansions of thermodynamic potentials. Indeed, the first complete calculations [2,21] of these expansions for nonzero mass M were done by this other method, which employs a Mellin transformation to resum series such as (1.1c) , (1.2c) to power series form [38] .

As in Sec. IV , consider the general series

$$F_{\pm}(x_1,\dots x_A) = \sum_{m=1}^{\infty} (\pm)^{m+1} \frac{1}{m^s} f_1(m x_1) \times \dots \times f_A(m x_A) . \qquad (C1)$$

Here the summand can be mellin-transformed [38,38]

$$\frac{1}{m^s} f_1(m x_1) \dots f_A(m x_A) \equiv \frac{1}{2\pi i} \int_{\mathrm{Re}\,z - i\infty}^{\mathrm{Re}\,z + i\infty} dz \frac{1}{m^z} G(z, x_i) , \qquad (C2)$$

$$G(z, x_i) = \int_0^{\infty} dm\, m^{z-1} \times \left[\frac{1}{m^s} f_1(m x_1) \dots f_A(m x_A) \right] , \qquad (C3)$$

where m is regarded as a complex variable, as is z , and Rez in Eq. (C2) is chosen such that this equation is well-defined (normally there exists a vertical strip in the z-plane for which both integrals converge). Inserting Eq. (C2) into (C1) gives

$$F_{\pm}(x_1,\dots x_A) = \frac{1}{2\pi i}\int_{\mathrm{Re}z-i\infty}^{\mathrm{Re}z+i\infty} dz\, \zeta^{\pm}(z)\, G(z,x_i) \qquad (C4)$$

where the ζ-functions $\zeta^{\pm}(z)$ are defined by Eq. (4.2). These functions are well-behaved at infinity. Assuming $G(z, x_i)$ is such that the contour at infinity contributes nothing, one closes the contour in Eq. (C4) at infinity. If C is this closed contour then

$$F_{\pm}(x_1,\dots x_A) = \text{sum of residues of poles within C} . \qquad (C5)$$

Eq. (C5) may not look like a high-T expansion, but it turns out to be one. It is interesting that the Riemann ζ-function is also involved in this calculation. However, this method and the ζ-function method differ very substantially. Here we need to know the Mellin transform G(z) of the entire summand in Eq. (C1). Then the Cauchy theorem can be used as in Eq. (C5). There exist books which list mellin transformation, and in many cases one can look up G(z). However, if one or more additional factors, $f_i(mx_i)$ are included in the summand, an entirely new Mellin transformation is needed. When one encounters a summand (C2) whose Mellin transform is unknown, this approach becomes difficult.

The ζ-function method, on the other hand, requires as input the power series expansions of the individual functions

$f_i(mx_i)$ in the summand. If the problem is changed by including some additional factors in the summand, nothing much happens. The method still works, only that the high-T expansion becomes more complicated. But this method relys on the power series for the $f_i(mx_i)$ not involving nonpower functions of mx_i other than $\ln mx_i$ (or powers of $\ln mx_i$) because it is not clear how one would handle these functions within the minicalculus of ζ-function regularization.

One final point: the ζ-function method is clearly simpler as long as it is applicable.

REFERENCES

1) A. Actor, Phys. Lett. 157B (1985) 53.

2) H. Haber and H. Weldon (a), Phys. Rev. D25 (1982) 502 ; (b) J. Math. Phys. 23 (1982) 1852.

3) R.P. Feynman, Phys. Rev. 91 (1953) 1251.

4) T. Matsubara, Prog. Theor. Phys. 14 (1955) 351.

5) C. Bernard, Phys. Rev. D9 (1974) 3312.

6) A.A. Abrikosov, L.P. Gorkov and I.E. Dzyaloshinski, "Methods of Quantum Theory in Statistical Mechanics", Prentice-Hall, NJ (1964).

7) A.L. Fetter and J.D. Walecka, "Quantum Theory of Many-Particle Systems", McGraw-Hill (1971).

8) T. Applequist and R.D. Pisarski, Phys. Rev. D23 (1981) 2305.

9) P. Ginsparg, Nucl. Phys. B170 (1980) 388.

10) B.A. Freedman and L.D. McLerran, Phys. Rev. D16 (1977) 1130.

11) J.I. Kapusta, Nucl. Phys. B148 (1979) 461.

12) J.I. Kapusta, Phys. Rev. D24 (1981) 426.

13) A. Actor, Ann. Phys. 159 (1985) 445.

14) M. Abramowitz and I. Stegun, "Handbook of Mathematical Functions", Dover (1970).

15) I.S. Gradshteyn and I.M. Ryzhik, "Table of Integrals, Series and Products", Academic Press (1965).

16) A. Actor, Phys. Rev. D27 (1983) 2548.

17) L.D. Landau and E.M. Lifshitz, "Statistical Physics", Pergamon, Oxford (1968).

18) K. Huang, "Statistical Mechanics", Wiley, New York (1963).

19) A. Actor, to appear in Nucl. Phys. B.

20) D. Gross, R. Pisarski and L. Yaffe, Rev. Mod. Phys.53 (1981)43.

21) H. Braden, Phys. Rev. D25 (1982) 1028 .

22) A. Niemi and G. Semenoff, Nucl Phys. B230 [FS10] (1984)181.

23) D. Kirzhnits and A. Linde, Phys. Lett. 42B (1972) 471.

24) L. Dolan and R. Jackiw, Phys. Rev. D9 (1974) 3320.

25) S. Weinberg, Phys. Rev. D9 (1974) 3357.

26) D. Kirzhnits and A. Linde, Zh. Eksp. Teor. Fiz. 67 (1974) 1263 (Sov. Phys. JETP 40 (1975) 628).

27) D. Kirzhnits and A.linde, Ann. Phys. 101 (1976) 195.

28) A. Linde, Rep. Prog. Phys. 42 (1979) 389.

29) E. Abers and B. lee, Physics Reports 9C (1973) 1.

30) G.'t Hooft, Nucl. Phys. B35 (1971) 167.

31) A. Polyakov, Phys. Lett. 72B (1978) 477.

32) L. Susskind, Phys. Rev. D20 (1979) 2610.

33) J. Honerkamp, Nucl. Phys. B48 (1972) 269.

34) E. Shuryak, Phys. Reports 61 (1980) 71.

35) A. Actor, in preparation.

36) L. McLerran and B.Svetitsky, Phys.Rev. D24 (1981) 450.

37) B. Svetitsky and L.Yaffe, Nucl. Phys. B210 (FS6) 423.

38) G. MacFarlane, Phil. Mag. 40 (1949) 188.

39) B. Davies, "Integral Transformations and Their Applications", Springer, New York, 1978.

40) H.A. Weldon,"Proof of zeta Function Regularization of High--Temperature Expansions, U. of Penn. preprint, July 1985.

IS ϕ^4 A FREE THEORY IN 4 DIMENSIONS?

I. G. HALLIDAY

Blackett Laboratory, Imperial College,
Prince Consort Road,
London SW7 2BZ, England

ABSTRACT

Analytic methods have failed to complete the expected proof of the triviality of ϕ^4 in 4 dimensions. We survey here the results of several Monte Carlo investigations. These are entirely consistent with the theory being trivial.

1. INTRODUCTION

It has been believed for many years, starting with ideas due to Landau, that the simple scalar field theory with interaction ϕ^4 in 4 dimensions does not exist. This in spite of the fact that it is a theory much loved by pedagogical text books. The problem arises from the renormalisation. This is well defined at any given order in perturbation theory. Thus given some ultraviolet cut-off Λ we can find bare values for the mass $m_o(\Lambda)$ and coupling $g_o(\Lambda)$ depending in the cut off Λ such that the physical mass m and coupling g are held fixed as $\Lambda \to \infty$. This works at fixed order in perturbation theory. The current belief is that this cannot be done in the theory defined non-perturbatively unless g is zero. In other words the only pure scalar theory is the free theory.

Recently analytic techniques showed that in dimension greater than four the same result holds. This is not unexpected naively. The theories are then not renormalisable even in perturbation theory. Thus it is no surprise that a proof can be found independently although the elegance of the proof is remarkable. The proof is outlined below in section 2 and the problems with extension to 4 dimensions mentioned. In section 3 the results of naive renormalisation studies using Monte Carlo on the lattice are outlined. In section 4 the use of the renormalisation group technology in numerical studies is surveyed.

In all these methods it is taken as given that the correct approach to constructing a quantum field theory is to approximate it by a lattice system in say a hypercubic lattice with a scalar field ϕ_s defined at the sites s. The lattice has a finite volume $V = L^d$ and lattice spacing a. Then a acts as an ultra violet cut-off. The functional integrals in Euclidean space are then well defined. The limits $\Lambda \to \infty$, $a \to 0$ must now be controlled. The Lagrangian density is, of course, approximated by some discrete form for space derivatives, etc. This is taken usually as a polynomial of low order.

2. ANALYTIC METHODS

We first outline a derivation of the polymer or random walk representation of ϕ^4. The action is

$$A = \sum_s (g_o \phi_s^4 + \alpha \phi_s^2) + \beta \sum_{s,s^1} \phi_s \phi_{s^1}$$
$$\alpha = 4 + \frac{m_o^2 a^2}{2} \tag{2.1}$$

where the sum over s is over all sites and s, s^1 are nearest neighbours only. The theory then is defined by expressions like

$$\langle \phi_a \phi_b \phi_c \phi_d \rangle = \frac{1}{Z} \int \prod_s d\phi_s (\phi_a \phi_b \phi_c \phi_d) e^{-A}$$
$$Z = \int \prod_s d\phi_s e^{-A} \tag{2.2}$$

where a, b, c, d are four sites and s runs over all sites on the lattice. Clearly e^{-A}/Z acts as a probability distribution.

The trick introduced by Brydges, Frohlich and Spencer [1] and Aizenmann[2] is to write expressions (2.2) in terms of a second scalar field theory and then use trivial input such as $\langle \phi_s^2 \rangle > 0$ for this second field theory. The results following immediately from $\langle \phi_s^2 \rangle$ acting as a probabilistic expectation value of a positive quantity.

Consider

$$\langle \phi_a \ \phi_b \rangle = \frac{1}{Z} \int \prod_s d \ \phi_s \ \phi_a \ \phi_b \ e^{-A} = G_2 \ (a, \ b) \tag{2.3}$$

Apart from the ϕ^4 terms in A the integral is Gaussian. Thus at each site write

$$f(\phi_s^2) = e^{-g_o \ \phi_s^4 + \alpha_s \ \phi_s^2} = \int d \ \alpha_s \ e^{-i \ \alpha_s \ \phi_s^2} \ \tilde{f} \ (\alpha_s) \tag{2.4}$$

The ϕ_s integrals are now Gaussian

$$\int \Pi \ d \ \phi_s \ e^{-i \ \alpha \ \phi_s^2} \ e^{\beta \sum \phi_s \ \phi_{s^1}} \ \phi_a \ \phi_b \tag{2.5}$$

$$= \int \Pi \ d \ \phi_s \ \ \phi_a \ \phi_b \ e^{-\frac{1}{2} \ \phi^T [2i\alpha + \beta J] . \phi}$$

where α is a diagonal matrix with entries α_s and J is an off diagonal matrix with entries $J_{ss^1} = 1$ where $_{s,s^1}$ are nearest neighbours otherwise zero. ϕ is now to be thought of as a column vector with entries ϕ_s.

Now any Gaussian integral

$$\int \Pi \, d \, \phi \, \phi_{1_1} \; - - \; \phi_{1_m} e^{-\phi.A.\phi}$$

$$= \sum_{\text{pairs}} (A^{-1}_{1_i 1_j} \; - - \; A^{-1}_{1_m 1_n}) \, . \int \Pi \, \phi \, \phi \, e^{-\phi.A.\phi} \tag{2.6}$$

where the A^{-1} indices are summed over all ways of pairing them off - shades of Wicks theorem. The crucial trick is not to do the final ϕ integral. Thus we need to know

$$A^{-1}{}_{ij} = (2i\alpha + \beta J)^{-1}_{ij} = (\Lambda + \beta J)^{-1}_{ij}$$

where Λ is diagonal, J is off-diagonal. This gives the Neumann expansion

$$(A^{-1})_{ij} = \Lambda^{-1}_{ij} + (\Lambda^{-1} \; \beta J \; \Lambda^{-1})_{ij} + (\Lambda^{-1} \; \beta J \; \Lambda^{-1} \; \beta J \; \Lambda^{-1})_{ij} + --$$

Now Λ is diagonal and J_{lm} is zero unless l,m are nearest neighbours. Thus a term like$(\Lambda^{-1}\beta J\Lambda^{-1}\beta J\Lambda^{-1})_{ij}$ is zero unless i, j can be joined by two nearest neighbour jumps by J and we sum over all such terms. Clearly we get a sum over all walks from i→j constructed by making nearest neighbour jumps.

There is a separate walk for each A^{-1}_{lm} term.

The power of α_i in (2.6) is clearly given by minus the sum of the number of times any walk hits site i. Thus the α_i integral is of the form

$$I = \int d\alpha_i \, . \, \tilde{f}\,(\alpha_i) e^{-i\alpha_i \phi_i^2} (\alpha_i)^{-N_i}$$

Write $(\alpha_i^{-N_i}) \;\alpha\; \int_o^\infty d\, t_i \; t_i^{N_i - 1} \; e^{-i\alpha_i t_i} / \; \Gamma(N_i)$

Then $I \;\alpha\; \int_o^\infty d\, t_i \dfrac{t_i^{N_i - 1}}{\Gamma(N_i)} \, f\,(\phi_i^2 + t_i)$. Thus finally, for example,

$$\langle \phi_a \; \phi_b \; \phi_c \; \phi_d \rangle = \sum_{\text{pairs}} \; \sum_{\substack{\text{paths} \\ a \to b \\ c \to d}} \Big[\int \prod_s d\,\phi_s \prod_s d\, t_s \frac{t_s^{N_s - 1}}{\Gamma(Ns)}$$

$$e^{-g_o(\phi_s^2 + t_s)^2} . e^{-\alpha\,(\phi_s^2 + t_s)} . e^{-\beta\,\phi\,\phi} \Big] / Z. \qquad (2.7)$$

The ϕ_s integrals at fixed paths and fixed t_s still give a funny field theory; crucially it still defines a probability distribution. The "mass" varies from site to site.

Defining the corrected 4 point function

$$G_4^c = \langle \phi_a \ \phi_b \ \phi_c \ \phi_d \rangle - \sum_{\text{pairs}} \langle \phi \ \phi \rangle \langle \phi \ \phi \rangle$$

we would like to prove the following inequalities

$$0 \geqslant G_4^c \ (x_1 \text{--} \ x_4) \geqslant - 3 \ \beta^2 \sum_Z \prod_{i=1}^{4} \ G_2 \ (x_i, Z) \qquad (2.8)$$

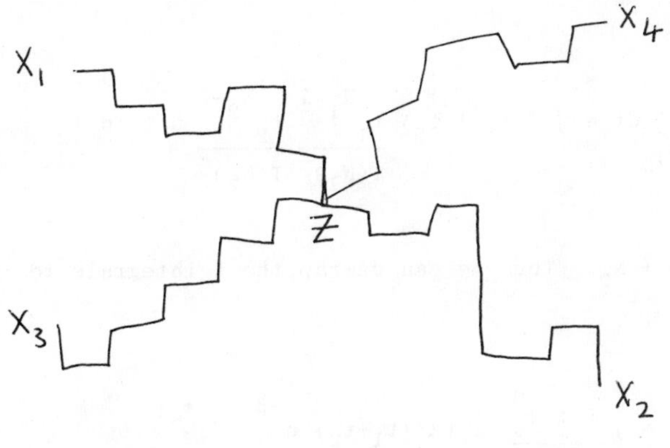

Fig. 1

The first due to Lebowitz the second to Frohlich, Aizenmann [3,2].

We first undo the sum over the walks and look at a given pairing.

$$F_4^c = \sum_{\substack{w_i : x_1 \to x_2 \\ w_2 : x_3 \to x_4}} [Z(w_1, \ w_2) - Z \ (w_1) \ Z(w_2)]$$

$$\geqslant \ - \sum_{w_1 \cap w_2 \neq 0} Z(w_1) \ Z(w_2) \qquad (2.9)$$

In words the connected four point function is greater than a negative term got by summing over intersecting walks only! To prove this we need only

$$Z(W_1,W_2) \geq Z(W_1)\, Z(W_2) \quad W_1 \cap W_2 = 0 \tag{2.10}$$
$$Z(W_1,W_2) \geq 0$$

Now undo the t integrals. At first site $Z(W_1,W_2)$ involves one t integral at each site and counts how often both walks visit site i to compute N_i while $Z(W_1)\, Z(W_2)$ involves 2 t_i integrals one in each term and each term only counts how often its walk visits each site. But the following identity is easily proven

$$\int f(t) \frac{t^{N-1}}{\Gamma(N)} dt = \int f(t_1 + t_2) \frac{t_1^{N_1-1}}{\Gamma(N_1)} \frac{t_2^{N_2-1}}{\Gamma(N_2)} d\,t_1\, d\,t_2$$

provided $N = N_1 + N_2$. Thus we can unwrap the t integrals to compare

$$Z(t_1 + t_2) = \int \frac{\Pi\, d\, \phi_s}{Z}\; g\, (\phi^2 + t_1 + t_2)\; e^{-\beta\, \phi_s\, \phi_{s1}}$$

with $Z(t_i)\, Z(t_2)$. Knock off the terms independent of ϕ^2

$$R(t) = Z(t) e^{+2\, g\, t^2 + \alpha\, t}$$

Then trivially

$$\frac{1}{R(xt + s)} \frac{\partial}{\partial x} R(xt + s) = -g_o \sum_i \langle t_i \phi_i^2 \rangle_{xt+s} < 0 \qquad (2.12)$$

Similarly

$$\frac{\partial}{\partial y} \left(\frac{1}{R(xt + sy)} \frac{\partial}{\partial x} R(xt + sy)\right) = -g \langle \phi^2, \phi^2 \rangle \leq 0 \qquad (2.13)$$

Note the rather minimal dynamical information which is being fed into the problem.

Now consider the identity

$$\ln R(s + t) = \int_0^1 dx \frac{1}{R(xt+s)} \frac{\partial}{\partial x} R(xt+s) + \ln R(s)$$

The argument of the integral is negative (2.12) and of negative gradient. Thus it is greater than the integral with s set equal to zero. This gives instantly

$$\ln R(s + t) \geq \ln R(t) + \ln R(s)$$

or $\quad Z(s + t)\, e^{-4g \sum_i t_i s_i} \geq Z(s)\, Z(t)$

Now if a site is not visited by a walk i.e. $N_i = 0$ then clearly $t_i = 0$. Thus if the two walks W_1 W_2 have no points in common

$$Z(s + t) \geq Z(s)\, Z(t)$$

which is (2.10). Hence (2.9) is proven.

Now we jump much tedious detail. We have

$$G_4^c \geq - \sum_{\text{pairs}} \sum_{w_1, w_2} \int \frac{dt\, t^{N-1}}{\Gamma(N)} \int \Pi\, d\phi\, f\, (\phi^2 + t)\, e^{-\beta H} \quad (2.14)$$

where W_1, W_2 intersect at least once at say Z. Now consider the 2 walks W_1, W_2 as 4 from $x_i \to Z$. We are avoiding the problems of multiple intersections (see 3 for details).

The powers of β don't quite match the numbers of visits per site and we get

$$0 \geq G_4^c \geq - \beta^2 \sum_Z \prod_i G_2\, (x_i, Z) \quad (2.15)$$

where we have cheated slightly since the Z for each walk may be one site apart. Not a problem as $a \to o$.

Now hold $G_2(x_1, x_2)$ finite for x_1, x_2 measured in metres. A known bound is

$$0 \leq \langle \phi_{n_1} \phi_{n_2} \rangle \leq \frac{C}{\beta\, |n_1 - n_2|^{d-2}}$$

for n_1 n_2 lattice sites and, d = space-time dimension and C an absolute constant. In metres as $a \to o$.

$$0 \leq \langle \phi_{n_1} \phi_{n_2} \rangle \leq \frac{c\, a^{d-2}}{\beta\, [a |n_1 - n_2|]^{d-2}} \quad (2.16)$$

$$a^{d-2} / \beta \geq \text{constant as } a \to o.$$

Stuffing this in (2.15) gives, for (x, z) in metres now, as $a \to o$.

$$0 > G_4^c (x_1 - - x_4) > - a^{d-4} . \frac{3\beta^2(a)}{a^{2(d-2)}} . \sum_Z a^d \{\prod_1^4 G_2 (x_i, z)\}$$

Then

$$|G_4^c| < a^{d-4} \text{ const. } \int d Z. F(Z)$$

where the Z integral is convergent due to (2.16). Thus if $d > 4$ $G_4^c \to 0$ as $a \to o$. The connected 4 point function going to zero of course implies $g = 0$ and a no-scattering Gaussian model.

In 4 dimensions the hope for a proof lies in the intersection properties of random walks. For example Lawler[(4)] has shown in 4 dimensions on Z^4 that simple random walks of length n starting at 0 and X such that $|X^2| \approx n$ intersect with probability $\sim 1/\log n$. Thus in a lattice for $X^2 = 1$ metre as $a \to o$ $n \to \infty$! This is the style of result which may kill the right hand side of our inequality and bring to fruition Symanziks[(5)] dream.

3. NAIVE RENORMALISATION

We have a theory with two free parameters $g_o(a)$, $m_o(a)$. Standard renormalisation theory would say that we pick two physical quantities, historically the renormalised or physical coupling and physical mass, and holding them fixed we let $a \to o$. The functions $g_o(a)$, $m_o(a)$ are adjusted so that this holds if possible.

The first calculation in this style was carried out by Freedman, Weingarten and Smolensky[(6)]. They varied a by setting $Na = 1$ where N is the cubic lattice size in lattice units i.e. the lattice has N^4 sites. They then varied N. The bare mass was adjusted to make $1/m_R \sim 1/3$ lattice. The renormalised coupling was then calculated by going to Fourier transform space. As shown in fig. 2 for all values of g_o, at fixed m_R, the renormalised coupling $g_R \to o$ as $a \to o$. It does not get too close to zero.

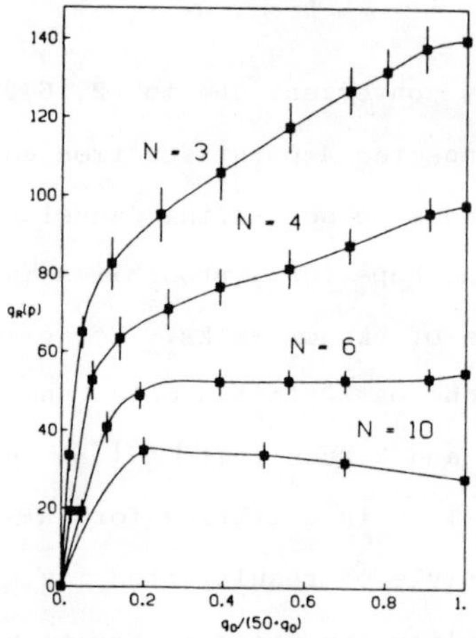

Fig. 2

The real statement is that there is no evidence for any fixed points, cross-overs etc. The calculation can be criticised for the use of small lattices and other authors have not been able to reproduce error bars of this size(7).

This style of calculation has been repeated by Fox & Halliday(8). Here an attempt was made to renormalise $g_o(a)$, $m_o(a)$ by fixing the 2 point function at 2 distances. This was foiled by the bizarre behaviour of the 2 point functions. Given the value of $\langle \phi^2(0) \rangle$ they then agreed to great accuracy with the free ($g_o = 0$) two point function. This is shown in fig. 3 which plots the free 2 point function against the interacting two point function for $\alpha = -2$ for $\xi = 4(o), 2(+), 1(x), .5(\cdot)$ with errors less than the data points.

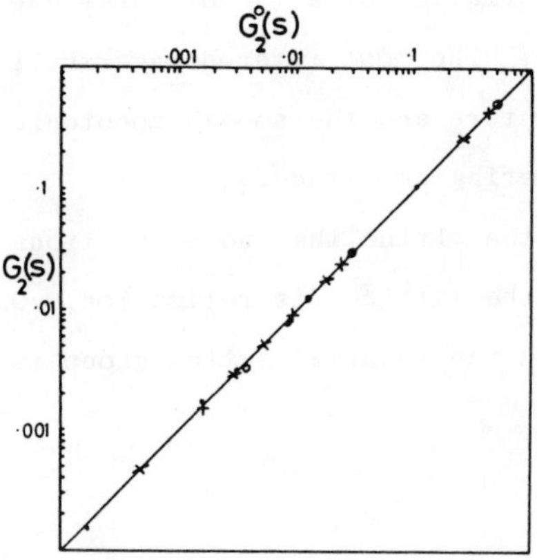

Fig. 3

Here, in a fixed 10^4 lattice, a was varied by letting the correlation length vary from 4, 2, 1, ·5 lattice spacings by varying α or m_o. The ratio R of the connected 4 point function to the disconnected 4 point function was then measured at a range of values of g_o. As $a \to o$ or ξ increasing the scattering amplitude decreased as shown in fig. 4.

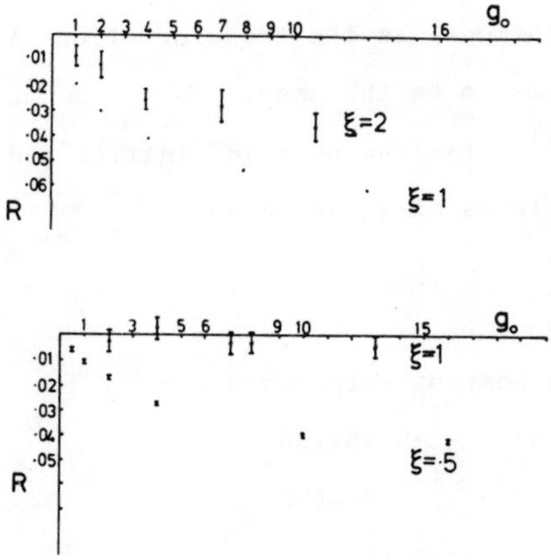

Fig. 4

These results and their error bars are the results of up to 960,000 iterations. Great care was taken over the estimates of the error bars.

Again the range of variation of a is not large due to the size limit in the lattice. The most interesting result is the non-existence of any structure and the smooth monotonic fall towards zero of the scattering amplitude.

These analyses have the virtue that no assumptions are made or approximations beyond the lattice discretisation. On the other hand no insight from the renormalisation group is used.

4. RENORMALISATION GROUP

The idea is implement Wilson's renormalisation group using the Monte Carlo technique as proposed by Ma, Swendson(9). Given a canonical form of the lattice action as in (2.1), with some choice of the ϕ scaling to fix say $\beta = 1$ or $\alpha = 1$ or whatever, a block spin transformation is carried out to find an action in a lattice with twice the spacing which gives the same physics. Thus $\phi \rightarrow \phi^1$ and $A(\phi) \rightarrow A^1(\phi^1)$. The new action A^1 may contain many couplings. The idea(10) is that the parameters in A should approach a free Gaussian model as the transformations $A \rightarrow A^1 \rightarrow A^{11} \rightarrow$ are repeated. This seems to be the case.

For example Lang(11) starting on a 16^4 lattice and blocking twice finds the coefficients of ϕ^2 going as

$\cdot 6 \rightarrow \cdot 9425 \rightarrow \cdot 9918$

$\cdot 2 \rightarrow \cdot 9172 \rightarrow \cdot 9955$

The coefficients of $\phi\ \phi$ nearest neighbours went

$- \cdot 2811 \rightarrow - \cdot 2973(8) \rightarrow - \cdot 3.457(30)$

$- \cdot 2844 \rightarrow - \cdot 301 \rightarrow - \cdot 3.487(29)$

while the ϕ^4 terms went like

$\cdot 2 \to \cdot 0288(3) \to \cdot 0041(8)$

$\cdot 4 \to \cdot 0414(3) \to \cdot 0023(41)$

Higher coupling terms behaved similarly. Thus the Gaussian terms are tending to limits while the non-Gaussian terms all tend to zero.

However if we ignore the higher coupling the renormalisation group has dumped us into an unstable region i.e. $g_o = 0$, $m_o^2 < 0$ in conventional language!

The crucial point is that there does not appear to be any sign of any other fixed points!

Clearly much work remains to be done.

4. CONCLUSIONS

We have seen that the numerical evidence supports the original arguments that ϕ^4 is free in 4 dimensions. However it is far from conclusive.

REFERENCES

1) Brydges, Frohlich and Spencer, Comm. Math. Phys. 83, 123(82).
2) M. Aizenmann, Comm. Math. Phys. 86, 1(82).
3) J. Frohlich, Nucl. Phys. B200. 281(82).
4) G. Lawler, Comm. Math. Phys. 86, 539(82).
5) K. Symanzik, Local Quantum Theory ed. R. Just (Academic Press, New York, 1969).
6) Freedman, Smolensky and Weingarten, Phys. Letts. 113B, 481(82).
7) J. Wheater (Private communication).
8) I. A. Fox and I. G. Halliday, Imperial Preprint 84-85/19.
9) S. K. Ma, Phys. Rev. Lett. 37, 461(76).
   R. H. Swendson, Phys. Rev. Lett. 42, 859(79).
10) K. G. Wilson and J. Kogut, Phys. Reports 12C no 2 (74).
11) C. B. Lang, Graz Preprint (85).
   D. J. E. Callaway and R. Petronzio, Nucl. Phus. B240, (FS12) 577(84).

FLUCTUATIONS OF CONSTANT POTENTIALS IN QCD AND THEIR CONTRIBUTION TO FINITE SIZE EFFECTS

C.P. Korthals Altes
Centre de Physique Théorique
CNRS - Luminy - Case 907
F - 13288 - MARSEILLE Cedex 9

ABSTRACT

Fluctuations of constant potentials give sizeable and sometimes dominant contributions to finite size effects in weak coupling. Apart from their practical implication for lattice gauge theory with periodic boundary conditions they present a theoretical curiosity : weak coupling with quartic, (not Gaussian) modes. The same mechanism applies to many multiply connected compact spaces.

I. Introduction

In this lecture I would like to discuss a theoretical curiosity and a phenomenological headache. It concerns the constant potentials and their fluctuations in weak coupling, when we study (lattice) QCD with periodic boundary conditions. Dynamical quarks will be left out, since the crux of the affair lies in the non-abelian gauge fields.[1,2)]

Why a theoretical curiosity ? We are used to weak coupling implying small fluctuations of the gauge potentials. The fluctuations are small once we have introduced a gauge fixing to eliminate the degeneracy of the saddle points due to gauge invariance. In the case of periodic boundary conditions there is an extra degeneracy of the saddlepoints that remains after gauge fixing. It is a continuous degeneracy and cannot be traced back to some symmetry (like translation and dilation symmetry in the case of instantons). It turns out that these saddlepoints (or "torons") are generally <u>not</u> physically equivalent. This obliges one to integrate over all of them ; thus we are faced with large fluctuations, which only under specific conditions -including the physically interesting case of $d = 4$ and $N = 3$ - can be easily integrated. This is done in the next section for the Wilsonloop and the glue ball-correlation.

In section 3,4 and 5, we justify the simplistic calculation in section two. The last section discusses some prospects.

2. A simplistic computation

The best way to understand the features of the constant modes is to make a simple lowest order computation. We will consider a box of size L^d in the continuum with periodic boundary conditions for the potentials

We will not need to specify the ultraviolet cut off. Momenta in the box have components

$$p_\mu = \ell_\mu \, 2\pi/L \quad , |\ell_\mu| = 0, 1, 2, \ldots \; ; \; \mu = 1, \ldots, d .$$

The action, together with a Feynman-gauge fixing, and a ghost term, can be Fourier analysed.

The result is :

$$\begin{aligned} S_{inv} + S_{g.f} = & \sum_{p,\mu,\nu} \mathrm{Tr}\, A_\mu(p) A_\nu(-p)\, p^2 \delta_{\mu\nu} \\ & + \sum_{p,\mu,\nu} \mathrm{Tr}\left[A_\mu(p), A_\nu(0)\right] p_\mu A_\nu(-p) \\ & - V \sum_{\mu,\nu} \mathrm{Tr}\left[A_\mu(0), A_\nu(0)\right]^2 + \cdots \end{aligned} \qquad 2.1$$

The ghost term plays no role in this lowest order calculation, so we omit it. Dots represent terms of order three with only non-constant potentials, terms of order four with at least two non-constant modes and anything of higher order.

Note that the gauge fixing does not fix the zero momentum modes and that the latter are absent from the quadratic part of the action. Quantization by the path integral leads to a partition function Z :

$$Z \equiv \int \prod_{\mu,a} dA_\mu^a(0) \prod_{\mu,p\neq 0,a} dA_\mu^a(p) \prod_p d(\text{ghost}) \exp -\frac{1}{g^2}\left(S_{inv} + S_{g.f} + S_{ghost}\right) \qquad 2.2$$

All non-constant modes $A_\mu(p)$, $p \neq 0$ will be of order g in weak coupling as follows from 1.1 and 1.2. However the constant modes seem to be of order one, unless we make an assumption :

$$Z_o = \int_{-\infty}^{\infty} \prod_{\mu,a} dA_\mu^a(o) \exp\left[\frac{V}{g^2}\sum_{\mu,\nu} \mathrm{Tr}\,[A_\mu(o), A_\nu(o)]^2\right] < \infty \qquad 2.3$$

This seems a natural assumption. If indeed true, we may set $A_\mu(o)$ of order $g^{1/2}$. Thus these modes fluctuate much more than the Gaussian ($p \neq o$) modes of order g. The reader can easily convince himself that after rescaling the potentials with the appropriate g-factors the leading terms (O(1)) in the action 2.1 are :

$$\sum_{p\neq o,\mu,\nu} \mathrm{Tr}\, A_\mu(p) A_\nu(-p) p^2 \delta_{\mu\nu} - V \sum_{\mu,\nu} \mathrm{Tr}\,[A_\mu(o), A_\nu(o)]^2 \qquad 2.4$$

whereas all the other terms are of order $g^{1/2}$ or higher.

Therefore, we can compute to lowest order the effect of the effect of the constant potentials with the commutator term in 2.4.

To this end consider first a Wilsonloop of size RxT in the 1-2 plane ; the contribution of the constant modes is :

$$W_{R,T} = \frac{1}{N} \mathrm{Tr}\, e^{iRA_1(o)} e^{iTA_2(o)} e^{-iRA_1(o)} e^{-iTA_2(o)}$$

$$+ \text{ non-constant modes.} \qquad 2.5$$

Since $A_{1,2}(o) = O(g^{1/2})$ we get after rescaling the A's in eqn 2.5 the following average to compute :

$$g^2 (RT)^2 \langle \mathrm{Tr}\,[A_1(o) A_2(o)]^2 \rangle \qquad 2.6$$

We do the average with respect to the commutator term in 2.4 ; therefore the average is independent of the the orientation of the loop. So we only need the value of the average of the commutator term in 2..4 which can be gotten by a trivial scaling argument ; we end up with :

$$\langle W_{R,T}\rangle = 1 - g^2 N \left\{ \frac{N^2-1}{N^2} \frac{1}{4(d-1)} \frac{(RT)^2}{V} + \left(\begin{matrix}\text{non constant}\\ \text{modes}\end{matrix}\right)\right\} + O(g^4) \quad 2.7$$

This result is also true for asymmetric boxes, since only the total volume V enters. In higher orders, this is no more true. Standard calculations show that the $\frac{1}{V}$ part of the non-constant modes do exactly cancel with the constant modes, when d = 2. This in accordance with the observation of Eguchi and Kawai [5), and is an example of the fact that the constant modes can be as important as the non-constant modes.

Another process of interest is the glueball correlation function. To lowest order (g^4) it is derived by the same scaling arguments as above. The result for the o^{++} glueball correlation is then[(9)] :

$$\Gamma_{o^{++}}(t) = \frac{1}{(2N)^2} \frac{1}{(d-1)^2(d-2)^2} \sum_{\substack{ijkl\\ \text{space}\\ \text{like}}} \frac{1}{V_s} \int d\vec{x} \langle \mathrm{Tr}\, F^2_{kl}(\vec{x}t) F^2_{ij}(0)\rangle_{conn}$$

$$\equiv \Gamma_{o^{++}}(p=o) + \Gamma_{o^{++}}(p \neq o, t) \quad 2.8$$

with the ratio in the last two terms of 1.8

$$\frac{\Gamma_{o^{++}}(p=o)}{\Gamma_{o^{++}}(p\neq o, t)} = \frac{1/V_s\, L^2}{\gamma(t)}$$

The symbol $\gamma(t)$ is the value of the Feynman graph shown in fig.1, apart from some normalization factors it has in common with the contribution $\Gamma_{o^{++}}$ (p = o). For dimensional reasons it behaves like

$$\gamma(t) \sim t^{-5}$$

and it can be small compared to the constant mode contribution.

in case of a box which is very elongated in the t-direction and thin in the space directions. For a more precise treatment, see ref.2.

Other interesting processes are the correlation between Polyakov-loops, energy density, etc. They need yet analysis.

In what follows, we will justify eqn. 2.3 and give a comprehensive way of computing the constant-mode effects, implying the conditions for 2.3 to be true.

3. Parametrization

This should bring out the special role played by the constant commuting modes. We will therefore split the anti-hermitean potentials as follows :

$$A_\mu(x) = Q_\mu(x) + B_\mu \qquad\qquad B_\mu \text{ constant diagonal}$$

The fieldstrength $F_{\mu\nu}$ becomes in terms of the new variables :

$$F_{\mu\nu} = D_\mu(B) A_\nu - \mu \leftrightarrow \nu + [Q_\mu, Q_\nu]$$

with $D_\mu(B)$ the covariant derivative

$$D_\mu(B) A_\nu \equiv \partial_\mu A_\nu + [B_\mu, A_\nu]$$

Integration over the Q-variables in 2.2 gives us a partition function Z(B), which has still to be integrated over the B variables. However, invariance under twisted gauge-transformations

$$\Omega_\mu = \exp i\, x_\mu L_\mu^{-1} H_\mu$$

with $\exp i H_\mu$ a center-group element , shows that Z(B) has a periodicity :

$$Z(B_\mu) = Z(B_\mu + L_\mu^{-1} H_\mu) \qquad 3.3$$

The integration range of the B fields can be restricted to the periodicity interval. (Z(B) should not be confused with the integrand in 2.3 !). This restriction is crucial for the gauge fixing.

4. The gauge fixing

This must be such that all the variables introduced in 3.1 are forced to go to zero, when the coupling goes to zero. If this can be done, then the variable B_μ just parametrizes the different inequivalent saddle points or torons.

4a. The non-constant modes.

Let us add a gauge term to S_{inv} :

$$S_{g.f.}(p\neq o) = \sum_{p\neq o, a} \left| \sum_\mu D_\mu^{ab}(B)\, Q_\mu^b(p) \right|^2$$

This gauge fixing 4.1 can be diagonalized in terms of the $Q_\mu^{ij}(p)$ components (i,j=1,...N) and the components B_μ^i of the diagonal matrix B_μ and reads

$$S_{g.f.}(p\neq o) = \sum_{p\neq o, i<j} \left| \sum_\mu \left(p_\mu - (B_\mu^i - B_\mu^j)\right) Q_\mu^{ij}(p) \right|^2 \qquad 4.2$$

The constraint on the limits of the B_μ variables mentioned in the previous section is enough to guarantee that $p_\mu - (B_\mu^i - B_\mu^i)$ never vanishes as long as $p \neq o$. Therefore the background field propagator $\left(\sum_\mu (p_\mu - (B_\mu^i - B_\mu^j))^2\right)^{-1}$ is well defined for any $p \neq o$ and all admissible values of B_μ. Thus the variables $Q_\mu^{ij}(p)$ $(p \neq o)$ emerge as genuine Gaussian fluctuations of O(g).

4(b). The constant off-diagonal modes

These modes are not entirely gauge fixed by the analogon of 4.2 :

$$S_{g.f.}(p=0) = \sum_{i<j} \left| \sum_\mu (B^i_\mu - B^j_\mu)\, Q^{ij}_\mu(0) \right|^2$$

Obviously if $B^i_\mu - B^j_\mu = O(1)$ then the off-diagonal modes are of order g :

$$Q^{ij}_\mu(o) = O(g) \qquad \text{I}$$

We will call this region of constant potential configuration space : region I.

Obviously there is a region, $B^i_\mu \cong B^j_\mu$, where the restoring force on the modes Q^{ij}_μ (o) disappears. They become of O(1)!. Therefore we have to mend the gauge condition, e.g. by :

$$S_{g.f.}(p=0) = \sum_{i<j} \left| \sum_\mu \left\{ (B^i_\mu - B^j_\mu)\, Q^{ij}_\mu(o) + Q^{ij}_\mu(o) \right\} \right|^2 \qquad 4.3$$

This gauge fixing makes that there appears a second regime :

$$B^i_\mu - B^j_\mu = O(g^{1/2}) \Rightarrow Q^{ij}_\mu(o) = O(g^{1/2}) \qquad \text{II}$$

Notice that we have only N(N-1) real gaugefixing conditions for constant modes. This corresponds to the fact that the N-1 constant diagonal gauge transformations do not transform at all the B-fields (themselves constant and diagonal). Therefore in constructing the ghostmatrix one can leave out these gauge-transformations. (One can also take N-1 additional gauge fixings and construct the ghostmatrix from all gauge variations. Then one has to live with a ghostmatrix that becomes singular - the determinant is zero - if all the Q^{ij}_μ (o) are set to zero, exactly for the reason explained above).

5. How to compute higher order effects

Integration of the Gaussian (i.e. non-constant) modes gives us the familiar background propagator $\left[\sum_\mu (p_\mu - (B^i_\mu - B^j_\mu))^2\right]^{-1}$.

The situation in the zero-momentum sector is less straightforward. Let us consider the effective action in that sector :

$$
\begin{aligned}
S_{eff}(p=o) &= \tilde{S}_{inv}(p=o) + S_{g.f.}(p=o) + S_{ghost}(p=o) \\
&= 1/2\, Q^{ij}_\rho(o)\left(\sum_\mu (B^i_\mu - B^j_\mu)^2\right) Q^{ji}_\rho(o) \\
&\quad + 1/2\, \bar{\eta}^{ij}(o)\left(\sum_\mu (B^i_\mu - B^j_\mu)^2\right) \omega^{ji}(o)
\end{aligned}
$$

+ (terms of order 4 in Q and B fields together, containing Q at least cubically).
+ (terms of order 4 in Q, B, $\bar{\eta}$ and ω fields, containing Q at least linearly). 5.1

$S_{g.f.}(p = o)$ is of course taken from eqn 4.3. In the first two terms we recognize the background propagators ; if we limit ourselves to region I ($B^{ij}_\mu \equiv B^i_\mu - B^j_\mu = O(1)$), then indeed the quadratic terms are dominant over the last two terms in 5.1, and the integration of the $Q^{ij}_\mu(o)$ and $\bar{\eta}^{ij}(o)$ ($\omega^{ij}(o)$) will give the background propagator. In region II both B^{ij} and Q^{ij} and the ghost are of the same order, so all terms in 5.1 are equally important (every field is now $O(g^{1/2})$, so all the quartic terms do contribute). The outcome of the integration will be :

$$
\begin{aligned}
Z_o(B) &\equiv \int \prod_{i,j,\mu} dQ^{ij}_\mu(o) \prod_{i,j} d\bar{\eta}^{ij}(o) \prod_{ij} d\omega^{ij}(o)\, \exp -\tfrac{1}{g^2} S_{eff}(p=o) \\
&= (g^2)^{\frac{1}{2}(d-1)N(N-1)} \prod_{i<j} \left(\sum_\mu (B^{ij}_\mu)^2\right)^{-(d-2)} F\left(\{B^{ij}/g^{1/2}\}\right) \quad 5.2
\end{aligned}
$$

The d powers come from the gauge boson degrees of freedom, the -2 from the ghost ; $F(\{x_\mu^{ij}\})$ is a formfactor and has the property

$$\lim_{x_\mu^{ij}\to\infty} F(\{x_\mu^{ij}\}) = 1$$

since in region I: $B_\mu^{ij}/g^{1/2} \equiv x_\mu^{ij} \to \infty$ as we discussed above. In region II stringent bounds 1,2) indicate that

$$\lim_{x_\mu^{ij}\to 0} F(\{x_\mu\}) = \prod_{i,j} (x_\mu^{ij})^{\alpha} \quad , \text{ with } \alpha > 2(d-2)$$

Thus we see that $Z_0(B)$ behaves always smoothly at $B_\mu^{ij} = 0$, due to the behaviour of the form factor. We can learn something interesting from eqn 5.2 : the form factor starts to play a role only in high enough dimensions. To determine this dimension d_c, we look at the behaviour of the background field propagator near $B_\mu^{ij} = 0$ (the $p \neq 0$ components of this propagator are smooth at $B_\mu^{ij} = 0$). By naive power counting, we see that 5.2 is integrable (without F !) up to a dimension d_c determined by :

$$d_c\,(N-1) = 2(d_c - 2)\; 1/2\; N(N-1)$$

$$\text{or} \qquad d_c = 2N/(N-1) \qquad 5.3$$

Then the integrand, for $d > d_c$, is well behaved for <u>large</u> values of B_μ^{ij} and therefore, upon rescaling with $g^{1/2}$, we have :

$$Z_0 \equiv \int \prod_{i,\mu} dB_\mu^i \prod_\mu \delta(\sum_i B_\mu^i)\, Z_0(B)$$

$$= (g^2)^{\frac14 d(N^2-1)} \int_{-L/g^{1/2}}^{L/g^{1/2}} \prod_{i,\mu} dx_\mu^i \prod_\mu \delta(\sum_\mu x_\mu^i)\, \Big(\sum_\mu (x_\mu^{ij})^2\Big)^{-(d-2)} F(\{x_\mu^{ij}\}) \qquad 5.4$$

For $d > d_c$ we can put the integration limits in 5.4 equal to ∞ to compute the leading order contribution to Z_0. Notice that for $d = d_c$ we get logarithms[1,2] in the coupling !

The justification of our simplistic computation in section 2 for $d > d_c$ lies in the following observations : the difference between the integrals in eqns 5.4 and 2.3 is that the former contains gauge fixing and ghost terms in addition to the invariant action, while the latter contains only the invariant action. It is straightforward to see that if 5.4 converges (with $L/g^{1/2}$ set to infinity) then so does 2.3. Therefore gauge fixing in p = 0 sector is then not necessary. It also means that for $d > d_c$ the B fields are of order $g^{1/2}$ and can be omitted in the background propagator, when we are only interested in lowest order effects[8].

There are many other interesting aspects such as the various limits of large volume, and how these constant modes influence these limits[2)3),4),7)]. One of the most prominent practical outcomes of this calculation is the insight that twisted boundary conditions do diminish the finite size effects[2], since, if the twist is well chosen, the constant modes are eliminated from the system[6].

REFERENCE

1) A. Gonzalez-Arroyo, J. Jurkiewicz and C.P. Korthals Altes, Proceedings of the Freiburg 11th Nato Summer institute, J. Honerkamp editors, Plenum Press, New York 1982.

2) A. Coste, A. Gonzales-Arroyo, J. Jurkiewicz and C.P. Korthals Altes, Marseille preprint CPT-85/P.1777.
A. Coste, A. Gonzales-Arroyo, C.P. Korthals Altes, B. Söderberg and A. Tarancon, Marseille preprint, 1985.

3) U. Heller, F. Karsch, CERN TH 3879 (1984)

4) B.E. Baaquie, Phys. Rev. D 16, 2612 (1977)

5) T. Eguchi, H. Kawai, Phys. Rev. Lett 48, 1063 (1982)

6) A. Gonzales-Arroyo, M. Okawa, Phys Rev D27 2397 (1983)

7) M. Lüscher, Nucl. Phys. B 219 (1983), 233

8) This caveat applies also to the absence of gauge fixing in the p = o sector. If we compute higher order effects then the higher moments of the integral 2.3 are going to diverge for high enough moments. Only for N = all moments do converge and are we allowed to omit gauge fixing in the p = o sector altogether.

9) V_s is understood to be the spatial volume of the box : $V_s = L^{d-1}$

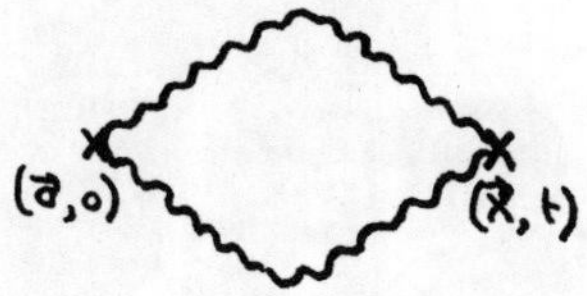

fig.1

Lowest order perturbative contribution to the correlation function between two plaquettes at points $(\vec{x},t)$ and $(\vec{o}\ ,\ o)$, see eqn. 2.8

LIST OF PARTICIPANTS

J. ABAD	Universidad de Zaragoza
A. ACTOR	Universidad de Salamanca
J.M. AGUIRREGABIRIA	Universidad del Pais Vasco
V. ALESSANDRINI	L.P.T.H.E. Orsay
J.L. ALONSO	Universidad de Zaragoza
R. ALVAREZ-ESTRADA	U.C. de Madrid
T. ALVAREZ-MOLINA	U.A. de Madrid
B. ALLES	U.C. de Barcelona
M. ASOREY	Universidad de Zaragoza
J.A. AZCARRAGA	Universidad de Valencia
V. AZCOITI	Universidad de Zaragoza
E. BAGAN	U.A. de Barcelona
M. BAIG	U.A. de Barcelona
B. BERG	Hamburg University
J. BERNABEU	Universidad de Valencia
C. BONA	Universidad de Palma de Mallorca
J.M. BORDES	Universidad de Valencia
L.J. BOYA	Universidad de Zaragoza
J. CARIÑENA	Universidad de Zaragoza
J. CASADO	Universidad de Santiago
J. CASAHORRAN	Universidad de Zaragoza
A. CASAS	U.A. de Madrid
J.M. CERVERO	Universidad de Salamanca

A. COSTE	C.N.R.S. Marseille
A. CRUZ	Universidad de Zaragoza
J.C. CUCHI	Universidad de Zaragoza
A. CHAMORRO	Universidad del Pais Vasco
J. CHINEA	U.C. de Madrid
J.W. DAREWICH	York University
A. DOBADO	Universidad de Santiago
V. ELIAS	University of Winnipeg
D. ESPRIU	Oxford University
J.G. ESTEVE	Universidad de Zaragoza
P. ESTEVEZ	Universidad de Salamanca
H. EVERTZ	I.T.P. Aachen
F. FALCETO	Universidad de Zaragoza
A. FERNANDEZ-PACHECO	Universidad de Zaragoza
X. FUSTERO	U.A. de Barcelona
J. GAITE	Universidad de Salamanca
A. GALINDO	U.C. de Madrid
S. GARCIA	Universidad de Valencia
L.M. GARRIDO	U. C. de Barcelona
E. GAVA	Universita di Trieste
P. GINSPARG	Harvard University
A. GONZALEZ-ARROYO	U.A. de Madrid
M.A. GOÑI	Universidad del Pais Vasco
A. GRIFOLS	U.A. de Barcelona
I. HALLIDAY	Imperial College
M.J. HERRERO	U.A. de Madrid
G.'t HOOFT	Utrecht University
L.A. IBORT	Université Paris VI

C. ITZYKSON	C.E.N. Saclay
K. JANSEN	I.T.P. Aachen
F. JIMENEZ	U.C. de Madrid
J. JORDAN DE URRIES	C.S.I.C.
C.P. KORTHALS-ALTES	C.N.R.S. Marseille
J.I. LATORRE	U.C. de Barcelona
J. LEON	C.S.I.C.
L. MARTINEZ ALONSO	U.C. de Madrid
A. MENDEZ	U.A. de Barcelona
J. MIRAMONTES	Universidad de Santiago
P.K. MITTER	Université Paris VI
A. MORALES	Universidad de Zaragoza
J.J. MORENO	Centro Astronomico de Salerno
T. MORGAN	M.I.T.
A. MUÑOZ SUDUPE	U.C. de Madrid
G. MUSSARDO	Universita di Trieste
J. NARGANES	U.C. de Madrid
J.L. NAVARRO	C.S.I.C.
J. NEGRO	C.S.I.C.
R. NUÑEZ-LAGOS	Universidad de Zaragoza
M.A. DEL OLMO	Université de Montreal
M. PARANJAPE	University of Toronto
P. PASCUAL	U.C. de Bárcelona
J.F. PASCUAL-SANCHEZ	Universidad de Valladolid
J. PEIRO	U.A. de Madrid
J. PEÑARROCHA	Universidad de Valencia
C. PEREZ-MARTIN	U.A. de Madrid
J. PEREZ - MERCADER	C.S.I.C.

J. PERIS	U.A. de Barcelona
M. QUIROS	C.S.I.C.
A. RAMALLO	Universidad de Santiago
M. RAMON	U.C. de Madrid
E. RAMOS	U.A. de Madrid
M.J. RODRIGUEZ	U.C. de Madrid
R. RODRIGUEZ TRIAS	Universidad de Zaragoza
F. RUIZ	U.C. de Madrid
M. RUIZ ALTABA	University of Florida
J. SANCHEZ GUILLEN	Universidad de Santiago
A. SANTAMARIA	Universidad de Valencia
M. SANTANDER	Universidad de Valladolid
A. SEGUI	Universidad de Zaragoza
J. SERRANO BLAZQUEZ	Universidad de Oviedo
G. SIERRA	U.C. de Madrid
J. SOLA	U.A. de Barcelona
R. STORA	L.A.P.P. Annecy
A. TARANCON	Universidad de Zaragoza
R. TARRACH	U.C. de Barcelona
T.F. TREML	University of Toronto
R. TRESGUERRES	C.S.I.C.
L. VALLE	Universidad del Pais Vasco
L. VAZQUEZ	U.C. de Madrid
J. VILLARROEL	Universidad de Salamanca
P. VINDEL	Universidad de Valencia
J.P. DE VRIES	Oxford University
J.F. WHEATER	Durham University
E. ZAS	Universidad de Santiago